SOMATOSTATIN AND ITS RECEPTORS

The Ciba Foundation is an international scientific and educational charity (Registered Charity No. 313574). It was established in 1947 by the Swiss chemical and pharmaceutical company of CIBA Limited—now Ciba-Geigy Limited. The Foundation operates independently in London under English trust law.

The Ciba Foundation exists to promote international cooperation in biological, medical and chemical research. It organizes about eight international multidisciplinary symposia each year on topics that seem ready for discussion by a small group of research workers. The papers and discussions are published in the Ciba Foundation symposium series. The Foundation also holds many shorter meetings (not published), organized by the Foundation itself or by outside scientific organizations. The staff always welcome suggestions for future meetings.

The Foundation's house at 41 Portland Place, London W1N 4BN, provides facilities for meetings of all kinds. Its Media Resource Service supplies information to journalists on all scientific and technological topics. The library, open five days a week to any graduate in science or medicine, also provides information on scientific meetings throughout the world and answers general enquiries on biomedical and chemical subjects. Scientists from any part of the world may stay in the house during working visits to London.

Ciba Foundation Symposium 190

SOMATOSTATIN AND ITS RECEPTORS

1995

JOHN WILEY & SONS

Chichester · New York · Brisbane · Toronto · Singapore

Published in 1995 by John Wiley & Sons Ltd
Baffins Lane, Chichester
West Sussex PO19 1UD, England
Telephone (+44) (1243) 77977

Other Wiley Editorial Offices

John Wiley & Sons, Inc., 605 Third Avenue,
New York, NY 10158-0012, USA

Jacaranda Wiley Ltd, G.P.O. Box 859, Brisbane,
Queensland 4001, Australia

John Wiley & Sons (Canada) Ltd, 22 Worcester Road,
Rexdale, Ontario M9W 1L1, Canada

John Wiley & Sons (SEA) Pte Ltd, 37 Jalan Pemimpin #05-04,
Block B, Union Industrial Building, Singapore 2057

Suggested series entry for library catalogues:
Ciba Foundation Symposia

Ciba Foundation Symposium 190
ix+274 pages, 46 figures, 11 tables

Library of Congress Cataloging-in-Publication Data
Somatostatin and its receptors.
      p. cm.—(Ciba Foundation symposium; 190)
      Includes bibliographical references and indexes.
      ISBN 0 471 95382 2
      1. Somatostatin—Congresses. 2. Somatostatin—Receptors—
Congresses. I. Chadwick, Derek. II. Cardew, Gail. III. Series.
[DNLM: 1. Somatostatin—physiology—congresses. 2. Receptors,
Somatostatin—physiology—congresses. W3 C161F v. 190 1995/WK
515 S6933 1995]
QP572.S59S664   1995
612.4'05—dc20
DNLM/DLC
for Library of Congress   94-43530
                                                     CIP

British Library Cataloguing in Publication Data
A catalogue record for this book is
available from the British Library

ISBN 0 471 95382 2

Phototypeset by Dobbie Typesetting Limited, Tavistock, Devon.
Printed and bound in Great Britain by Biddles Ltd, Guildford.

Contents

Participants

A. Beaudet Department of Neurology and Neurosurgery, McGill University, Montreal Neurological Institute, 3801 University Street, Montreal, Quebec, Canada H3A 2B4

G. I. Bell Howard Hughes Medical Institute, Departments of Biochemistry & Molecular Biology & Medicine, University of Chicago, 5841 S Maryland Avenue, MC 1028, Chicago, IL 60637, USA

M. Berelowitz Division of Endocrinology, SUNY Health Sciences Center, Level 15 Rm 060, Stony Brook, NY 11794-8154, USA

Ch. Bruns Preclinical Research, Bldg 386/646, Sandoz Pharma AG, CH-4002 Basle, Switzerland

M.-F. Chesselet Department of Pharmacology, University of Pennsylvania, 3620 Hamilton Walk, Philadelphia, PA 19104-6084, USA

P. Cohen Université Pierre et Marie Curie, CNRS URA 1682, 96 Blvd Raspail, F-75006 Paris, France

D. H. Coy Department of Medicine, Tulane University School of Medicine, 1430 Tulane Avenue, New Orleans, LA 70112-2699, USA

J. P. Epelbaum Unité de Dynamique des Systèms Neuroendocriniens, INSERM U159, Centre Paul Broca, 2 ter rue d'Alesia, F-75014 Paris, France

C. M. Eppler Dept of Animal Industry Discovery, American Cyanamid Co. ARD, PO Box 400, Princeton, NJ 08543, USA

R. Goodman The Vollum Institute for Advanced Biomedical Research, Oregon Health Sciences University L474, 3181 SW Sam Jackson Park Road, Portland, OR 97201, USA

P. P. A. Humphrey Glaxo Institute of Applied Pharmacology, Department of Pharmacology, University of Cambridge, Tennis Court Road, Cambridge CB2 1QJ, UK

C. Kleuss Department of Pharmacology, Southwestern Medical Center, University of Texas, 5323 Harry Hines Blvd, Dallas, TX 75235-9041, USA

S. W. J. Lamberts Department of Medicine, Erasmus University, 40 Dr. Molewaterplein, NL-3015 GD Rotterdam, The Netherlands

P. Leroux Unité de Neuroendocrinologie Cellulaire et Moléculaire INSERM U413, UA CNRS, Faculté des Sciences, Université de Rouen, F 76821 Mont-St-Aignan Cedex, France

S. L. Lightman Department of Medicine, University of Bristol, Bristol Royal Infirmary, Marlborough Street, Bristol BS2 8HW, UK

W. Meyerhof Deutsches Institut für Ernährungsforschung, Universität Potsdam, Arthur-Schennert-Allee 114-116, D-14558 Potsdam-Rehbrücke, Germany

M. R. Montminy The Clayton Foundation Laboratories for Peptide Biology, The Salk Institute, 10010 North Torrey Pines Road, La Jolla, CA 92037-1099, USA

A.-M. O'Carroll Laboratory of Cell Biology, Building 36 Room 3A-17, NIMH 9000 Rockville Pike, Bethesda, MD 20892, USA

Y. C. Patel Department of Medicine, Neurology and Neurosurgery, McGill University, Royal Victoria Hospital and the Montreal Neurological Institute, Rm M3-15 687 Pine Avenue West, Montreal, Quebec, Canada H3A 1A1

S. Reichlin *(Chairman)* Division of Endocrinology, Metabolism, Diabetes and Molecular Medicine, Tufts University, New England Medical Center, 750 Washington Street, Boston, MA 02111, USA

T. Reisine Department of Pharmacology, University of Pennsylvania, School of Medicine, 36th Street and Hamilton Walk, Philadelphia, PA 19104-6084, USA

R. J. Robbins Endocrinology Section, Memorial Sloan-Kettering Cancer Center, 1275 York Avenue, New York, NY 10021, USA

A. Schonbrunn Department of Pharmacology, University of Texas Medical School, PO Box 20708, Houston, TX 77225, USA

R. M. Señaris Rodriguez *(Bursar)* Departamento de Fisiologia, Facultad de Medicina, Universidad de Santiago de Compostela, c/San Francisco S/n, E-15705 Santiago de Compostela, Spain

P. J. S. Stork The Vollum Institute for Advanced Biomedical Research, Oregon Health Sciences University L474, 3181 SW Sam Jackson Park Road, Portland, OR 97201, USA

C. Susini INSERM U151, CHU Rangueil Bâtiment L3, 1 Avenue Jean Poulhès, F-31054 Toulouse, France

J. E. Taylor Biomeasure Inc., 27 Maple Street, Milford, MA 01757, USA

Introduction

Seymour Reichlin

Division of Endocrinology, Tufts University, New England Medical Center, Box 268, 750 Washington Street, Boston, MA 02111, USA

This symposium was designed to summarize current knowledge of the mechanisms of control of somatostatin synthesis and secretion and how the recent discovery of the structure of the somatostatin receptor family has illuminated our understanding of somatostatin regulation of cell secretion and growth. We will also consider the insight that these findings have given in brain development, neurological disorders and clinical diagnosis and therapy. In the tradition of Ciba symposia, an active and free discussion among experts will clarify areas of controversy and important questions for future work will be identified.

To establish a context for this dialogue, I will retrace some of the steps that led to the serendipitous discovery of somatostatin and the subsequent elucidation of its biological importance (for reviews see Reichlin 1983, 1987, Patel & Srikant 1985, Patel 1992, 1994). This history is a paradigm for the development of our understanding of neuroendocrinology, from the early days of whole-animal physiology and bioassays, through peptide structure determination by amino acid sequencing, to elucidation at the molecular level of mechanisms of neuropeptide synthesis, secretion and action.

Growth hormone (GH) secretion inhibitory factors were first discovered by Krulich et al (1968) who were attempting to demonstrate GH-releasing activity in extracts of rat hypothalamus. Unlike previous work, which had utilized whole hypothalamic extracts, they assayed extracts of various regions of the hypothalamus, finding some that inhibited GH release and some that stimulated GH release. They made the novel proposal that GH secretion was governed by the interaction of hypothalamic stimulatory and inhibitory factors. At about the same time, Hellman & Lernmark (1969) found a component of whole pancreatic extracts that inhibited insulin secretion and suggested that it might be under local inhibitory control.

At first, the view of Krulich et al was not readily accepted, but became more credible when a group of workers in Roger Guillemin's laboratory determined the chemical structure of a GH release inhibitory factor during the search for GH-releasing activity in whole hypothalamic extracts (Brazeau et al 1973). It is an ironic footnote to this history, recounted by Guillemin many years later

(Guillemin 1992), that he did not acknowledge the prior report of Krulich et al (1968) in his somatostatin paper (Brazeau et al 1973) because he did not find their bioassay results convincing. It is also ironic that the structure of somatostatin, whose existence had been suspected for only a few years, was elucidated within a few months of its discovery, whereas rat GH-releasing hormone was sequenced in 1983 (Spiess et al 1983), a gap of more than 20 years from the first suggestion of its existence (Reichlin 1960).

Once somatostatin was shown to be a 14 amino acid peptide which could be synthesized by classical chemical methods, a flood of new clinical and physiological insights followed. Trials in humans showed that, as in rats, somatostatin inhibited GH secretion, thus opening a potential new mode of therapy for acromegaly (Yen et al 1974). Somatostatin was also shown to inhibit both insulin secretion (Alberti et al 1973) and thyroid-stimulating hormone (TSH) secretion in humans (Siler et al 1974), the latter suggesting that it was part of a dual control system of TSH regulation in conjunction with TSH-releasing hormone. Somatostatin was then shown to inhibit glucagon secretion in humans (Gerich et al 1974) and to inhibit gastric acid secretion, even when it was stimulated by pentagastrin (Arnold & Cretzfeldt 1975).

Initially, the inhibitory actions of somatostatin on structures outside the pituitary gland were considered to be pharmacological effects without physiological significance. However, with the discovery that immunoreactive somatostatin was present in a specific population of pancreatic islet cells (Dubois 1975, Hökfelt et al 1975), it became clear that it was part of an intrinsic paracrine control system. It was also present in a selective population of gastrointestinal secretory cells and gut neurons, suggesting that it was an intrinsic paracrine regulator of all internal and external secretions of the gastrointestinal tract. By 1978 the wide distribution of somatostatin in the central and peripheral nervous systems (Epelbaum 1986) and its regulation by depolarizing agents and several neurotransmitters (Robbins et al 1982a,b) suggested that somatostatin was a neurotransmitter or neuromodulator.

At this point, only modest progress had been made in understanding the synthesis of somatostatin, although it was clear from the demonstration of large molecular weight forms of somatostatin in various tissue extracts (Patel & Reichlin 1978, Lauber et al 1979) that it was synthesized as part of a larger prohormone, similarly to insulin and vasopressin. Somatostatin$_{1-28}$ was then discovered (Pradayrol et al 1980), demonstrating that there was more than one active form of somatostatin in tissues. At this symposium, Yogesh Patel will present results clarifying the mechanisms of somatostatin prohormone processing.

During the 1980s physiological studies of somatostatin action were initiated, leading to the demonstration that somatostatin receptors are present in a number of tissues (Patel 1992) and that its inhibitory actions include suppression of both cAMP synthesis and action and inhibition of K^+ transport (Patel et al 1990,

Schonbrunn et al 1985, Koch & Schonbrunn 1988). Agnes Schonbrunn, a pioneer in this field, will summarize additional work on the control of somatostatin secretion in the light of the newly isolated somatostatin receptors.

By 1980 the powerful tools of molecular biology were applied to the study of somatostatin for the first time. Between 1982 and 1983 the prohormone sequence of somatostatin in several species had been identified (Goodman et al 1982), and between 1983 and 1984 the sequence of both the rat (Funckes et al 1983, Montminy et al 1984) and human somatostatin gene (Shen & Rutter 1984) had been elucidated. Using immunological and molecular biological methods, it was demonstrated that as many as seven or eight different somatostatin genes were present in the animal kingdom. These findings, together with the observation that somatostatin could be detected in unicellular organisms (Berelowitz et al 1981), suggested that it was highly conserved amongst eukaryotes.

Between 1985 and 1992 important advances were made in understanding the mechanism of control of somatostatin synthesis. The discovery of the cAMP response element (CRE) in the somatostatin gene and the protein that binds to this sequence (CREB protein) by Montminy & Bilezikjian (1987) revealed the molecular basis by which signals arising in the membrane could activate the genome, a finding of general biological importance. At this symposium additional intranuclear binding factors that modulate somatostatin regulation of cAMP will be described by Marc Montminy and discussed by Richard Goodman.

The availability of various somatostatin analogues including somatostatin-14, somatostatin-28, the synthetic peptide octreotide and a series of compounds synthesized by David Coy and colleagues confirmed, by pharmacological and binding studies, that there was more than one type of somatostatin receptor (Srikant & Patel 1987, Patel et al 1990). The explanation of the differences in receptor activity was also revealed when the molecular sequence of a somatostatin receptor was reported from Graeme Bell's laboratory (Reisine & Bell 1993, Yamada et al 1992), followed by the description of four additional sequences and two additional variants of one of those sequences.

All of these receptors are G protein-coupled, seven membrane-spanning structures. At this symposium a number of participants, including Graeme Bell, Christian Bruns, Yogesh Patel and Terry Reisine, will clarify the relationship between receptor sequence and pharmacological specificity.

Knowledge of the structure of the somatostatin receptors and their coupling to G proteins illuminates a number of actions of somatostatin on cells, and will be reviewed by Christiane Kleuss and Agnes Schonbrunn. In addition to G protein coupling, with its relevance to cAMP-mediated control, Christiane Susini will expand earlier work on the influence of somatostatin on tyrosine phosphatases to show that somatostatin receptors are bound to a tyrosine phosphatase. This important claim may explain the antiproliferative action of somatostatin and its analogues. Indeed, much research is now being directed at somatostatin as a possible antineoplastic peptide.

At this symposium Marie-François Chesselet will review her work on the role of somatostatin in basal ganglion function, Philippe Leroux will describe his work on the ontogeny of somatostatin and somatostatin receptors in the brain and Alain Beaudet will show how the new knowledge of somatostatin receptors has clarified neuroendocrine mechanisms of GH regulation.

For the clinician, the development of novel somatostatin analogues has provided new tools for diagnosis and treatment. The introduction of the somatostatin analogue octreotide established a practical use of somatostatin for the treatment of hypersecretory states of the pituitary (GH, TSH), of gastrointestinal and neuroendocrine tumours (vasoactive intestinal peptide carcinomas, carcinoids, gastrinomas and glucagonomas) and of diarrhoeal states (Reichlin 1983, Gorden et al 1989). The newest clinical innovation, the use of octreotide tagged with radioactive indium, was introduced in 1990 (Lamberts et al 1990) and has now been used successfully to image neuroendocrine tumours and other cells that express somatostatin receptors such as Hodgkin's lymphoma and inflammatory cells. Its use will be described at this symposium by Steven Lamberts, a pioneer in this field.

The story of somatostatin is convoluted and confusing. From a serendipitous discovery, it has become a powerful tool for cell biologists, neurobiologists and clinicians. The presentations over the next three days will summarize how far we have come and will define the main issues for future investigation.

References

Alberti KG, Christensen MJ, Christensen SE et al 1973 Inhibition of insulin secretion by somatostatin. Lancet II:1299–1301

Arnold R, Creutzfeldt W 1975 Inhibition of pentagastrin-induced gastric-acid secretion by somatostatin in man [in German] Deutsche Medizin Wochenschr 100:1014–1016

Berelowitz M, LeRoith D, Von Schenk H et al 1981 Somatostatin-like immunoactivity and biological activity is present in *Tetrahymena pyriformis*, a ciliated protozoan. Endocrinology 110:1939–1944

Brazeau P, Vale WW, Burgus R et al 1973 A hypothalamic peptide that inhibits the secretion of immunoreactive pituitary growth hormone. Science 179:77–79

Dubois MP 1975 Immunoreactive somatostatin is present in discrete cells of the endocrine pancreas. Proc Natl Acad Sci USA 72:1340–1343

Epelbaum J 1986 Somatostatin in the central nervous system: physiology and pathological modifications. Prog Neurobiol 27:63–100

Funckes CL, Minth CD, Deschenes R et al 1983 Cloning and characterization of a mRNA encoding rat preprosomatostatin. J Biol Chem 258:18781–87871

Gerich JE, Lorenzi M, Schneider V et al 1974 Inhibition of pancreatic glucagon responses to arginine by somatostatin in normal man and insulin-dependent diabetics. Diabetes 23:876–880

Goodman RH, Aron DC, Roos BA 1982 Rat preprosomatostatin: structure and processing by microsomal membranes. J Biol Chem 258:5570–5573

Gorden P, Comi RJ, Maton PN, Go VLW 1989 Somatostatin and somatostatin analogue (SMS 201-995) in treatment of hormone-secreting tumours of the pituitary and gastrointestinal tract and non-neoplastic diseases of the gut. Ann Intern Med 110:35–50

Guillemin R 1992 Somatostatin: the early days. Metabolism (suppl 2) 41:2–4

Hellman B, Lernmark A 1969 Inhibition of the *in vitro* secretion of insulin by an extract of pancreatic α_1 cells. Endocrinology 84:1484–1487

Hökfelt T, Efendic S, Hellerstrom C, Johansson O, Luft R, Arimura A 1975 Cellular localization of somatostatin in endocrine-like cells and neurons of rat with special references to A_1-cells of the pancreatic islets and to hypothalamus. Acta Endocrinol (suppl 200) 80:5–41

Koch BD, Schonbrunn A 1988 Characterization of the cAMP-independent actions of somatostatin in growth hormone (GH) cells. II. An increase in potassium conductance initiates somatostatin-induced inhibition of prolactin secretion. J Biol Chem 263:226–234

Krulich L, Dhariwal APS, McCann SM 1968 Stimulatory and inhibitory effects of purified hypothalamic extracts on growth hormone release from rat pituitary *in vitro*. Endocrinology 83:783–790

Lamberts SWJ, Bakker WH, Reubi J-C, Krenning EP 1990 Somatostatin receptor imaging in the localization of endocrine tumours. N Engl J Med 323:1246–1249

Lauber M, Camier M, Cohen P 1979 Higher molecular weight forms of immunoreactive somatostatin in mouse hypothalamic extracts: evidence of processing *in vitro*. Proc Natl Acad Sci USA 76:6004–6008

Montminy MR, Bilezikjian LM 1987 Binding of a nuclear protein to the cAMP response element of the somatostatin gene. Nature 328:175–178

Montminy MR, Goodman RH, Horovitch SJ, Habener JF 1984 Primary structure of the gene encoding rat preprosomatostatin. Proc Natl Acad Sci USA 81:3337–3340

Patel YC 1992 General aspects of the biology and function of somatostatin. In: Thorner MO, Muller EE (eds) Basic and clinical aspects of neuroscience. vol 4, Springer-Verlag, Berlin p 1–16

Patel YC 1994 Somatostatin. In Becker KL (ed) Principles and practise of endocrinology and metabolism. Lippincott, Philadelphia, PA, in press

Patel YC, Reichlin S 1978 Somatostatin in hypothalamus, extrahypothalamic brain and peripheral tissues of the rat. Endocrinology 102:523–530

Patel YC, Srikant CB 1985 Somatostatin mediation of adenohypophysial secretion. Annu Rev Physiol 48:551–567

Patel YC, Murthy KK, Escher E, Banville D, Spleiss J, Srikant CB 1990 Mechanism of action of somatostatin: an overview of receptor function and studies of the molecular characterization and purification of somatostatin receptor proteins. Metabolism (suppl 2) 39:63–69

Pradayrol L, Jornvall H, Mutt V, Ribet A 1980 N-terminally extended somatostatin: the primary structure of somatostatin-28. FEBS Lett 109:55–58

Reichlin S 1960 Growth and the hypothalamus. Endocrinology 67:760–773

Reichlin S 1983 Somatostatin. N Engl J Med 309:1495–1501, 1556–1563

Reichlin S 1987 Somatostatin: basic and clinical status. Plenum, New York

Reisine T, Bell G 1993 Molecular biology of somatostatin receptors. Trends Neurosci 16:34–38

Robbins RJ, Sutton RE, Reichlin S 1982a Sodium- and calcium-dependent somatostatin release from dissociated cerebral cortical cells in culture. Endocrinology 100:496–499

Robbins RJ, Sutton RE, Reichlin S 1982b Effects of neurotransmitters and cyclic AMP on somatostatin release from cultured cerebral cortical cells. Brain Res 234:377–386

Schonbrunn A, Dorflinger LJ, Koch BD 1985 Mechanisms of somatostatin action in pituitary cells. Adv Exp Med Biol 188:305–324

Shen LP, Rutter WJ 1984 Sequence of the human somatostatin I gene. Science 224:168–171

Siler TM, Yen SSC, Vale W, Guillemin R 1974 Inhibition by somatostatin on the release of TSH induced in man by thyrotropin-releasing factor. J Clin Endocrinol & Metab 38:742–745

Spiess J, Rivier J, Vale W 1983 Characterization of rat hypothalamic growth hormone-releasing factor. Nature 303:532–535

Srikant CB, Patel YC 1987 Somatostatin receptors: evidence for functional and structural heterogeneity. In: Reichlin S (ed) Proceedings of the international conference on somatostatin. Plenum, New York, p89–102

Yamada Y, Post SR, Wang K, Tager HS, Bell GI, Seino S 1992 Cloning and functional characterization of a family of human and mouse somatostatin receptors expressed in brain, gastrointestinal tract and kidney. Proc Natl Acad Sci USA 89:251–255

Yen SS, Siler TM, DeVane GW 1974 Effect of somatostatin in patients with acromegaly: suppression of growth hormone, prolactin, insulin and glucose levels. N Engl J Med 29:935–938

Regulation of somatostatin gene transcription by cAMP

M. Montminy, P. Brindle, J. Arias, K. Ferreri and R. Armstrong

The Clayton Foundation Laboratories for Peptide Biology, The Salk Institute, 10010 North Torrey Pines Road, La Jolla, CA 92037, USA

Abstract. A number of hormones and growth factors stimulate target cells through receptors which are coupled to second messenger pathways. The second messenger cAMP, for example, mediates a wide variety of cellular responses to hormonal signals, including changes in intermediary metabolism, cellular proliferation and cellular motility. In mammalian cells, all of these biological responses are triggered by the activation of the cAMP-dependent protein kinase A, a heterotetramer consisting of paired catalytic and regulatory subunits. Upon hormonal stimulation, cAMP binds tightly to the regulatory subunits, thereby liberating catalytic subunits and promoting the phosphorylation of cellular substrates. In the liver, cAMP functions as a starvation state signal, mediating hormonal cues from the pancreas and adrenal gland to stimulate glucose production. cAMP stimulates glucose production, in part, by regulating transcription of the gene for phosphoenolpyruvate carboxykinase (PEPCK), a rate-limiting enzyme in gluconeogenesis. Following hormonal stimulation, cAMP induces PEPCK gene expression 10-fold within 20–30 min. This induction appears to be independent of new protein synthesis.

1995 Somatostatin and its receptors. Wiley, Chichester (Ciba Foundation Symposium 190) p 7–25

Characterization of the cAMP response unit

Previous studies (Montminy et al 1986, Comb et al 1986, Short et al 1986) have identified a consensus cAMP response element (CRE), usually positioned within 100 nucleotides of the transcriptional initiation site, which mediates transcriptional induction of phosphoenolpyruvate carboxykinase (PEPCK) and other cAMP responsive genes. Characterized by a core motif which often contains the palindromic sequence 5′-TGACGTCA-3′, the CRE can confer cAMP inducibility when placed upstream of a heterologous promoter and can function in both distance- and orientation-independent contexts. Remarkably, mutant cell lines deficient in the cAMP-dependent protein kinase A (PKA) cannot support cAMP-responsive transcription from a CRE reporter gene (Montminy et al 1986), suggesting that this kinase phosphorylates proteins which bind to the CRE and thereby activates transcription of cAMP responsive genes. Towards this end, we purified a 43 kDa CRE-binding (CREB) protein from rat adrenal phaeochromocytoma PC12 cells and brain nuclear extracts using CRE–oligonucleotide affinity chromatography (Montminy & Bilezikjian 1987, Yamamoto et al 1988). The purified CREB protein was highly phosphorylated by PKA *in vitro,* and primary sequencing of a CREB tryptic phosphopeptide revealed a consensus PKA site: (R)-R-P-S-Y-R. Using primary sequence information provided by other tryptic fragments of CREB protein, we obtained cDNAs encoding a 341 amino acid protein which contained the PKA phosphoacceptor site at Ser133 (Gonzalez et al 1989). cDNAs for the CREB protein were also characterized independently by Hoeffler et al (1988).

Previous work showing that most, if not all, of the biological effects of cAMP are mediated by PKA, prompted us to test whether CREB protein activity in response to cAMP might be dependent on phosphorylation by PKA. We found that CREB protein is indeed phosphorylated by PKA at a single serine phospho-acceptor site Ser133 (Gonzalez & Montminy 1989) and this phosphorylation is induced by hormonal stimuli which signal through the cAMP pathway. Remarkably, mutant cell lines which are deficient in PKA activity cannot stimulate CREB protein phosphorylation at Ser133, nor can they induce CRE-dependent transcription. To determine whether PKA phosphorylation of CREB protein at Ser133 was critical for cAMP-responsive transcription *in vivo*, we compared wild-type and mutant forms of CREB protein containing substitutions at the PKA phosphoacceptor site (Gonzalez & Montminy 1989). When transfected into mouse F9 teratocarcinoma cells, a wild-type CREB protein expression plasmid can stimulate CRE–CAT (chloramphenicol acetyltransferase) reporter activity greater than 200-fold in the presence of PKA. By contrast, a mutant CREB protein plasmid containing a Ser133 to Ala133 substitution was completely unable to respond to PKA induction. Acidic substitution mutants (Ser133 to Asp133 or Glu133) were also inactive, suggesting that the negative charge provided by phosphorylation is not sufficient to stimulate transcription.

The presence of phosphorylation sites for other kinases, most notably for casein kinase II (CKII), prompted us to test whether these might also regulate CREB protein activity (Gonzalez et al 1991, Hagiwara et al 1992). When treated with CKII, recombinant CREB protein was phosphorylated at several serine phosphoacceptor sites, including Ser111, 114, 117 and 156, as determined by primary sequencing of tryptic peptides (not shown). In contrast with the Ser133 phosphoacceptor site, however, mutagenesis of any or all of the CKII sites did not reduce but rather increased CREB protein activity somewhat in the presence of cAMP, suggesting that a single phosphorylation event may be both necessary and sufficient to induce transcription of cAMP-responsive genes. To test this hypothesis, we performed microinjection experiments with purified recombinant CREB protein (Alberts et al 1994). When co-injected with a CRE–lacZ reporter plasmid, unphosphorylated CREB protein had no effect on lacZ reporter activity in microinjected rat embryo fibroblasts. In contrast, phosphorylated CREB protein markedly stimulated CRE-dependent transcription, suggesting that PKA-mediated phosphorylation at Ser133 might indeed be sufficient to induce expression of cAMP regulated genes.

Kinetic profile of the transcriptional response to cAMP

Like other signalling pathways, cAMP stimulates transcription of somatostatin and other target genes with burst–attenuation kinetics: transcription usually peaks within 20–30 min of induction and gradually declines over the next 4–8 h (Fig. 1A). The kinetics of PKA-dependent CREB protein phosphorylation parallel the changes in transcription of cAMP-responsive genes by run-on assay (Fig. 1B). Nuclear translocation of PKA, visualized by microinjection of fluorescently labelled PKA holoenzyme, appears to represent the rate-limiting step in CREB protein phosphorylation and transcriptional activation (Hagiwara et al 1992, Hagiwara et al 1993).

To follow the stoichiometry of CREB protein phosphorylation in response to hormonal stimulation, we developed a phosphorylated Ser133-specific CREB protein antiserum (Hagiwara et al 1993). Maximal activation of the cAMP pathway with forskolin (10 μM) causes 40% of total CREB protein to be phosphorylated. By varying the strength of hormonal stimulus, we found that the degree of CREB protein phosphorylation and CRE-dependent transcription were directly proportional to the extent of PKA activation. Rather than being simply turned on or off, peak levels of cAMP-responsive transcription are proportional to signal strength.

The second phase in cAMP-responsive transcription is an attenuation phase during which CREB protein phosphorylation and CRE-dependent transcription decrease in parallel over a four hour period. During that time, levels of total CREB protein remain unchanged, suggesting that a CREB protein phosphatase may be critical in down-regulating cAMP-induced transcription (Hagiwara et al 1992). Indeed, CREB phosphorylation at Ser133 and CRE–CAT reporter activity in mouse NIH/3T3 fibroblasts were markedly augmented by the phosphatase inhibitors okadaic acid ($ID_{50} = 20$ nM) and the inhibitor-1 (I-1) protein 4 h after cAMP induction. Of the four serine threonine phosphatases characterized to date, only protein phosphatase 1 (PP1) can be inhibited by both okadaic acid and I-1 protein. When transfected into F9 cells, an expression vector for PP1, but not the protein phosphatase 2A (PP2A) catalytic subunit, could inhibit cAMP-dependent transcription from a CRE–CAT reporter plasmid. Furthermore, microinjection of purified PP1 but not PP2 protein severely diminished CRE–lacZ reporter expression in rat embryo fibroblasts, suggesting that cAMP-responsive transcription may depend not only on the level of PKA activation, but also on cellular levels of PP1. Indeed, microinjection of constitutively active I-1 protein not only increases

A

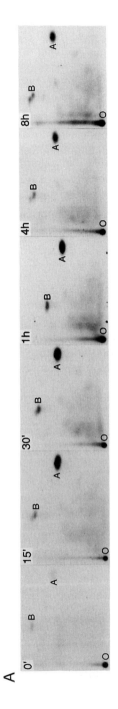
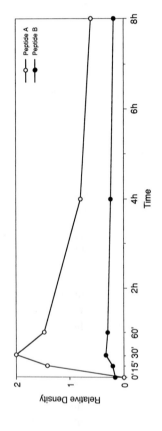

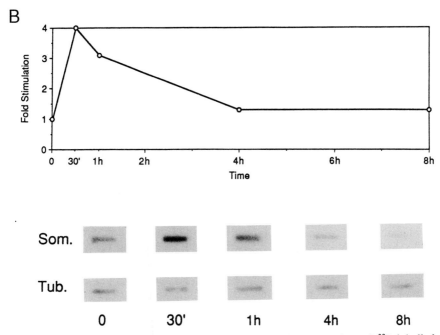

FIG. 1. (A) (*Opposite*) Two-dimensional phosphotryptic mapping of ^{32}P-labelled CREB protein in PC12 cells after treatment with forskolin. Spot A contains PKA phosphorylation site (Ser133). Spot B contains casein kinase II (Ser156) phosphorylation site. (B) Run-on transcription of the cAMP-responsive somatostatin gene in stable lines of NIH/3T3 cells transfected with the somatostatin (Som) gene. Tub., tubulin mRNA transcription.

CRE-dependent transcription, but also stimulates CREB protein phosphorylation *in situ*. In contrast, the SV40 T antigen, which binds and specifically inhibits PP2A, has no effect on either CRE-inducible transcription or on CREB protein phosphorylation.

Mechanism of phosphorylation-dependent induction

Phosphorylation may regulate nuclear factors at several levels, including DNA binding and nuclear targeting activities. To determine whether phosphorylation might induce CREB protein DNA-binding activity, we performed gel mobility shift and footprinting assays using recombinant purified CREB protein. These assays, combined with filter binding studies, revealed that CREB protein and PKA-phosphorylated CREB protein both bind to the CRE with a K_d of 10^{-9} M (Gonzalez & Montminy 1989, Hagiwara et al 1993). If PKA phosphorylation promotes CREB protein DNA-binding activity, then we reasoned that CREB fusion proteins containing a heterologous binding domain (e.g. the yeast Gal4

DNA-binding domain) may not respond to PKA induction. When transfected into F9 or PC12 cells, Gal4–CREB protein nevertheless induced CAT reporter activity 50–100 fold in response to PKA (Hagiwara et al 1992). Moreover, mutagenesis of the Ser133 phosphoacceptor site completely abrogated PKA-inducible activity, suggesting that the Gal4–CREB protein may function like the wild-type CREB protein on a heterologous DNA-binding domain.

To determine whether phosphorylation may affect nuclear entry of CREB protein, we performed immunocytochemical studies with CREB protein-specific antisera. We found that endogenous CREB protein was nuclear in both unstimulated as well as forskolin-treated cells (Gonzalez & Montminy 1989). Moreover, mutant forms of CREB protein lacking the Ser133 phosphoacceptor site were also targeted correctly to nuclei of transfected cells, suggesting that PKA-mediated phosphorylation may not induce nuclear transport of CREB protein.

Having eliminated CRE binding activity, dimerization or nuclear translocation as plausible mechanisms by which PKA might stimulate CREB activity, we hypothesized that phosphorylation might directly regulate the transactivation potential of CREB protein. In this regard, we have recently determined that a 70 amino acid kinase-inducible domain (KID) spanning amino acids 90–160 in the CREB protein can modulate transcription of target genes in response to cAMP (Brindle et al 1993). The KID does not function alone, but is paired with a constitutive activator to modulate transcription. In this regard, the CREB protein contains a glutamine rich activator termed Q2 with properties similar to glutamine-rich regions in the nuclear factor SP1. As deletion of Q2 completely abrogates CREB protein activity in response to cAMP, we have speculated that the KID must cooperate with Q2 to stimulate transcription in response to PKA phosphorylation.

To determine whether the KID and Q2 could function as independent activators, we attached these regions to the Gal4 DNA-binding domain and examined their activities on a 5X Gal4 CAT reporter plasmid following transfection into F9 cells (Brindle et al 1993) (Fig. 2A). The Q2–Gal4 fusion protein stimulated reporter activity 15-fold but did not respond to PKA induction. A KID–Gal4 effector plasmid showed low constitutive activity which was stimulated threefold in response to PKA. When co-transfected together (Fig. 2B), Q2–Gal4 and KID–Gal4 effectors could reconstitute full PKA inducible transcription, suggesting that Q2 and KID are indeed independent activators which synergize to stimulate transcription in response to cAMP.

In order to characterize factors that interact functionally with CREB protein, we employed an *in vitro* transcription assay system using purified wild-type or mutant CREB polypeptides (Ferreri et al 1994). When added to crude HeLa nuclear extracts, the unphosphorylated CREB protein stimulated transcription from the CRE-containing somatostatin promoter but not a control α-globin template. In contrast, a mutant CREB polypeptide, lacking the glutamine-rich Q2 region, was transcriptionally inactive despite having wild-type DNA-binding

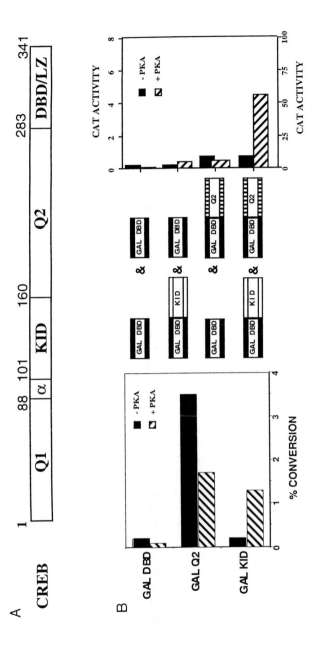

FIG. 2. (A) The transcription factor, CREB protein, is organized in modular fashion. Q1 and Q2, hydrophobic, glutamine-rich domains; α, alternately spliced 14 amino acid sequence which forms alpha helical structure *in vitro*; KID, kinase-inducible domain containing PKI phosphoacceptor site at SER133; DBD/LZ, basic DNA-binding/leucine zipper dimerization domain. (B) The KID and Q2 domains in the CREB protein function independently as modulatory and constitutive activation domains, respectively. Left, CAT activity derived from F9 cells following co-transfection with GAL4 DNA binding domain (Gal DBD), GAL4–Q2, or GAL4–KID fusion protein expression vectors as listed, PKA (− or + as indicated) plus GAL4–CAT reporter plasmid. Right, the KID and Q2 activators can synergize in *trans* to stimulate transcription of cAMP responsive genes. F9 cells were co-transfected with expression vectors as indicated, PKA (− or + as indicated) plus GAL4–CAT reporter plasmid.

affinity *in vitro*. To test whether the Q2 domain was sufficient to stimulate transcription, we expressed and purified a fusion protein containing the Q2 region attached to the Gal4 DNA-binding domain. When evaluated on a somatostatin promoter template containing two Gal4 recognition sites, the Gal4–Q2 protein stimulated somatostatin promoter activity in a dose-dependent manner, but the Gal4 DNA-binding domain alone was minimally active, suggesting that the Q2 domain does indeed function as a constitutive activator both *in vitro* and *in vivo* (Brindle et al 1993).

Interactions between CREB protein and the transcriptional apparatus

On the basis of the mutagenesis results presented above, we hypothesized that CREB protein would form both constitutive and inducible interactions with proteins that are associated with the transcriptional apparatus. To characterize constitutive interactions between Q2 and the basal transcription complex, we prepared partially purified fractions of the basal transcription factors and monitored their ability to complement Q2 activity (Ferreri et al 1994). Transcription reactions reconstituted with all basal factors, except the transcription factor (TF) IID fraction, showed minimal activity from either the somatostatin or control adenovirus 2 major late promoters. Addition of recombinant TBP (TATA-binding protein) restored basal but not Gal4–Q2 activator-dependent activity to both templates. Only the holo–TFIID fraction could fully reconstitute Gal4 Q2 transactivation, suggesting that some factor in this fraction, in addition to TBP, is required for Q2 function. Within the TFIID fraction, TBP is found associated with a number of factors (TAFs) which appear to have co-activator activity (Dynlacht et al 1991, Hoey et al 1993).

Inspection of sequences within Q2 revealed an intriguing resemblance to the hydrophobic and glutamine rich Sp1 activation domain B (Gill et al 1994). As mutational analysis indicated that this region was important for interaction with the *Drosophila* dTAF$_{II}$110 protein, we tested for similar interactions between Q2 and dTAF$_{II}$110. CREB protein stimulates transcription of target genes comparably in *Drosophila* and mammalian nuclear extracts (M. Montminy, unpublished observations), thereby permitting us to evaluate interactions between Q2 and the recently cloned *Drosophila* TAFs. Towards this end, we prepared CRE–oligonucleotide affinity resins bound by purified CREB protein or mutant ΔQ2 proteins. We observed that ^{35}S-labelled dTAF$_{II}$110 protein from reticulocyte lysates programmed with dTAF$_{II}$110 RNA could indeed bind to CRE resin containing wild-type CREB protein, albeit with low efficiency (10%). No such binding was observed between dTAF$_{II}$110 protein and ΔQ2 mutant CREB protein, suggesting that the interaction we detected was specific for the Q2 domain. Indeed, dTAF$_{II}$110 was also retained on resins containing Gal4–Q2 protein but not Gal4 DBD (DNA-binding domain) alone.

To confirm these *in vitro* assays and begin to map the important sequences in Q2 that mediate interaction with $dTAF_{II}110$, we employed the yeast two hybrid system (Chien et al 1991) (Fig. 3). A yeast expression plasmid encoding the Q2 domain fused to the acidic activation domain of GAL4 was co-transformed into a yeast strain with a second vector expressing $dTAF_{II}110$–Gal4 DBD fusion protein. Interaction of the CREB protein Q2 domain with $dTAF_{II}110$, monitored by the recruitment of the acidic activation domain onto the promoter of a Gal4–lacZ reporter gene, was readily observed. In similar assays, the CREB protein Q2 domain did not appear to interact with other components of the TFIID complex such as TBP, TAF40 and TAF80.

Based on sequence similarity with Sp1 (amino acid 375–378) (Gill et al 1994), we postulated that the putative $dTAF_{II}110$ interaction motif in CREB protein is located at amino acid 204–208. This motif is absent from the CREM (cAMP response element modulator) α, β and ϵ repressors, which lack sequences in Q2 extending from amino acid 183 to 244 of CREB protein. Similar deletion mutants of CREB protein (Δ204–214 and Δ204–224) exhibit far lower constitutive and PKA-dependent activity than the wild-type protein in co-transfection assays using a somatostatin–CAT reporter vector (Brindle et al 1993, Ferreri et al 1994). Consistently, $dTAF_{II}110$ association with these and other inactive CREB protein deletion mutants was markedly decreased compared to the wild-type Q2 domain (Fig. 3), demonstrating that Q2 activity was indeed correlated with its ability to bind $dTAF_{II}110$.

Dr Goodman's group has recently characterized a CREB protein binding protein (CBP) which specifically recognizes sequences within the Ser133 phosphorylated form of CREB protein (Chrivia et al 1993). CBP does not regulate the DNA binding, dimerization or nuclear targeting properties of CREB protein. Rather CBP binds selectively to the KID of CREB, a 60 amino acid transactivation domain which is critical for PKA inducible transcription. To characterize the functional properties of CBP, we developed an antiserum directed against amino acids 634–648 within the CREB protein binding domain of CBP (Arias et al 1994). Using this antiserum, we detected a 265 kDa polypeptide on Western blots (Fig. 4A, lane 1) as predicted from the cDNA (Chrivia et al 1993), which coincided with the predominant phospho-CREB protein binding activity in HeLa nuclear extracts by 'far-Western' blot assay (Fig. 4A, lane 2). An identical phospho-CREB binding activity was also found in NIH/3T3 cells (not shown). This phospho-CREB protein binding protein appeared to be specific for Ser133 phosphorylated CREB protein because no such band was detected with CREB protein labelled to the same specific activity at a non-regulatory phosphoacceptor site (Ser156) by CKII (Hagiwara et al 1992) (Fig. 4B). To demonstrate that the major phospho-CREB protein binding protein in HeLa and NIH/3T3 cells is bound specifically by the anti-CBP antibody, we prepared immunoprecipitates from crude nuclear extracts using the CBP antiserum. Far-Western analysis of these immunoprecipitates

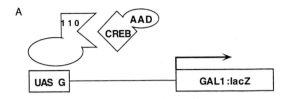

B		Fusion to GAL4 AAD	Fusion to GAL4 DBD	β-galactosidase Activity
	NO		dTAF$_{11}$ 110	<1
Sp1 B			dTAF$_{11}$ 110	51
Q2			dTAF$_{11}$ 110	19
Q2[Δ204-213]			dTAF$_{11}$ 110	4
Q2[Δ204-224]			dTAF$_{11}$ 110	<1
Q2[Δ204-256]			dTAF$_{11}$ 110	<1
Q2[Δ185-256]			dTAF$_{11}$ 110	<1
Q2			NO	<1

FIG. 3. Interaction between dTAF110 and the CREB Q22 requires sequences in Q2 which are homologous to SP1. (A) Diagram of yeast two hybrid assay showing that recruitment of GAL4 acidic activator requires sequences in the CREB protein which interact with GAL4–dTAF110 and promote transcription from a GAL4–lacZ reporter. (B) β-galactosidase activity obtained from wild-type and mutant Q2 regions shown with SP1 activity included for comparison.

(Fig. 4C) revealed a 265 kDa band in samples incubated with CBP antiserum but not with control IgG. To examine whether the phosphorylation-dependent interaction between CREB protein and CBP was critical for cAMP-responsive transcription, we employed a microinjection assay using CBP antiserum which would be predicted to impair formation of a CREB–CBP complex (Fig. 5). Following microinjection into nuclei of NIH/3T3 cells, a CRE–lacZ reporter was markedly induced by treatment with 8-Br cAMP (8-bromoadenosine 3':5'-cyclic monophosphate) plus IBMX (3-isobutyl-1-methylxanthine phosphodiesterase inhibitor). Co-injection of CBP antiserum with the CRE–lacZ

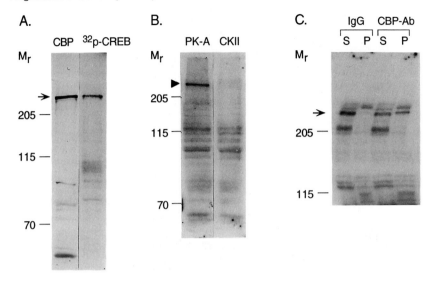

FIG. 4.   CBP is the predominant phospho-Ser133 CREB protein binding activity in nuclear extracts. (A) Western (CBP) and far-Western ([^{32}P]CREB protein) blot analysis of HeLa nuclear extract following SDS-PAGE and transfer to nitrocellulose. (B) Far-Western blot of crude HeLa nuclear extract using [^{32}P]CREB protein phosphorylated with PKA or casein kinase II (CKII). In A, B and C, arrow points to the major 265 kDa band. Mr, relative mass in kilodaltons. (C) Far-Western blot analysis of immunoprecipitates prepared from HeLa nuclear extracts with control IgG or affinity purified CBP antiserum. CREB protein binding activity was detected with [^{32}P]CREB phosphorylated with PKA. S,P, supernatant and pellet fractions, respectively.

plasmid inhibited cAMP-dependent activity in a dose-dependent manner (Fig. 5A), but control IgG had no effect on this response (Fig. 5B).

To determine whether CBP antiserum inhibited cAMP responsive transcription by binding specifically to CBP, we performed peptide blocking experiments (Arias et al 1994) (Fig. 5B). CBP antiserum, preincubated with synthetic CBP peptide, was unable to recognize the 265 kDa CBP product on a Western blot (not shown) and could not inhibit CRE–lacZ reporter activity upon microinjection into NIH/3T3 cells (Fig. 5B, top). Antiserum treated with an unrelated synthetic peptide (ILS) retained full activity in both Western blots and microinjection assays, suggesting that the ability of the antiserum to bind CBP was critical for its inhibitory effect on cAMP-dependent transcription.

To determine whether CBP activity is restricted to a subset of promoters, we tested several constitutively active reporter constructs (Arias et al 1994): cytomegalovirus, Rous sarcoma virus and SV40. When examined in NIH/3T3 cells by transient transfection assay, each of these constructs had comparable basal activity relative to the cAMP-stimulated CRE reporter plasmid (not

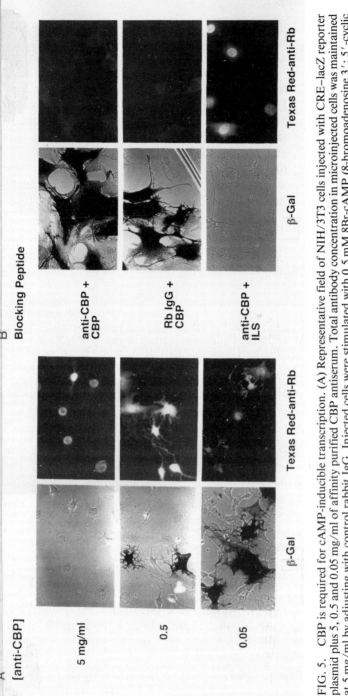

FIG. 5. CBP is required for cAMP-inducible transcription. (A) Representative field of NIH/3T3 cells injected with CRE–lacZ reporter plasmid plus 5, 0.5 and 0.05 mg/ml of affinity purified CBP antiserum. Total antibody concentration in microinjected cells was maintained at 5 mg/ml by adjusting with control rabbit IgG. Injected cells were stimulated with 0.5 mM 8Br-cAMP (8-bromoadenosine 3′:5′-cyclic monophosphate) plus IBMX (3-isobutyl-1-methylxanthine phosphodiesterase inhibitor) for 4 h, then fixed and assayed for lacZ activity (β-GAL) as well as antibody content (Texas Red anti-Rb). (B) Effect of CBP antiserum on CRE–lacZ reporter activity following pretreatment of CBP antiserum with synthetic CBP peptide (anti-CBP + CBP) or unrelated peptide (anti-CBP + ILS). Rb IgG + CBP, control rabbit IgG pretreated with CBP peptide. NIH/3T3 cells were injected with CRE–lacZ reporter plus various CBP antisera, stimulated with 0.5 mM 8Br cAMP plus IBMX for 4 h and assayed for lacZ activity. Cells expressing the lacZ gene product form a blue precipitate which quenches immunofluorescent detection of the injected antibody.

shown), thereby permitting us to compare the effects of CBP antiserum on these reporters directly. Although co-injected CBP antiserum could block cAMP-stimulated activity from a CRE–lacZ reporter in contemporaneous assays, no inhibition was observed on basal expression from any of the constitutive promoter constructs we tested, even when 10-fold lower amounts of reporter plasmid were employed. These results suggest that CBP can indeed discriminate between basal and signal-dependent activities *in vivo*.

With these factors in hand, we can now begin the process of reconstituting PKA-dependent transcription *in vitro*, using well characterized proteins such as CREB protein, TAF110 and CBP. The assembly of such factors on cAMP-regulated promoters like somatostatin may thereby permit responsiveness to a variety of hormonal stimuli which employ cAMP as their second messenger.

Acknowledgements

This work was supported by an NIH grant no. GM37828 and the Foundation for Medical Research, Inc. Marc Montminy is a Foundation for Medical Research investigator.

References

Alberts A, Arias J, Hagiwara M, Montminy M, Feramisco J 1994 Recombinant cyclic AMP response element binding protein (CREB) phosphorylated on Ser133 is transcriptionally active upon its introduction into fibroblast nuclei. J Biol Chem 269: 7623–7630

Arias J, Alberts A, Hua X et al 1994 Activation of cAMP and mitogen responsive genes relies on a common nuclear factor. Nature 370:226–229

Brindle P, Linke S, Montminy M 1993 Analysis of a PKA dependent activator in CREB reveals a new role for the CREM family of repressors. Nature 365:855–858

Chien C, Bartel P, Sternglanz R, Fields S 1991 The two hybrid system: a method to identify and clone genes for proteins that interact with a protein of interest. Proc Natl Acad Sci USA 88:9578–9582

Chrivia JC, Kwok RP, Lamb N, Hagiwara M, Montminy MR, Goodman RH 1993 Phosphorylated CREB binds specifically to the nuclear protein CBP. Nature 365:855–859

Comb M, Burnberg NC, Seascholtz A, Herbert E, Goodman HM 1986 A cyclic-AMP-and phorbol ester-inducible DNA element. Nature 323:353–356

Dynlacht B, Hoey T, Tjian R 1991 Isolation of co-activators associated with the TATA-binding protein that mediate transcriptional activation. Cell 66:563–576

Ferreri K, Gill G, Montminy M 1994 The cAMP regulated transcription factor CREB interacts with a component of the TFIID complex. Proc Natl Acad Sci USA 91:1210–1213

Gill G, Pascal E, Tseng Z, Tjian R 1994 A glutamine hydrophobic patch in transcription factor Sp1 contacts the dTAF$_{II}$110 component of the *Drosophila* TFIID complex and mediates transcriptional activation. Proc Natl Acad Sci USA 91:192–196

Gonzalez GA, Montminy MR 1989 Cyclic AMP stimulates somatostatin gene transcription by phosphorylation of CREB at serine 133. Cell 59:675–680

Gonzalez GA, Yamamoto KK, Fischer WH et al 1989 A cluster of phosphorylation sites on the cyclic AMP-regulated nuclear factor CREB predicted by its sequence. Nature 337:749–752

Gonzalez GA, Menzel P, Leonard J, Fischer WH, Montminy MR 1991 Characterization of motifs which are critical for activity of the cyclic AMP-responsive transcription factor CREB. Mol Cell Biol 11:1306–1312

Hagiwara M, Alberts A, Brindle P et al 1992 Transcriptional attenuation following cAMP induction requires PP-1-mediated dephosphorylation of CREB. Cell 70:105–113

Hagiwara M, Brindle P, Harootunian A et al 1993 Coupling of hormonal stimulation and transcription via cyclic AMP-responsive factor CREB is rate limited by nuclear entry of protein kinase A. Mol Cell Biol 13:4852–4859

Hoeffler JP, Meyer TE, Yun Y, Jameson JL, Habener JF 1988 Cyclic-AMP-responsive DNA-binding protein: structure based on a cloned placental cDNA. Science 242:1430–1432

Hoey T, Weinzieri R, Gill G, Chen J, Dynlacht B, Tjian R 1993 Molecular cloning and functional analysis of Drosophila TAF110 reveal properties expected of coactivators. Cell 72:247–260

Montminy MR, Bilezikjian LM 1987 Binding of a nuclear protein to the cyclic-AMP response element of the somatostatin gene. Nature 328:175–178

Montminy MR, Sevarino KA, Wagner JA, Mandel F, Goodman RH 1986 Identification of a cyclic-AMP responsive element within the rat somatostatin gene. Proc Natl Acad Sci USA 83:6682–6686

Short JM, Wynshaw-Boris A, Short HP, Hanson RW 1986 Characterization of the phosphoenolpyruvate carboxykinase (GTP) promoter-regulatory region. II. Identification of cAMP and glucocorticoid regulatory domains. J Biol Chem 261:9721–9726

Yamamoto KK, Gonzalez GA, Biggs WH IIIrd, Montminy MR 1988 Phosphorylation-induced binding and transcriptional efficacy of nuclear factor CREB. Nature 334:494–498

DISCUSSION

Goodman: So much of what is understood about cAMP regulation of genes has come from studies of the somatostatin promoter and is probably out of proportion to the importance of cAMP in somatostatin regulation. It's become the most clear-cut pathway for activation of transcription. Our lab is in agreement with the studies that Marc Montminy has presented here, with the exception of some minor details.

The idea that the Q2 region of the CREB (cAMP response element-binding) protein works synergistically with the KID (kinase-inducible domain) is very appealing. The results from both Marc Montminy's lab and our lab (Arias et al 1994, Kwok et al 1994) suggest that the primary step involved in CREB protein activation involves CBP (CREB binding protein). These papers indicate that CBP potentiates CREB protein action and show that antibodies against the CREB protein binding domain of CBP block the cAMP response.

The issue on which we have differed is whether phosphorylation induces an allosteric change in the CREB protein, resulting in the presentation of the Q2 region to basal transcription factors. Three lines of evidence have been used to support this allosteric model. Firstly, Marc Montminy's lab have shown that phosphorylation of the CREB protein changes the pattern of tryptic peptide fragments (Montminy et al 1995, this volume). This is probably due to some local change of the phosphate group itself which blocks a local tryptic site; it doesn't necessarily indicate an allosteric change.

Secondly, Nichols et al (1992) have shown that phosphorylation of the CREB protein increases its binding to a low affinity CRE site. This is an interesting finding because the phosphorylation site is in the N-terminus, but the binding domain is in the C-terminus. This would suggest that some structural change has to be transduced in this molecule to change the binding affinity. We've developed a new assay using fluorescence polarization to look at binding of proteins to DNA and proteins to each other under equilibrium conditions (J. Richards, J. R. Lundblad, R. Brennan, H. D. Bachinger, R. Goodman, unpublished results). Using these techniques, we can show binding to low affinity sites, but phosphorylation has no effect on affinity. These results contradict those of Nichols et al (1992), possibly undermining one of the strongest arguments for the allosteric model.

The final support for allosteric changes in CREB protein comes from studies of CREM (CRE modulator), which is a negative regulator of the CRE site and does not have a Q2 region. CREM doesn't work very well as a transcriptional activator and it is not known what model CREM supports.

Direct biophysical studies of CREB protein phosphorylation are just beginning. Analysis of secondary structure using circular dichroism and studies of tertiary structure using equilibrium sedimentation have been performed and neither showed allosteric changes in CREB protein after phosphorylation (R. Goodman, unpublished results). Joe Habener found no changes in the KID region of CREB after phosphorylation (J. Habener, unpublished results).

I think that the most parsimonious explanation is that the activation of CREB protein by phosphorylation occurs through the recruitment of CBP, which is probably independent of anything other than a very local allosteric change. The Q2 region may be required to modulate that interaction or to potentiate the activity of CBP.

Montminy: I agree, with some reservation. When we proposed an allosteric mechanism, there were two proteins, isocitrate dehydrogenase and glycogen phosphorylase, that were known to be regulated by phosphorylation and whose structures had been elucidated at the crystal level. Glycogen phosphorylase is regulated by an allosteric mechanism in response to phosphorylation, whereas isocitrate dehydrogenase is not. Mutagenesis of isocitrate dehydrogenase, replacing the serine phosphoacceptor with aspartate, has the same effect on the activity of the protein as phosphorylation. This is not the case for glycogen phosphorylase, where mutations to aspartate do not substitute for phosphorylation. We did the same experiment with the CREB protein, changing the serine phosphoacceptor to either aspartate or to glutamate, and found that the protein remained inactive. This suggested that the stereochemistry of the phosphate group is very important in the regulation of CREB protein activity and it prompted us to look at its structure by tryptic analysis. We have also performed other enzymatic assays, for instance using elastase which does not cleave at a specific sequence, and we still see some changes, although I would say that those changes are limited to the KID.

What really proves that the allosteric mechanism is not operative is that when the KID and Q2 region are introduced together as heterologous binding domains, they work synergistically to the same extent as when they're within the same molecule. This ability of the two domains to work in *trans* conclusively shows that an allosteric mechanism is not involved. We then modified the model, to suggest that the changes were restricted to the KID; however, the question remains as to whether there are structural changes in that part of the protein which allow it to bind CBP. I don't know whether acidic substitutions in the phosphoacceptor site allow it to bind. Dick Goodman, have you looked at this?

Goodman: Aspartate doesn't allow binding as detected by the fluorescence assay, so there's something special about the phosphate group. We haven't studied this extensively.

Patel: There's little doubt that the somatostatin gene is a prototype for transcriptional regulation by the cAMP/CREB protein pathway; however, there are other promotor interactions which we have not mentioned. For instance, glucocorticoids, GH (growth hormone), cytokines and NMDA (N–methyl–D–aspartate) receptor agonists are all known to activate the somatostatin gene. Even chronic exposure of brain cell cultures with depolarizing concentrations of K^+ lead to increases in somatostatin mRNA levels via a Ca^{2+}-dependent pathway (Tolon et al 1994). Is the cAMP/CREB protein–DNA interaction the final common pathway for transcriptional control of the somatostatin gene by all of these agents or do you see other promoter interactions?

Montminy: K^+ seems to be more cell type restricted than Ca^{2+} in normal cell types. This appears to act through a similar mechanism (through cAMP) but perhaps using a different kinase, possibly the calmodulin-dependent kinase system. These studies have been done very elegantly by a number of groups such as Michael Greenberg's, although he hasn't looked at somatostatin (Sheng et al 1991). Does the glucocorticiod receptor bind to sequences in the somatostatin promotor?

Patel: No. We reported that glucocorticoids activate somatostatin gene transcription via a cooperative interaction with the cAMP signalling pathway (Liu et al 1994).

Montminy: Was that through the CRE?

Patel: Yes. Glucocorticoid-induced transactivation appears to be mediated via protein–protein interactions between the glucocorticoid receptor and the CREB protein.

Reisine: Marc Montminy, you showed an action potential in relation to the regulation of gene expression and mentioned both the subsequent refractory period and also the refractory period to further stimulation (Fig. 1). What do you think is causing this? Does it represent phosphorylation or protein interactions?

Montminy: There are at least four groups of serine phosphatases, called serine PP1, PP2A, B and C, that are distinguished by their ability to be inhibited by okadaic acid. We found that PP1 seemed to have a primary role during the attenuation phase. If PP1 activity is inhibited, maximal levels of CREB protein phosphorylation are maintained and CREB protein activity is potentiated correspondingly. The refractory period has been more difficult to characterize. It seems to be more cell specific than the mechanism used to maintain refractoriness. We have investigated this in the rat thyroid cell line FRTL-5 which is dependent on cAMP for proliferation and differentiation (M. Montminy & R. Armstrong, unpublished results). When these cells are stimulated with forskolin, CREB protein activity increases via the same mechanism as in PC12 cells. If these cells are maintained for up to 20 h in forskolin and re-stimulated with forskolin after removing the medium, a peak or similar rise is obtained that is the reverse of cAMP accumulation. Phosphorylation of CREB protein is not observed in response to that stimulation but somatostatin transcription is. We have been able to trace this down to a loss of a catalytic subunit protein. Western blotting indicates that the levels of catalytic subunit decrease progressively and after about five hours are about five to sevenfold lower than they are in unstimulated cells. This loss reflects a decrease in the translation of the protein which we have determined by pulse–chase studies and suggests that there is some translational repression of protein synthesis during that period. When forskolin is removed, a recovery period ensues which can last about three days. During this time, the levels of catalytic subunit protein accumulate and prevent the cell responding. I don't know how important this is biologically in terms of shutting down the cells because these are maximal signals to the cell, whereas physiologically most signals are transient. It is possible, however, that the shape of the curve is affected by these signals. For instance, high levels of TSH (thyroid-stimulating hormone) caused by chronic TSH stimulation, result in a loss of responsiveness to further stimulation. This may occur not only at the receptor level but also at the nuclear level.

Goodman: So if the catalytic subunit is injected into these cells, a refractory period is not observed.

Montminy: Exactly. The refractory period is bypassed either by injecting catalytic subunit or by transfecting an expression plasmid encoding the subunit without the sequences that are important for translational repression.

Stork: Is there a corresponding refractory period that limits the release of synthesized somatostatin?

Montminy: We haven't checked what happens to somatostatin release. It is possible that the cytoplasmic events have a different sensitivity because the amount of protein kinase A (PKA) that's required to enter the nucleus is different from the amount that is present in the cytoplasm. cAMP stimulation of somatostatin release may be less sensitive, so it may not show as pronounced a refractory period.

Reichlin: How does this relate to growth stimulation by cAMP? The markers that you're following are the synthesis rate and secretion rate. Is there an effect on cell proliferation and hyperplasia through the cAMP mechanism? If so, how does it work?

Montminy: That's more complicated. In cells such as fibroblasts, proliferation is inhibited through a MAPK (microtubule-associated protein kinase) inhibitory mechanism: PKA directly phosphorylates MAPK which inhibits the ability of the signalling system to induce proliferation. In thyrotrophs and somatotrophs, however, the situation is different. cAMP is known to regulate the proliferation of these cell types and the CREB protein may also be involved. Evidence supporting this includes the observation that GH levels are low when a mutant isoform of CREB protein, containing a serine to alanine substitution at the phosphorylation site, is introduced into somatotroph cells of transgenic mice and overexpressed using a GH promoter (Struthers et al 1991). The number of somatotrophs is also reduced to about one tenth of normal. This is probably not a toxic effect because this mutant CREB protein can also be expressed in lactotrophs without affecting their numbers. The effect is therefore very specific for cells which are known to be regulated by cAMP. This suggests that the CREB protein may have a biochemical intermediate not only in the transcription pathway but also in the proliferative pathway.

Stork: cAMP can both stimulate and inhibit proliferation in different cells; it can also both inhibit and stimulate MAPK activity, suggesting that MAPK activation may be both stimulatory and inhibitory. For example, in fibroblasts cAMP has been shown to inhibit MAPK activation and thereby inhibit growth, whereas in PC12 cells cAMP stimulates MAPK activation, yet inhibits growth. It is possible that the duration of MAPK activation may dictate the growth effects, i.e. sustained activation of MAPK may be inhibitory. In somatotrophs, at different stages of development or different stages of malignancy, cAMP may have opposite effects. For example, it inhibits the growth of GH_4C_1 rat pituitary cells. This is in contrast to its actions in experiments performed by Marc Montminy using developing somatotrophs as well as the transgenic model. In addition, natural mutations of $G_{s\alpha}$ have been identified in pituitary adenomas suggesting stimulatory effects of cAMP production on growth. These effects through MAPK are probably independent of the CREB protein and may involve phosphorylation of kinases upstream of MAPK in the cascade. Each cell may have a different programme for responding to cAMP to meet the needs of that cell.

Lamberts: Koper et al (1992) demonstrated that meningiomas, which are tumours of the human blood–brain barrier are also under control of cAMP. High cAMP levels in cultured human meningioma cells are correlated with low mitotic activity. Human meningiomas also express a high number of somatostatin receptors. Exposure of cultured human meningioma cells to somatostatin results in inhibition of adenylyl cyclase activity, a decrease in intracellular cAMP levels and stimulation of growth. This is unexpected because inhibition of intracellular cAMP levels by somatostatin in human endocrine tumours is, in most instances, followed by a decrease in hormone secretion and in mitotic activity (Hofland et al 1992). This suggests that different somatostatin receptor-positive tumours respond differently to somatostatin with regard to the effect on growth rate. This puzzling situation is further illustrated by recent studies on the antiproliferative effects of somatostatin analogues in a number of human thyroid carcinoma cell lines (Ain & Taylor 1994). In these studies, biphasic dose–response curves were demonstrated on the growth of a number of tumour cell lines by different somatostatin analogues, suggesting that the presence of distinct somatostatin receptors might result in differential stimulation and inhibition of tumour growth.

References

Ain KB, Taylor KD 1994 Somatostatin analogs affect proliferation of human thyroid carcinoma cell lines in vitro. J Clin Endocrinol Metab 78:1097–1102

Arias J, Alberts AS, Brindle P et al 1994 Activation of cAMP and mitogen responsive genes relies on a common nuclear factor. Nature 370:226–229

Hofland LJ, van Koetsfeld PM, Wouters N, Waaijers M, Reubi J-C, Lamberts SWJ 1992 Dissociation of antiproliferative and antihormonal effects of the somatostatin analog octreotide on 7315b pituitary tumor cells. Endocrinology 131:571–577

Koper JW, Markstein R, Kohler C et al 1992 Somatostatin inhibits the activity of adenylate cyclase in cultured human meningioma cells and stimulates their growth. J Clin Endocrinol Metab 74:543–547

Kwok RPS, Lundblad JR, Chrivia JC et al 1994 Nuclear protein CBP is a coactivator for the transcription factor CREB. Nature 370:223–226

Liu J-L, Papachristou DN, Patel YC 1994 Glucocorticoids activate somatostatin gene transcription through cooperative interaction with the cyclic AMP signaling pathway. Biochem J 301:863–869

Nichols M, Weih F, Schmid W (1992) Phosphorylation of CREB affects its binding to high and low affinity sites—implications for cAMP-induced gene transcription. EMBO J 11:3337–3346

Sheng M, Thompson MA, Greenberg ME 1991 CREB: a Ca^{2+}-regulated transcription factor phosphorylated by calmodulin-dependent kinases. Science 252:1427–1430

Struthers RS, Vale WW, Arias C, Sawchenko PE, Montminy MR 1991 Somatotroph hypoplasia and dwarfism in transgenic mice expressing a non-phosphorylatable CREB mutant. Nature 350:622–624

Tolon RM, Sanchez-Franco F, de los Frailes MT, Lorenzo MJ, Cacicedo L 1994 The effect of potassium-induced depolarization on somatostatin gene expression in cultured fetal rat cerebrocortical cells. J Neurosci 14:1053–1059

Processing and intracellular targeting of prosomatostatin-derived peptides: the role of mammalian endoproteases

Yogesh C. Patel and Aristea Galanopoulou

Fraser Laboratories, Departments of Medicine, Neurology and Neurosurgery, McGill University, Royal Victoria Hospital and Montreal Neurological Institute, Montreal, Quebec, H3A 1A1, Canada

Abstract. Prosomatostatin is cleaved at dibasic and monobasic sites to produce somatostatin-14 and somatostatin-28 respectively. The mammalian pro-protein convertases comprising furin, PACE4 and PC1–6 have recently been identified and are believed to mediate endoproteolysis of prohormone precursors such as prosomatostatin. Furin is membrane bound, localized to the Golgi and mediates constitutive processing. PC1 and PC2 are soluble and are expressed solely in endocrine and neuroendocrine tissues suggesting a key role in prohormone processing. We have investigated the endogenous and heterologous synthesis and processing of rat prosomatostatin in $1027B_2$ rat islet somatostatinoma cells and in constitutive (COS-7, PC-12) and regulated (AtT-20, GH_3/GH_4C_1) secretory cells. We have correlated processing efficiency with: secretion through the constitutive or regulated pathways; endogenous expression of furin, PC1 and PC2; and expression or overexpression of furin, PC1 and PC2. Pulse–chase studies showed that prosomatostatin is rapidly and independently processed to somatostatin-14 and somatostatin-28. Furin is capable of monobasic processing of prosomatostatin and is a candidate somatostatin-28 convertase. PC1 and PC2 both effect dibasic processing of prosomatostatin and qualify as putative somatostatin-14 convertases. PC1 is active in constitutive and regulated secretory cells, has a broader specificity and is overall more potent than PC2. Efficient processing of prosomatostatin begins in a Golgi or pre Golgi compartment. It requires the milieu of the secretory cell but not the secretory granule.

1995 Somatostatin and its receptors. Wiley, Chichester (Ciba Foundation Symposium 190) p 26–50

Somatostatin, a multifunctional peptide, is produced in neurons and secretory cells in a variety of tissues notably brain, pancreatic islets, gut and thyroid. It regulates key cellular processes such as cell secretion, neurotransmission, smooth muscle contractility and cell proliferation (Patel 1992). There are two biologically active forms of somatostatin—somatostatin-14 and somatostatin-28. In mammals, these two products are generated by endoproteolytic processing of prosomatostatin at two distinct regions at the C-terminal segment of the molecule (Fig. 1). Cleavage at a dibasic Arg–Lys site produces somatostatin-14 and a 8 kDa peptide whereas cleavage at a monobasic Arg site releases somatostatin-28 and a 7 kDa form. In addition, there is a second monobasic Lys processing site at the N-terminal segment of the precursor which generates the decapeptide antrin, or prosomatostatin$_{[1-10]}$, a molecule without any known biological activity. In lower vertebrates such as fish, there are two distinct somatostatin genes expressed in separate subpopulations of cells which encode for separate somatostatin-14 and somatostatin-28 related products (reviewed in Patel 1992). Mammals, on the other hand, are characterized by a single prosomatostatin molecule which undergoes differential processing to give rise to tissue-specific amounts of somatostatin-14 and somatostatin-28. For instance, the hypothalamus and cerebral cortex synthesize both somatostatin-14 and somatostatin-28 in an approximate molar ratio of 4:1; whereas the stomach, pancreatic islets, enteric neurons and retina produce almost entirely somatostatin-14 (Patel et al 1981, 1985). In contrast, intestinal mucosal cells synthesize somatostatin-28 as the principal terminal product. The sequence of somatostatin-14 and prosomatostatin$_{[1-10]}$ is conserved between fish and mammals indicating a remarkable structural conservation at both the C- and N-termini of prosomatostatin.

Little is known of the role of prohormone converting enzymes and subcellular compartmentalization of processing events—two key steps in prosomatostatin maturation. Secretory proteins like somatostatin are synthesized as precursors on ribosomes, translocated into the lumen of the endoplasmic reticulum and transported through budding coated vesicles through the Golgi stacks to the trans Golgi network (Rothman & Orci 1992). Here the protein is sorted via clathrin-coated vesicles into a regulated pathway consisting of secretory granules that are stored pending release, or via nonclathrin-coated vesicles into a constitutive, or bulk flow, pathway. These nonclathrin-coated vesicles then migrate to the plasma membrane (Rothman & Orci 1992). The recent cloning of several members of an expanding family of subtilisin-related serine convertases (SPCs) has shed some light on their role in pro-protein processing (Halban & Irminger 1994, Seidah et al 1993, Steiner et al 1992). Six members of this family have been identified in mammals: furin (PACE); PC1/PC3; PC2; PACE4; PC4; and PC5/PC6. Furin is a membrane-bound enzyme that is expressed in most tissues and cells (Halban & Irminger 1994). It is a resident of the trans Golgi compartment and mediates processing of constitutively secreted proteins and membrane glycoproteins typically at an RXK/RR site (Hasaka et al 1991, Molloy et al 1994, Watanabe et al 1992). In contrast, PC1 and PC2 are not membrane bound and are targeted to secretory vesicles. They cleave at dibasic residues and are expressed solely in endocrine and neuroendocrine tissues including the brain, suggesting a key role in the maturation of neuropeptide and prohormone precursors (Halban & Irminger

1994, Schafer et al 1993, Seidah et al 1993, Zhou et al 1993). PC4 has so far been identified only in germ cells in the testis, whereas PC5 and PC6 are expressed broadly in many tissues. Their role in prohormone processing has not been determined (Halban & Irminger 1994).

Methods

The following cell lines were obtained and cultured as previously described (Galanopoulou et al 1993, Patel et al 1991): $1027B_2$, somatostatin-producing cells derived from a transplantable rat islet cell tumour; COS-7, African green monkey kidney cells; PC-12, rat adrenal phaeochromocytoma cells; AtT-20, mouse pituitary tumour cells; GH_3/GH_4C_1, rat pituitary mammosomatotroph cells.

The following expression vectors were used: pKS5, rat preprosomatostatin cDNA under the control of a cytomegalovirus (CMV) promoter (from K. Sevarino); pPES, rat preprosomatostatin cDNA in pRC/CMV; pmPC1, mouse PC1 cDNA in pRC/CMV; pmPC2, mouse PC2 cDNA in pRC/CMV; pMJ601: rPSS, rat prosomatostatin cDNA under the control of a late vaccinia virus (vv) promoter; W:wt, wild type vv; W:hfurin, vv expressing human furin cDNA.

Cells were grown in monolayer culture, transiently transfected by calcium phosphate precipitation and studied after 72 h. Stable GH_4C_1 cell lines were established following Lipofectin (Gibco–BRL) transfection with pPES, pKS5 and pmPC1, or pKS5 and pmPC2. They were selected with 300 μg/ml genticin (G418, Gibco–BRL) starting on the third post-transfection day. Vaccinia infections were carried out in COS-7 cells infected with W:wt (2.4 or 12 plaque-forming units [pfu]/cell) or W:hfurin (0.8 or 2.4 pfu/cell) followed by transfection with pMJ601:rPSS.

$1027B_2$ cells, transfected cells or vaccinia-infected cells were incubated for 4 h in serum free media containing phenylmethylsulphonyl fluoride (PMSF) and pepstatin A. Regulated secretion was tested by stimulation with 20 μM forskolin and 10 mM dibutyryl cyclic AMP (dbcAMP). Cell extracts in 1 M acetic acid and media were harvested for analysis of somatostatin-like immunoreactivity (somatostatin-LI) by high performance liquid chromatography (HPLC) and radioimmunoassay (RIA). For measurement of cellular PC1 and PC2 immunoreactivities (PC1-LI, PC2-LI), cells were extracted in 50 mM Tris–HCl, 150 mM NaCl, 0.05% SDS and 1% NP40 buffer containing protease inhibitors.

Somatostatin RIAs were performed with the following antisera: R149, directed against somatostatin-14 and detects somatostatin-14, somatostatin-28 and prosomatostatin equally (Galanopoulou et al 1993); R203, directed against prosomatostatin$_{[1-10]}$ and detects prosomatostatin$_{[1-10]}$ and its C-terminally extended forms (prosomatostatin, 8 kDa, 7 kDa peptides) (Galanopoulou et al 1993).

PC1 RIA was performed using a rabbit antibody against synthetic PC1$_{[84-100]}$ peptide (I. Lindberg), $[^{125}I]$Tyr PC1$_{[84-98]}$radioligand and

PC1$_{[84-100]}$standards (Galanopoulou 1993). The minimum detectable dose was ≤1.9 pg.

PC2-LI was measured with the antibody 1159 (gift of Drs Mains and Eipper) which was raised against the N-terminal tyrosinated mouse PC2$_{[627-636]}$peptide. This peptide was also used for iodination and PC2 standards. The minimum detectable dose was ≤8 pg standard.

HPLC was performed using a C18 μBondapak column (Waters, Milford, MA). Material was eluted with 12–55% CH$_3$CN (acetonitrile), 0.2% HFBA (heptafluorobutyric acid). Aliquots of the eluate were evaporated and analysed by RIA (Galanopoulou et al 1993).

Northern blots were performed using 20 μg total RNA and cRNA probes for prosomatostatin, furin, PC1 and PC2 as described (Galanopoulou et al 1993).

Prosomatostatin is rapidly and independently processed to somatostatin-14 and somatostatin-28

The kinetics of somatostatin-14 synthesis were first studied in hypothalamic and cerebrocortical neurons and in islet cells by pulse–chase experiments (Patel et al 1985, Robbins & Reichlin 1983, Zingg & Patel 1982). Somatostatin-14 synthesis begins within five minutes of pulse labelling and proceeds through direct cleavage of prosomatostatin, the main route for somatostatin-14 synthesis in these tissues. Labelling of somatostatin-28 in the normal tissues studied in these experiments was not sufficient to draw any firm conclusions about its synthesis from prosomatostatin.

The recent availability of 1027B$_2$ islet somatostatin tumour cells has allowed us to elucidate the kinetic relationship between prosomatostatin, somatostatin-28 and somatostatin-14. Cultured 1027B$_2$ cells were incubated with [^{3}H]Pro, [^{3}H]Leu, and [^{3}H]Phe (100 μCi/ml each). Cell extracts were purified with C18 silica cartridges and labelled prosomatostatin peptides isolated by separate affinity columns using R149 and R203 antisera. These were then fractionated by reverse phase HPLC. The size of the labelled peptide peaks was expressed as total counts/min (c.p.m.) per number of labelled amino acids. Following a 15 min pulse incubation, prosomatostatin was the principal labelled peak accounting for 87% of total somatostatin-14-LI synthesis. Small but definite conversion to somatostatin-14 (8%) and somatostatin-28 (5%) was already evident at this time. During subsequent chase, extensive conversion of prosomatostatin to both somatostatin-14 (59%) and somatostatin-28 (34%) occurred in parallel. Simultaneous analysis of the labelled products for 7 kDa and 8 kDa forms (Fig. 1) showed that conversion of prosomatostatin to 8 kDa and 7 kDa peptides occurred concomitant with the synthesis of somatostatin-14 and somatostatin-28.

These results suggest that prosomatostatin is cleaved at its dibasic site to generate somatostatin-14 and the 8 kDa peptide, and at the C-terminal

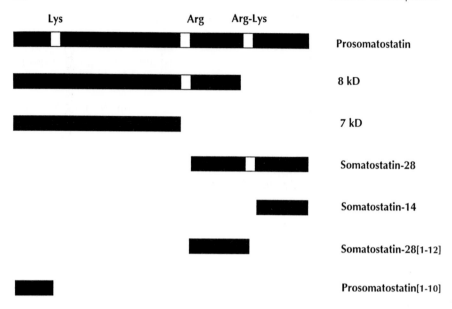

FIG. 1.   Schematic depiction of mammalian prosomatostatin, its dibasic and monobasic processing sites, and known cleavage products.

monobasic cleavage site to yield somatostatin-28 and the 7 kD peptide. Furthermore, these two processes occur independently in parallel. The rapid time course of somatostatin-14 and somatostatin-28 synthesis suggests that conversion of both peptides is initiated early, probably in a Golgi or pre Golgi compartment.

Heterologous processing of prosomatostatin in constitutive and regulated cells

As a first step towards elucidating a putative role of the SPCs in prosomatostatin maturation, we compared the heterologous processing of prosomatostatin in both endocrine (GH$_3$, AtT-20) and constitutively secreting (COS-7, PC-12) tumour cells using cDNA transfection experiments (Galanopoulou et al 1993). Processing at dibasic and monobasic sites was correlated with endogenous expression of mRNA for furin, PC1 and PC2. A 4.5 kb furin mRNA transcript was found in all four tumour cell lines, but with a high level of expression in PC-12 cells (Galanopoulou et al 1993). In the case of PC1, two transcripts of 5 and 3 kb were observed only in AtT-20. Northern blots of PC2 mRNA revealed a 2.8 kb transcript in AtT-20 and GH$_3$ cells.

COS-7 cells secreted more somatostatin-LI than they stored under basal conditions, showed no response to standard secretagogues (forskolin, dbcAMP)

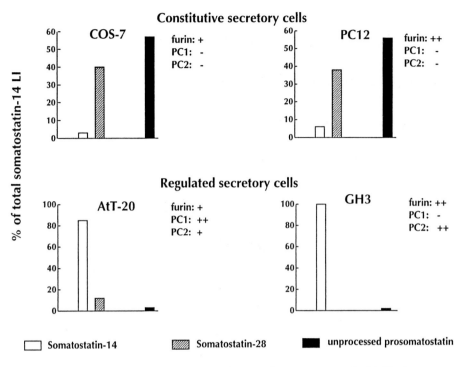

FIG. 2. Pattern of processing of prosomatostatin to somatostatin-14-like immunoreactivity (somatostatin-14 LI) forms in COS-7, PC-12, AtT-20 and GH₃ cells transfected transiently with rat prosomatostatin cDNA. Per cent processing to somatostatin-14, somatostatin-28 and unprocessed prosomatostatin is calculated from representative HPLC runs of cell extracts. Expression of furin, PC1 and PC2 in the four cell lines is based on Northern analysis of corresponding mRNAs.

and processed rat prosomatostatin mainly to somatostatin-28 (40% of total somatostatin-14-LI in cells, 10% in medium) and small but detectable amounts of somatostatin-14 (2% in cells, 7% in medium). Processing, however, was inefficient based on the presence of high concentrations of unprocessed prosomatostatin (about 60% of total somatostatin-14-LI) (Fig. 2).

PC-12 cells were similar to COS-7 cells in that they exhibited inefficient constitutive processing of prosomatostatin to predominantly somatostatin-28 (40% of total somatostatin-14-LI in cell extracts, 50% in medium) and a small per cent of somatostatin-14 (4% of somatostatin-14-LI in cells, 3% in medium) (Fig. 2).

In contrast to COS-7 and PC-12 cells, AtT-20 cells not only targeted prosomatostatin-derived peptides to the regulated secretory pathway but also processed prosomatostatin efficiently to both somatostatin-14 and somatostatin-28 (85% and 12% of total somatostatin-14-LI respectively in cell extracts) (Fig. 2). Unprocessed prosomatostatin represented only 3% of total cellular somatostatin-14-LI.

GH$_3$ cells synthesized and stored somatostatin-14-LI and responded to secretagogues with stimulated release. Rat prosomatostatin was almost entirely (about 97%) processed to somatostatin-14 in these cells (Fig.2). Analysis of N-terminal prosomatostatin immunoreactivity showed significant conversion of prosomatostatin to prosomatostatin$_{[1-10]}$ (27 and 29% respectively in COS-7 and PC-12 cells) and the 7 kDa peptide (prosomatostatin$_{[1-64]}$) comprising 16% and 14% of total immunoreactivity in COS-7 and PC-12 cells respectively.

These results provide clear evidence of endoproteolytic activity in the constitutive pathway of COS-7 and PC-12 cells capable of processing prosomatostatin at its monobasic cleavage sites at Arg^{64} and Lys^{13} to release somatostatin-28 and prosomatostatin$_{[1-10]}$. The finding of significant amounts of the 7 kDa peptide co-processed in these cells gives further support for monobasic processing at the Arg^{64} residue since cleavage at this locus would yield both somatostatin-28 and the 7 kDa peptide (Fig.1). Such monobasic processing could be effected by furin, the only known SPC expressed in these cells. In contrast, the endogenous expression of PC1 and PC2 correlates with efficient dibasic processing in At-20 and GH$_3$ cells. The observations of a high level of PC2 expression and virtually complete dibasic processing of prosomatostatin to somatostatin-14 in both GH$_3$ and AtT-20 cells suggests that PC2 is a candidate somatostatin-14 convertase. In addition, AtT-20 cells coexpress high levels of PC1 mRNA and protein, raising the possibility that PC1 also mediates somatostatin-14 conversion.

Furin is a putative somatostatin-28 convertase

Furin cleaves pro-proteins typically at R–X–K/R–R or R–X–X–R motifs (Halban & Irminger 1994, Hosaka et al 1991, Watanabe et al 1992). Structure–function studies have revealed an even broader specificity conforming to the following rule: Arg at position -1 is essential and, in addition, at least two out of three basic residues at positions -2, -4 and -6 are required (Watanabe et al 1992). Such a specificity profile shows considerable overlap with the structural requirements for processing at monobasic cleavage sites proposed by Devi (1991). These criteria predict the monobasic cleavage motifs Arg–Leu–Glu–Leu–Gln–Arg and Arg–Gln–Phe–Leu–Gln–Lys at the site of somatostatin-28 and prosomatostatin$_{[1-10]}$ processing respectively. Except for the presence of Lys rather than Arg at the site of prosomatostatin$_{[1-10]}$ conversion, both these sequences possess the necessary substrate specificity for catalysis by a furin-like enzyme.

To determine the role of furin in monobasic prosomatostatin processing, we compared the heterologous processing of rat prosomatostatin in COS-7 cells overexpressing human furin using vaccinia furin infection experiments. COS-7 cells were infected with different doses of either vv:wt (2.4 or 12 pfu/cell) or vv:hfurin (0.8 or 2.4 pfu/cell) and then transfected with the plasmid pMJ 601:rPSS. After 20 h, cells were incubated for 4 h in secretion media, then cells and media were collected for HPLC analysis of somatostatin-14, somatostatin-28 and unprocessed prosomatostatin. Expression of human furin was confirmed by Northern blots hybridized with the rat furin probe and showed only endogenous furin mRNA in control cells whereas W:hfurin infected cells expressed dose-dependent amounts of a second 3 kb mRNA transcript corresponding to human furin mRNA. Coexpression of W:wt and prosomatostatin resulted in a comparable pattern of prosomatostatin processing to that observed in our earlier transfection studies (Fig. 2), i.e. processing mainly to somatostatin-28 (30%), small amounts of somatostatin-14 (5%) and 60% unprocessed prosomatostatin. This excludes artifacts produced by viral infection. Infection with human furin at 0.8 pfu/cell resulted in a dose-dependent increase in somatostatin-28 conversion (42% and 41% of somatostatin-14-LI in cells and media respectively). Infection at 2.4 pfu/cell also resulted in a dose-dependent increase in somatostatin-28 conversion (58% and 32% of somatostatin-14-LI in cells and media respectively). These were accompanied by corresponding decreases in the amount of unprocessed prosomatostatin, especially in cell extracts where it dropped from 70% in controls to 22% at the higher 2.4 pfu/cell infection. The higher level of furin infection also produced a significant increase in somatostatin-14 conversion up to 20%. There was no increase in prosomatostatin$_{[1-10]}$ conversion.

These results indicate that furin can mediate monobasic cleavage of prosomatostatin to somatostatin-28 and thus qualifies as a somatostatin-28 converting enzyme. Furin does not increase prosomatostatin$_{[1-10]}$ conversion suggesting that another endoprotease with preference for Lys rather than Arg at position −1 may be responsible for N-terminal monobasic prosomatostatin processing. The requirement of furin for Arg at position −1 is further confirmed by the low enzyme affinity for the somatostatin-14 cleavage domain (R–X–R–K), a typical furin motif except for Lys substituting for Arg at position −1.

PC1 but not PC2 effects dibasic prosomatostatin processing in constitutive cells

Since furin was the only SPC expressed in COS-7 cells, the effect of PC1 and PC2 cDNA on prosomatostatin processing was determined in separate co-transfection experiments with expression vectors for prosomatostatin and either PC1 or PC2 (Galanopoulou et al 1993). The level of expression of the two endoprotease genes was determined by Northern analysis for PC1 and PC2 mRNA (Fig. 3A). Immunoblot analysis of cell extracts with PC2 antibody revealed a single 75 kDa protein band corresponding to the precursor form of PC2. The level of expression of PC1 was determined by RIA of cell extracts

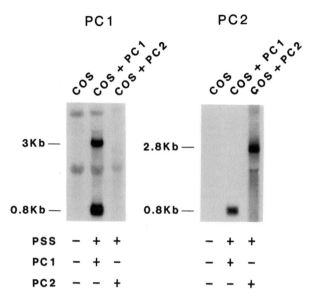

FIG. 3A. Northern blot analysis of total RNA from COS-7 cells co-transfected with cDNAs for prosomatostatin (PSS) and either PC1 or PC2. Separate autoradiograms were obtained with cRNA probes for PC1 (left hand side lanes) and PC2 (right hand side lanes). PC1 or PC2 transfected COS-7 cells show single specific mRNA bands of 3 kb (PC1 mRNA) and 2.8 kb (PC2 mRNA), respectively. The additional 0.8 kb band observed in PC1 transfected cells is detectable with both enzyme probes and appears to be non-specific.

and media and compared with endogenous PC1 expression in AtT-20 cells. PC1 transfected COS-7 cells showed undetectable levels of PC1-LI in cell extracts but secreted 272 ± 42 pg/dish during the post-transfection period. In comparison, the same number of AtT-20 cells stored 546 ± 94 pg of PC1-LI and secreted 1720 ± 120 pg of PC1-LI. HPLC analysis of cell extracts and media (schematically depicted in Fig. 3B) showed a significant increase in the per cent of somatostatin-14 processed from $3.5 \pm 2\%$ to $13.7 \pm 3\%$ in cell extracts and from 7 ± 1 to $31 \pm 5\%$ in media as a result of co-transfection with prosomatostatin and PC1 cDNAs compared to the control (prosomatostatin alone). There was no net change in somatostatin-28 processing as a result of PC1 transfection. In contrast to PC1, co-transfection of COS-7 cells with PC2 and prosomatostatin cDNAs did not influence the efficiency nor pattern of processing of prosomatostatin to somatostatin-14 and somatostatin-28 compared to the control (Fig. 3B).

These results indicate that PC1 is capable of dibasic processing of prosomatostatin and is a candidate somatostatin-14 convertase. Furthermore, because COS-7 cells are constitutively secreting, these findings suggest that proteolytic

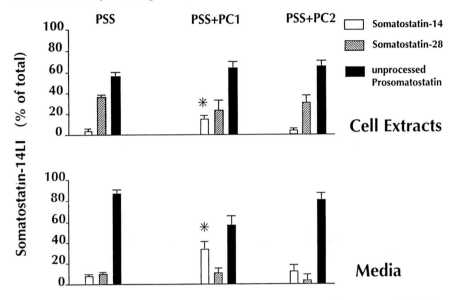

FIG. 3B.   Comparison of per cent processed somatostatin-14 and somatostatin-28 with unprocessed prosomatostatin (PSS) in COS-7 cells transfected with prosomatostatin alone, co-transfected with prosomatostatin and mouse PC1 or co-transfected with prosomatostatin and mouse PC2. SS-14LI, somatostatin-14-like immunoreactivity. Mean data ± SE (n = 3). *p < 0.05 compared to prosomatostatin control. Coexpression of PC1 with prosomatostatin results in a significant increase in the per cent of somatostatin-14 processed in cell extracts and secretion media. Per cent somatostatin-28 processed is unaffected by PC1 co-transfection. PC2 is without effect on the per cent of somatostatin-14 or somatostatin-28 processed compared to control.

cleavage by PC1 begins in a Golgi or pre Golgi compartment proximal to the formation of secretory granules corroborating our kinetic data from pulse–chase analyses.

Expression of PC1 mRNA in rat brain by *in situ* hybridization correlates with the regional brain distribution of somatostatin especially in areas rich in somatostatin neurons such as the hypothalamic periventricular nucleus, lending support to a physiological role of PC1 in prosomatostatin processing in neural cells (Schafer et al 1993). In the case of PC2, several indirect lines of evidence predict that this enzyme might also qualify as a somatostatin-14 convertase. GH$_3$ cells and rat pancreatic islet D cells express PC2 rather than PC1 and possess heterologous or endogenous prosomatostatin almost entirely to somatostatin-14 (Fig. 2) (Marcinkiewicz et al 1994, Neerman-Arbez et al 1994, Stoller & Shields 1988). A somatostatin-14 convertase isolated from anglerfish islets appears to be the fish homologue of mammalian PC2 (Mackin et al 1991). Recent studies of the biosynthesis of SPCs have shown that PC2 (and PC1) undergo extensive post-translational modifications, resulting in several molecular forms with differential activity, intracellular storage and secretion (Benjannet et al 1993, Shen et al 1993, Vindrola & Lindberg 1992). Our inability to demonstrate somatostatin-14 conversion by PC2 in COS-7 cells can probably be explained by the inability of constitutive cells such as COS-7 cells to process PC2 from the inactive 75 kDa proPC2 form to the active 68 kDa form found in regulated secretory cells (Benjannet et al 1993). Furthermore, PC2 is known to be activated relatively slowly in secretory cells requiring 1–2 h from the time of synthesis compared to the more rapid 30 min activation time for PC1 (Benjannet et al 1993). Based on these results, it is probable that the inability of PC2 to effect dibasic processing of prosomatostatin in COS-7 cells is due to enzyme inactivity which might be restored in regulated cells.

PC2 mediates dibasic prosomatostatin processing in regulated cells

We next investigated processing of prosomatostatin coexpressed with PC2 in stably transfected GH_4C_1 cells. Although related to GH_3 cells, GH_4C_1 cells express 20-fold lower levels of PC2 mRNA compared to GH_3 cells. They also exhibit poor granularity, but respond to hormonal treatment with 10 nM epidermal growth factor (EGF), 1 nM 17β-oestradiol and 300 nM insulin with a 50-fold increase in granule number (Scammell et al 1986).

We first determined the effect of overexpression of PC2 on somatostatin-14 conversion in these cells. GH_4C_1:rPSS stable transfectants showed high basal secretion of somatostatin-14-LI (42% of cell content/four hours) and processed prosomatostatin relatively inefficiently to both somatostatin-14 and somatostatin-28. Stable co-transfection of GH_4C_1 cells with prosomatostatin and PC2 led to increased somatostatin-14 conversion especially in cell extracts. Somatostatin-14 conversion increased from 36% of total somatostatin-14-LI to 53% concomitant with a reduction in the level of unprocessed prosomatostatin (from 39% to 26%). This experiment demonstrates that although PC2 is inactive in COS-7 cells, it can certainly mediate dibasic cleavage of prosomatostatin to somatostatin-14 in an endocrine cell such as GH_4C_1. The overall efficiency of this event, however, is much less than that found in GH_3 cells suggesting that either higher levels of PC2 or induction of secretion granules is a further requirement.

Granule induction fails to improve prosomatostatin processing

We next investigated the effect of hormone-induced granulation in GH_4C_1:rPSS:PC2 stable transfectants on prosomatostatin processing efficiency. Cells were treated with 10 nM EGF, 1 nM 17β-oestradiol or 300 mM insulin for four days and studied on day five when four hour secretion media and cell extracts were harvested for analysis of somatostatin-14-LI and PC2-LI. This hormone treatment led to a 10-fold increase in intracellular somatostatin-14-LI.

In the case of PC2-LI, there was a small (25%), but not significant, increase in cell content of this enzyme. This difference in the storage pattern of somatostatin-14-LI and PC2-LI can be explained by recent findings characterizing the biosynthesis of PC2 in GH_4C_1 cells (Benjannet et al 1993). Most of the immunologically detected intracellular PC2-LI consists of proPC2 forms which reside in the endoplasmic reticulum and the trans Golgi network. Only a small per cent of proPC2 is processed and targeted to the regulated pathway as the mature PC2 form. This is in contrast to prosomatostatin and its cleavage products which are typically targeted to secretion granules in regulated cells.

Basal secretion of somatostatin-14-LI increased fourfold following granule induction and by a further 40% in response to forskolin. Likewise, there was a 50% stimulation of PC2-LI release in response to forskolin. These results demonstrate the existence of a co-regulated storage compartment of both somatostatin-14-LI and PC2-LI.

HPLC analysis of cell extracts and secretion media before and after granule induction showed no change in the pattern of prosomatostatin processing. Somatostatin-14 accounted for 53% and 47% of total somatostatin-14-LI in treated and untreated cell extracts. In secretion media, somatostatin-14 represented only 25% of total somatostatin-14-LI following treatment with hormones compared to 37% in control cells. These results indicate that although the activity of PC2 requires the secretory cell, there is no absolute requirement for secretory granules.

Further evidence excluding a role of secretory granules in prosomato-statin processing was obtained from studies of $1027B_2$ cells. Electron microscopy shows that these cells are devoid of secretory granules and immunogold labelling with somatostatin antisera fails to label any specific intracellular structures. Their ultrastructural morphology correlates with their secretory capability—they show high basal secretion and no stimulation with cAMP, theophylline or phorbol esters. They have a poorly developed secretory apparatus and are unresponsive to cAMP. They express high levels of PC1 and PC2 mRNA and protein, and co-secrete both enzymes and somatostatin-14-LI constitutively.

Characterization of prosomatostatin products in cell extracts and media showed 48% and 40% conversion of prosomatostatin to somatostatin-14 and somatostatin-28 respectively. Treatment with 1 μM monensin had little effect on prosomatostatin processing efficiency.

These results clearly demonstrate efficient dibasic and monobasic conversion via the constitutive pathway in $1027B_2$ cells. They are consistent with both the localization of mature products of somatostatin processing to various Golgi subfractions of neural cells (Lepage-Lezin et al 1991) and recent observations that prosomatostatin processing in GH_3 cells begins in the trans Golgi network before the generation of secretory vesicles (Xu & Shields 1993).

PC1 is more potent than PC2 in effecting dibasic processing of prosomatostatin

Although PC1 was effective as a somatostatin-14 convertase in the constitutive compartment of COS-7 cells and although the activity of PC2 was restored in secretory cells, the question of the relative potency of the two proteases in the same regulated cell remained. To investigate this, we produced stable GH_4C_1 cells coexpressing rat prosomatostatin and either mouse PC1 or PC2. PC1 transfectants produced significantly greater somatostatin-14 conversion (62% in cells, 66% in media) compared to PC2 transfectants (53% in cells, 47% in media). The relative amounts of somatostatin-28 were comparable in PC1 and PC2 transfected cells; however, there was a significantly lower level of unprocessed prosomatostatin in PC1 than in PC2 transfectants. Since the level of somatostatin-14-LI expression was comparable between the two cell lines and the conditions of enzyme expression were similar, these results indicate that PC1 is a more efficient somatostatin-14 convertase than PC2.

Conclusions

The experiments outlined here provide a framework for understanding specificity of the enzyme, substrate and intracellular environment required for prosomatostatin maturation. Kinetic studies show that prosomatostatin is rapidly and independently processed to somatostatin-14 and somatostatin-28. Furin is capable of monobasic processing of prosomatostatin and is a candidate somatostatin-28 convertase. PC1 and PC2 both effect dibasic processing of prosomatostatin and qualify as putative somatostatin-14 convertases. PC1 is active in constitutive and regulated secretory cells, has a broader specificity and is more potent than PC2. Efficient processing of prosomatostatin occurs within a Golgi or pre Golgi compartment. It requires the milieu of the secretory cell but not the secretory granule.

Acknowledgements

This study was supported by a grant from the Canadian Medical Research Council (MT6196). We acknowledge the following individuals who have collaborated on various aspects of the work included in this article: Drs L. Orci, M. Amherdt, N. Seidah, S. Rabbani and Mrs G. Kent. We are grateful to Mrs O. Dembinska for technical assistance and to Mrs M. Correia for secretarial support.

References

Benjannet S, Rondeau N, Paquet L et al 1993 Comparative biosynthesis, covalent post-translational modifications and efficiency of prosegment cleavage of the prohormone convertases PC1 and PC2: glycosylation, sulphation and identification of the intracellular site of prosegment cleavage of PC1 and PC2. Biochem J 294:735–743

Devi L 1991 Consensus sequence for processing of peptide precursors at monobasic sites. FEBS Lett 280:189–194

Galanopoulou A, Kent G, Rabbani SN, Seidah NG, Patel YC 1993 Heterologous processing of prosomatostatin in constitutive and regulated secretory pathways. J Biol Chem 268:6041–6049

Halban PA, Irminger J-C 1994 Sorting and processing of secretory proteins. Biochem J 299: 1–18

Hosaka M, Nagahama M, Kim W-S et al 1991 Arg-X-Lys/Arg-Arg motif as a signal for precursor cleavage catalyzed by furin within the constitutive secretory pathway. J Biol Chem 266:12127–12130

Lepage-Lezin A, Joseph-Bravo P, Devilliers G et al 1991 Prosomatostatin is processed in the Golgi apparatus of rat neural cells. J Biol Chem 266:1679–1688

Mackin RB, Noe BD, Spiess J 1991 Identification of a somatostatin-14 generating propeptide converting enzyme as a member of the Kex 2/furin/PC family. Endocrinology 129:2263–2265

Marcinkiewicz M, Rarnla D, Seidah NG, Chrétien M 1994 Developmental expression of the prohormone convertases PC1 and PC2 in mouse pancreatic islets. Endocrinology 135:1651–1660

Molloy SS, Thomas L, Van Slyke JK, Stenberg PE, Thomas G 1994 Intracellular trafficking and activation of the furin proprotein convertase: localization to the TGN and recycling from the cell surface. EMBO J 13:18–33

Neerman-Arbez M, Cirulli V, Halban PA 1994 Levels of the conversion endoproteases PC1 (PC3) and PC2 distinguish between insulin-producing pancreatic islet β cells and non-β cells. Biochem J 300:57–61

Patel YC 1992 General aspects of the biology and function of somatostatin. In: Muller EE, Thorner MO, Weil C (eds) Basic and clinical aspects of neuroscience, vol 4: Somatostatin. Springer-Verlag, Berlin, p1–16

Patel YC, Wheatley T, Ning C 1981 Multiple forms of immunoreactive somatostatin: comparison of distribution in neural and non-neural tissues and portal plasma of the rat. Endocrinology 109: 1943–1949

Patel YC, Zingg HH, Srikant CB 1985 Somatostatin-14 like immunoreactive forms in the rat: characterization, distribution, and biosynthesis. In: Patel YC, Tannenbaum GS (eds) Advances in experimental biology and medicine, vol 188: Somatostatin. Plenum Press, New York, p71–87

Patel YC, Papachristou DN, Zingg HH, Farkas EM 1991 Regulation of islet somatostatin secretion and gene transcription: selective effects of cAMP and phorbol esters in normal islets and in a somatostatin producing rat islet clonal cell line (1027B$_2$). Endocrinology 128:1754–1762

Robbins RJ, Reichlin S 1983 Somatostatin biosynthesis by cerebral cortical cells in monolayer culture. Endocrinology 113:574–581

Rothman JE, Orci L 1992 Molecular dissection of the secretory pathway. Nature 355:409–415

Scammell JG, Burrage TG, Dannies PS 1986 Hormonal induction of secretory granules in a pituitary tumour cell line. Endocrinology 119:1543–1548

Schafer MK-H, Day R, Cullinan WE, Chretien M, Seidah NG, Watson SJ 1993 Gene expression of prohormone and proprotein convertases in the rat CNS: a comparative in situ hybridization analysis. J Neurosci 13:1258–1279

Seidah NG, Day R, Chretien M 1993 The family of pro-hormone and pro-protein convertases. Biochem Soc Trans 21:685–691

Shen F-S, Seidah NG, Lindberg I 1993 Biosynthesis of the prohormone convertase PC2 in Chinese hamster ovary cells and in rat insulinoma cells. J Biol Chem 268:24910–24915

Steiner DF, Smeekens SP, Ohagi S, Chan SJ 1992 The new enzymology of precursor processing endoproteases. J Biol Chem 267:23435–23438

Stoller TJ, Shields D 1988 Retrovirus mediated expression of preprosomatostatin: posttranslational processing, intracellular storage and secretion in GH$_3$ pituitary cells. J Cell Biol 107:2087–2095

Vindrola O, Lindberg I 1992 Biosynthesis of the prohormone convertase mPC1 in AtT-20 cells. Mol Endocrinol 6:1088–1094

Watanabe T, Nakagawa T, Ikemizu J, Nagahama M, Murakami K, Nakayama K 1992 Sequence requirements for precursor cleavage within the constitutive secretory pathway. J Biol Chem 267:8270–8274

Xu H, Shields D 1993 Prohormone processing in the trans-Golgi network: endoproteolytic cleavage of prosomatostatin and formation of nascent secretory vesicles in permeabilized cells. J Cell Biol 122:1169–1184

Zhou A, Bloomquist BT, Mains RE 1993 The prohormone convertases PC1 and PC2 mediate distinct endoproteolytic cleavages in a strict temporal order during proopiomelanocortin biosynthetic processing. J Biol Chem 268:1763–1769

Zingg HH, Patel YC 1982 Biosynthesis of immunoreactive somatostatin by hypothalamic neurons in culture. J Clin Invest 70:1101–1109

DISCUSSION

Cohen: The field of prohormone processing is not well documented, except in the case of proinsulin where processing of one of the two potential proteolytic processing sites occurs during budding of the granules in the trans Golgi network (Orci et al 1987). For a long time it was thought that the processing of prohormones occurs either during translocation of the mature granule from the Golgi apparatus to the apical region of the pancreatic cell or during axonal transport. The kinetic data presented here by Yogesh Patel provide evidence for prohormone processing within the Golgi, i.e. monensin doesn't block processing and the induction of granule formation does not affect the efficiency of prosomatostatin processing to somatostatin-14 or somatostatin-28.

We examined prosomatostatin processing in pancreatic islets of anglerfish (Bourdais et al 1990) and in rat brain cortex cells (Lepage-Lezin et al 1991) by using a combination of sub-cellular fractionation studies, temperature and drug blocks and immunolocalization with antibodies directed against epitopes that are upstream (N-terminal) from monobasic (Arg) and dibasic (Arg–Lys) cleavage sites. The results for both cell types were compatible with those of Yogesh Patel's. Immunolabelling showed that both monobasic and dibasic processing occurs in the trans Golgi network and seems to be associated with a membrane compartment (Cohen et al 1994).

Nature has evolved several ways to cleave peptide bonds. Serine proteases, such as trypsin, chymotrypsin and the enzymes in the blood clotting system are ubiquitous; but there are also metallopeptidases, thiol proteases and aspartyl proteases. So, why are serine proteases considered to be the only family of enzymes that cleave dibasic bonds?

A few years ago, we characterized a complex that we called operationally the *somatostatin-28 convertase*, although it is not necessarily the genuine somatostatin-28 converting enzyme (Gluschankof et al 1987). It is able to convert somatostatin-28 into two fragments and remove the extrabasic residue from N-terminally extended somatostatin-14. It also cleaves at the N-terminal end of dibasic sites in several other peptides, including atrial naturietic factor and dynorphin A. Cleavage is efficient with a K_m in the micromolar range, but requires the integrity of dibasic moieties and the presence of an Arg residue.

In situ hybridization has shown that convertase mRNA is present in discrete regions of the rat cortex, as well as in the premamillary body, habenula and substantia nigra. Although it is present in the cortex, it is 12 times more abundant in the testis (Cohen et al 1995). We therefore used this testicular convertase to clone the gene (Pierotti et al 1994). From this, we deduced its amino acid sequence and found that it is not an aminopeptidase or a carboxypeptidase. An aminopeptidase B, which trims out Lys and Arg from the N-terminus of peptides, has also been cloned.

The convertase is 1161 amino acids long and has a 19–21 putative signal peptide at the N-terminus. It is a metalloenzyme with a Zn^{2+}-binding site. It contains six potential glycosylation sites and, surprisingly, a 71 amino acid sequence that is 80% glutamate and aspartate. This feature is also found in Kex1, a carboxypeptidase B in *Saccharomyces cerevisiae* (Dmochowska et al 1987). It has a C-terminal cysteine that is, on the basis of sequence comparison with homologous endoproteases, a putative component of the catalytic reaction. It is roughly 50% homologous to the insulin-degrading enzyme, insulysin (insulinase, E.C.3.4.99.45) which cleaves at the N-terminus of hydrophobic, acidic and basic residues and has less substrate specificity than N–Arginine dibasic (NRD) convertase (nardilysin, E.C.3.4.24.61). We reported the sequence of the latter enzyme recently (Pierotti et al 1994).

Rat, human and *Drosophila* insulin-degrading enzymes belong to the pitrilysin family. They are probably derived from an ancestral gene, related to the *Escherichia coli* protease III (pitrilysin, E.C.3.4.99.44). The amino acid sequence of protease III is 45% homologous to that of the NRD convertase.

The NRD convertase is the endoproteolytic component of the so-called somatostatin convertase and contains the Asp/Glu-rich region. It is present in large amounts in the male germline as the protein convertase PC4 and expressed in a short burst at a late stage of germline maturation (stage 7–8), when the spermatids are elongating. The enzyme is probably involved in some essential event related to spermiogenesis that we have not identified so far. Interestingly, PC4 is expressed in the same place and at the same time as ACE (angiotensin-converting enzyme). The endogenous substrate of this enzyme is not known— we are not claiming that it is the natural processing enzyme for somatostatin-28, but *in vitro* it does cleave this precursor correctly.

One of the most difficult things in the field of proteolytic processing of propeptides is to establish an unequivocal relationship between the substrate and the enzyme and to determine if a given species of enzyme is the enzyme that acts physiologically. The observation that PCs are related to the subtilisin-like enzyme family is a very powerful tool to investigate prohormone processing. Do you believe that the PC/subtilisin-like, furin-related, thiol reagent-sensitive enzyme family is a unique family that cleaves all monobasic and dibasic residues involved in the processing of hundreds of proproteins and propeptide hormones?

Patel: I don't think we need another processing enzyme! I have narrowed the candidates involved in prosomatostatin processing down to three convertases. PC4 is found only in the testis in germ cells and can therefore be excluded as a candidate protease. There is evidence, based on its expression in neuroendocrine cells, that PC5 may be involved in neuropeptide processing, so we are planning to look at its role specifically in prosomatostatin processing.

Not only are we dealing with several different processing enzymes, but also post-translationally modified forms of these enzymes, e.g. glycosylated, sulphated and enzymatically cleaved forms. These forms are differentially produced along the secretory pathway and may or may not be secreted. Their relative activities and precise roles in prosomatostatin processing are also uncertain.

Therefore, the subtilisin-related serine convertases characterized by furin and PCs constitute a complicated system. If you have found another processing enzyme, it will certainly add to this complexity.

Epelbaum: What is the localization of PCs and furin in cells that process prosomatostatin to somatostatin-14, such as nervous or pancreatic cells, compared with cells that process it to somatostatin-28, such as intestinal cells?

Patel: This has only been studied in islets and needs to be studied in other tissues. *In situ* hybridization has shown that PC1 and PC2 mRNAs are distributed widely throughout the CNS (central nervous system) (Schafer et al 1993). I was interested in the localization of these mRNAs in the anterior hypothalamic periventricular neurons in this study because these neurons comprise the majority of somatostatin-synthesizing cells in the hypothalamus. They contain mostly PC1 suggesting that it is important in prosomatostatin processing. Immunocytochemical co-localization shows that only PC2, and not PC1, is present in δ cells of rat islets (Marcinkiewicz et al 1994). These reports suggest a differential tissue-specific localization of PC1 and PC2 in different somatostatin-producing cells and, along with our direct studies, demonstrate a role for both enzymes in prosomatostatin processing. No one has yet looked at intestinal somatostatin-28-producing cells to see which enzyme is localized there.

Meyerhof: Does the distribution of the two peptides in different tissues correspond to the potency of each particular peptide in a given tissue?

Patel: Somatostatin-28 was originally isolated from the porcine intestine. Its actions overlap with somatostatin-14 but it also has distinct properties. For instance, somatostatin-28 is a more potent inhibitor of both GH (growth hormone) secretion from the pituitary and insulin secretion from islet β cells. Conversely, somatostatin-14 is a more potent inhibitor of glucagon secretion from islet α cells.

Meyerhof: If you use rat GH_3 or GH_4C_1 pituitary tumour cells, which express somatostatin receptors endogenously, would you expect that these receptors are regulated by the somatostatin transfected into those cells? Could the exogenous peptide bind to endogenous receptors and by signal transduction mechanisms feed back into the peptide-processing pathway?

Patel: It is possible, but I can't envision how such autoinhibition of a heterologous cell might affect the pattern or efficiency of processing.

Reichlin: Is there any evidence that somatostatin-28 is converted to somatostatin-14?

Patel: Not under normal conditions. In our furin vaccinia experiments, we saw dose-dependent conversion of prosomatostatin to somatostatin-28 with no synthesis of somatostatin-14 until we expressed furin to a high level of 2.4 pfu (plaque-forming units)/cell infection, when we started to see somatostatin-14. Somatostatin$_{[1-12]}$, the N-terminal segment of somatostatin-28, increased simultaneously with somatostatin-14, suggesting direct conversion of somatostatin-28 to somatostatin-14 and somatostatin-28$_{[1-12]}$. This is not surprising because the so-called 'dibasic' cleavage site for somatostatin-14 is really a multibasic site (RXRK) which qualifies as a weak furin substrate. However, this is an artificial system. Normally, pulse–chase experiments show that somatostatin-14 and somatostatin-28 are synthesized independently from prosomatostatin.

Reichlin: Is there a sequence in the somatostatin gene, that may or may not be expressed, that determines this alternative processing?

Goodman: Bloomquist et al (1991) have used antisense oligonucleotides to look at specific sequences involved in processing in a different system. Could you bring us up to date on this?

Patel: We've spent the last year making antisense *in vitro* knockouts of PC1 and PC2 in stable AtT-20 mouse pituitary tumour cells and GH_4C_1 cells respectively. These cells express large amounts of endogenous enzyme mRNA, so it has been difficult to select cells with sufficient amounts of mRNA knocked out. A reduction to about 50–60% is insufficient to show any effect, you need about 90% knockout. This was also a problem in the studies of Bloomquist et al (1991). Our studies are in progress, but I don't have any results to show you yet.

Sevarino et al (1989) showed that signals in the proregion of prosomastostatin are required for targeting into either the regulated or the constitutive secretory pathway. Lower vertebrates, such as the anglerfish, have two genes for somatostatin—one that encodes somatostatin-14 and one somatostatin-28. These genes are expressed in separate cells, producing two different somatostatin cell types. In contrast, mammals have one somatostatin gene and one somatostatin cell type. It is possible that there is a requirement for anatomical compartmentalization of somatostatin-14 and somatostatin-28 processing, so we looked at secretion granule heterogeneity using immunogold labelling with specific somatostatin-14 and somatostatin-28 antisera. In collaboration with L. Orci and M. Ravazzola (unpublished results), we studied fetal human pancreas which makes somatostatin-14 and somatostatin-28, and found two populations of granules—one that contains only somatostatin-14 and the other only somatostatin-28. This indicates that there are probably two separate pathways for targeting somatostatin-28 and somatostatin-14 in mammalian cells. It is not known what determines this targeting, but the N-terminal segment of prosomatostatin may be involved.

Cohen: Bersani et al (1988) and our lab (Bourdais et al 1990) have found large quantities of the N-terminal proregion of the somatostatin precursor (i.e. the domain situated at the N-terminus of the single Arg residue that constitutes the monobasic maturation site) in secretory granules of the pig and anglerfish. It looks like this proregion has to reach the trans Golgi compartment and is then packaged in the secretory granule. It is possible, therefore, that it plays a role in targeting the prohormone within the secretory machinery of the producing cells.

Reichlin: Is there a conserved sequence?

Cohen: Yes, we found N-terminal and C-terminal sequences indicating integrity of the pro-peptide. We also used antibodies against each terminus. Stoller & Shields (1989) also made a chimeric construction using the proregion of somatostatin attached to globin and found that this globin was sorted into the secretory pathway in transfected cells.

Stork: We have used rat prosomatostatin sequences to identify what we presume to be an address peptide, i.e. one that would target the proregion to the granules (Sevarino et al 1989). Progressive deletions of the proregion encoded by the rat cDNA were ineffective in disrupting targeting of the pro-peptide. The only deletion that really knocked out targeting encompassed the processing sites themselves (Sevarino & Stork 1991). Since any receptor-type molecule must be targeted and recognize sequences common to many proregions, the processing enzyme itself might serve a targeting function, bringing the molecules into the granules. Is that possible or does cleavage of the prohormone in or before the Golgi preclude continued association with the processing enzyme that, as Yogesh Patel pointed out, is itself targeted to the secretory granule? Is there any reason why these processing enzymes are targeted to the regulatory pathway if their function is prior to that?

Cohen: If the processing enzyme does play the role of the receptor, then gene expression constructs should include sequences coding for the structural components that are necessary for processing. In Dennis Shields' initial experiments, the dibasic moieties of prosomatostatin were left in the chimeric construct containing α-globin attached to the pro-peptide (Stoller & Shields 1989). It would be interesting to delete the dibasic site and see if this affects sorting. Philip Stork, did you do this experiment?

Stork: No. If we do that, we eliminate our ability to assay for the processing— the release of processed material was our assay for targeting.

Patel: The molecular mechanism involved in precursor targeting is a mystery. The most compelling evidence for the involvement of the proregion is reported by Sevarino et al (1989). Other theories propose condensation or aggregation of secretory proteins or association with secretogranins in the trans Golgi network, leading to the budding of a clathrin-coated vesicle by a process similar to receptor-mediated endocytosis.

Our results show that extensive processing to somatostatin-14 and somatostatin-28 occurs prior to the trans Golgi network and before the products are targeted to the secretory granules. This raises several questions. How are the separate products targeted to the secretory granules? Do they associate with the different molecular forms of the PC enzymes? How are the enzyme and precursor products packaged? These are complex issues which will take time to resolve.

Stork: Why would these enzymes be targeted? And are they efficiently recycled, as the yeast Kex enzyme is?

Patel: We all naively thought that because these enzymes work intracellularly, they would be recycled and not secreted out of the cell. So, it came as a surprise to find that PC1 and PC2 are secreted, often in large amounts. Even furin, which has a transmembrane domain and is a resident protein of the trans Golgi network, is capable of being cleaved and transported to the plasma membrane where it remains as a cell surface-bound enzyme. It does not appear to be secreted normally.

Montminy: One possible explanation for how somatostatin is processed to somatostatin-28 and somatostatin-14 is that it reflects cellular limitations in the regulated pathway. Basically, if there's not enough enzyme around to process prosomatostatin to somatostatin-14, then somatostatin-28 is formed by the default pathway. Then, if somatostatin biosynthesis is increased by different signals, the question is, to what extent would that shunt somatostatin processing to the somatostatin-28 pathway rather than the somatostatin-14 pathway?

Patel: That's an interesting point. Can you manipulate the ratio of somatostatin-14:somatostatin-28 by altering the level of secretory activity? In acute experiments, the ratio released is identical to the ratio of the two proteins stored intracellularly. When somatostatin transcription is increased in cultured islets with forskolin or cAMP, there is no difference between the somatostatin-14:somatostatin-28 ratio in cell extracts and secretion media. Indeed, in islet cells, somatostatin-14 is present almost exclusively and there is no significant increase in somatostatin-28.

Montminy: Are intestinal cells geared to granular secretion?

Patel: Intestinal mucosal somatostatin cells are the only mammalian cells that I know of which synthesize only somatostatin-28. Therefore, it is important to characterize processing in these cells, but they haven't been studied adequately yet.

Berelowitz: We looked at the proportion of somatostatin-28 and somatostatin-14 by HPLC (high performance liquid chromatography) in extracts of ventromedial nucleus, arcuate nucleus and the median eminence in animals that were either hypophysectomized, treated with excess GH or normal (Berelowitz et al 1983). Extracts from each area contain a particular ratio of somatostatin-28 to somatostatin-14 and we found no change in this ratio with the manipulations, despite the fact that treatment with GH produced a large increase in total somatostatin, whereas hypophysectomy caused a mild decrease in the amount of somatostatin.

Goodman: I am curious about the discrepancy between the amount of somatostatin-14 and somatostatin-28 in the medium secreted by transfected African green monkey COS-7 kidney cells versus that in the cell extract. More somatostatin-14 was secreted whereas the cells contain more somatostatin-28.

Patel: COS-7 cells are a constitutively secreting cell line. When they are transfected with prosomatostatin cDNA, they secrete more immunoreactive somatostatin (predominately somatostatin-28) than they store. In general, there is no difference in the relative amount of somatostatin-28 stored or secreted in these cells either under basal conditions or when infected with vaccinia furin. In these experiments, somatostatin-14 synthesis also increases with high furin expression. You are correct in noticing that this increase is predominantly intracellular rather than secreted somatostatin-14.

Eppler: What's the ratio of somatostatin-14 to somatostatin-28 in serum? Also, could COS cells secrete an enzyme that further processes somatostatin-28?

Patel: In serum there is preferential accumulation of somatostatin-28 because it's more stable and has a longer half-life than somatostatin-14 (Patel & Wheatley 1983). There is some conversion to somatostatin-14, but it is relatively slow.

No one has yet addressed the possible role of processing enzymes in the circulation. These enzymes are released from secretory cells and must therefore achieve significant levels in the blood, but it is not known what forms they are, what their levels are or what their role might be.

Schonbrunn: Is there a difference in processing when you compare constitutive secretion and stimulated secretion in AtT-20 and GH₃ cells that express prosomatostatin?

Patel: We have not looked at that. I showed you schematically the relative amounts of somatostatin-14 and somatostatin-28 in the regulated cell lines AtT-20 and GH₃ (Fig. 2). When stimulated with forskolin, the relative amounts released are identical to those found in cell extracts.

Schonbrunn: Is that release under basal conditions?

Patel: Basal secretion from these cells has been difficult to characterize, especially following acute transfections. Acute transfection of prosomatostatin cDNA in GH_3 cells results in basally secreted immunoreactivity of predominantly somatostatin-14. Stable transfection of GH_4C_1 cells, however, results in basally secreted immunoreactivity consisting of somatostatin-14, somatostatin-28 and large amounts of unprocessed precursor. We have not looked at the pattern of release of these forms following acute stimulation with forskolin, but I agree that this would be important to do.

Reichlin: The classical model of intragranular processing of neuropeptides as outlined by Sachs (1964) seems to be overturned by your findings. The early work, comparing chromatographs of extracts of the hypothalamus with the posterior pituitary, showed a change in the ratio of what we now recognize as provasopressin to vasopressin, as the granules were transported from the hypothalamus to the posterior pituitary. The belief was that the nascent protein was processed during transport. It takes a long time for that transport to occur, even at a rate of 200 mm/day. Are you arguing that processing occurs in the cytoplasm before the nascent protein is transported?

Patel: In those experiments, processing of vasopressin occurred during axonal transport, but it was not dependent on it because processing still occurred when transport was blocked with colchicine.

The concept that processing occurs in secretory granules comes from studies on the maturation of proinsulin. Orci et al (1987) found that clathrin-coated immature secretory granules, budding off the trans Golgi network, contain proinsulin which is rapidly processed to insulin as the immature secretion vesicles shed their clathrin coats and become mature secretory granules. Processing of prosomatostatin appears to be different. I have shown you four lines of evidence that prosomatostatin processing is underway in the trans Golgi network before the formation of secretory granules: (i) kinetic evidence for rapid cleavage of prosomatostatin to somatostatin-14 in 10–15 min; (ii) conversion of prosomatostatin to somatostatin-14 in constitutively secreting COS-7 cells co-transfected with prosomatostatin and PC1; (iii) failure of granule induction to influence prosomatostatin processing in GH_4C_1 cells; and (iv) efficient processing of prosomatostatin to somatostatin-14 and somatostatin-28 through the constitutive pathway in 1027 B_2 rat islet somatostatinoma cells and the failure of monensin to affect such processing. Paul Cohen has a fifth line of evidence showing the localization of mature products of prosomatostatin processing to various Golgi subfractions of neural cells by the immunogold labelling technique (Lepage-Lezin et al 1991). Finally, Xu & Shields (1993) have done experiments in permeabilized GH_3 cells using a low temperature (20 °C) to block the secretory pathway. They uncoupled the secretory pathway from the trans Golgi network and showed that prohormone processing occurs in the trans Golgi network. This collective evidence suggests that efficient processing of prosomatostatin does occur within a Golgi or pre Golgi compartment before targeting to the secretory granules.

Cohen: Another line of evidence was reported by Schnabel et al (1989). They performed elegant studies with a set of antibodies directed against various epitopes of the preopiomelanocortin (POMC) fragments generated during processing. They showed that the C-terminal amidated fragment of this precursor, i.e. adrenocorticotropic hormone (ACTH), occurred in the trans Golgi network. If amidation is occurring there, then proteolytic processing must have taken place beforehand (Cohen et al 1994).

Beaudet: I would like to add a note of caution regarding extrapolating results from the processing of other peptides. In collaboration with Patrick Kitabgi, we have looked at processing of the neurotensin precursor in the brain (Woulfe et al 1994). Like somatostatin, neurotensin and neuromedin N are processed very early, perhaps whilst still within the Golgi apparatus. However, further processing of the precursor molecule occurs *en route* within the axon. This processing appears to take place in the granules because we have shown, using confocal microscopy, the co-localization of immunoreactive neurotensin with other maturation products in the same granules. Some of the maturation products cannot be detected in the cytoplasm, only at the axon terminals.

As for the somatostatin convertases, there is a wide distribution of PC1 and PC2 throughout the CNS; however, they are not present in all cells and they are differentially distributed. It is important to determine whether PC1, PC2 or both co-localize with somatostatin in neurons in the CNS.

Patel: A co-localization study has not been done. All that's been reported is *in situ* hybridization to map PC1 and PC2 expression in different anatomical regions of the brain.

Stork: There was a model in favour a number of years ago that newly synthesized granules were preferentially secreted with respect to old granules. This suggests that there are distinct populations of granules. Does this relate to what Marc Montminy found in his refractory period (Montminy et al 1995, this volume), i.e. is transcription rate-limiting for secretion or is there sufficient surplus product around that the transcriptional refractory period has no effect on secretion? If transcription was rate-limiting, different agents that stimulate synthesis might have different refractory periods and different effects on release. Aside from the processing itself, is there evidence for distinct granule populations that respond to distinct secretagogues?

Patel: There are some kinetic data using pulse–chase labelling techniques which indicated that there are two storage pools in cells such as lactotrophs. During acute stimulation of lactotrophs with agents such as TRH (thyroid-stimulating hormone releasing hormone), the newly formed granules are released preferentially compared to hormone stored in older secretory granules (Walker & Farquhar 1980).

Stork: Marc Montminy, is the refractory period that you observed involved in limiting the release of synthesized somatostatin?

Montminy: We haven't checked what happens to somatostatin release. I would imagine that the cytoplasmic events might have different sensitivity, because the amount of protein kinase A that's required to get into the nucleus is different from the amount that is present in the cytoplasm. cAMP stimulation of release of somatostatin peptide may be less sensitive, so it may not show as much of a refractory period.

Reichlin: Is there a gross excess of processing enzyme or are the levels of these processing enzymes regulated?

Patel: There is little evidence of co-regulation of precursor and processing enzymes. Somatostatin synthesis is stimulated by many different agents, but there is no evidence that these same stimuli also increase the levels of processing enzymes.

References

Berelowitz M, DiTirro FJ, Thominet JL, Ting N-C, Pollack J, Frohman LA 1983 Growth hormone (GH) feedback effects on regional distribution and molecular forms of hypothalamic somatostatin. Clin Res 31:469A

Bersani M, Thim L, Furio GAB, Holst JH 1989 Prosomatostatin 1-64 is a major product of somatostatin gene expression in pancreas and gut. J Biol Chem 264:10633–10636

Bloomquist BT, Eipper BA, Mains RE 1991 Prohormone-converting enzymes—regulation and evaluation of function using antisense RNA. Mol Endocrinol 5:2014–2024

Bourdais J, Devilliers G, Girard R, Morel A, Benedetti L, Cohen P 1990 Prosomatostatin II processing is initiated in the trans-Golgi network of anglerfish pancreatic cells. Biochem Biophys Res Commun 170:1263–1272

Cohen P, Rholam M, Boussetta H 1994 Methods for the identification of neuropeptide processing pathways. In: Smith IA (ed) Methods in neurosciences. Acad Press Publishers, in press

Cohen P, Pierotti A, Chesneau V, Prat A, Foulon T 1995 N–Arginine dibasic convertase. In: Barrett AJ (ed) Methods Enzymol 148, in press

Dmochowska A, Dignard D, Henning D, Thomas DY, Bussey H 1987 Yeast KEX1 gene encodes a putative protease with a carboxypeptidase B-like function involved in killer toxin and alpha-factor precursor processing. Cell 50:573–584

Gluschankof P, Gomez S, Morel A, Cohen P 1987 Enzymes that process somatostatin precursors: a novel endoprotease that cleaves before the Arg Lys doublet is involved in somatostatin-28 convertase activity of rat brain cortex. J Biol Chem 262:9615–9620

Lepage-Lezin A, Joseph-Bravo P, Devilliers G et al 1991 Prosomatostatin is processed in the Golgi apparatus of rat neuronal cells. J Biol Chem 266:1679–1688

Marcinkiewicz M, Ramla D, Seidah NG, Chrétien M 1994 Development expression of the prohormone convertases PC1 and PC2 in mouse pancreatic islets. Endocrinology 135:1651–1660

Montminy M, Brindle P, Arias J, Ferreri K 1995 Regulation of somatostatin gene transcription by cAMP. In: Somatostatin and its receptors. Wiley, Chichester (Ciba Found Symp 190) p 7–25

Orci L, Ravazzola M, Storch MJ, Anderson RGW, Vasalli JD, Perrelet A 1987 Proteolytic maturation of insulin is a post-Golgi event which occurs in acidifying clathrin-coated vesicles. Cell 49:865–868

Patel YC, Wheatley T 1983 *In vitro* and *in vivo* plasma disappearance and metabolism of somatostatin-28 and somatostatin-14 in the rat. Endocrinology 112:220–225

Pierotti A, Prat A, Chesneau V et al 1994 N-Arginine dibasic convertase (NRD convertase), a novel metalloendopeptidase as a prototype of a class of processing enzymes. Proc Natl Acad Sci USA 91:6078–6082

Sachs H, Takabatake Y 1964 Evidence for a precursor in vasopressin synthesis. Endocrinology 75:943–948

Schafer MKH, Day R, Cullinans M, Chretien M, Seidah NG, Watson SJ 1993 Gene expression of prohormone and proprotein convertases in the rat CNS: a comparative *in situ* hybridization analysis. J Neurosci 13:1258–1279

Schnabel E, Mains RE, Farquhar MG 1989 Proteolytic processing of pro-ACTH/endorphin begins in the Golgi complex of pituitary corticotropes and AtT20 cells. Mol Endocrinol 3:1223–1235

Sevarino K, Stork PJ, Ventimiglia R, Mandel G, Goodman RH 1989 Amino-terminal sequences of prosomatostatin direct intracellular targeting but not processing specificity. Cell 57:11–19

Sevarino K, Stork P 1991 Independent signals mediate sorting of somatostatin to distinct regulated pathways. J Biol Chem 266:18507–18513

Stoller TJ, Shields D 1989 The propeptide of preprosomatostatin mediates intracellular transport and secretion of α-globin from mammalian cells. J Cell Biol 108:1647–1655

Walker AM, Farquhar MG 1980 Preferential release of newly synthesized prolactin granules is the result of functional heterogeneity among mammotrophs. Endocrinology 107:1095–1104

Woulfe J, Lafortune L, de Nadai F, Kitabgi P, Beaudet A 1994 Post-translational processing of the neurotensin/neuromedin N precursor in the central nervous system of the rat. II. Immunohistochemical localization of maturation products. Neuroscience 60:167–181

Xu H, Shields D 1993 Prohormone processing in the trans Golgi network: endoproteolytic cleavage of prosomatostatin and formation of nascent secretory vesicles in permeabilized cells. J Cell Biol 122:1169–1184

Anatomical localization and regulation of somatostatin gene expression in the basal ganglia and its clinical implications

Marie-Françoise Chesselet, Jean-Jacques Soghomonian and Pascal Salin

Department of Pharmacology, University of Pennsylvania, School of Medicine, 36th Street and Hamilton Walk, Philadelphia, PA19104-6084 USA

Abstract. The distribution of somatostatin in both the human and rat brain suggests that it is involved in numerous functions, including endocrine regulation, cognition and memory, autonomic regulation and motor activity. We have examined the regulation of somatostatin mRNA in the striatum, a brain region involved in motor and cognitive behaviour. Somatostatin and its mRNA are expressed in this region in interneurons which are resistant to ischaemia, excitotoxicity and Huntington's disease, possibly because they express high levels of superoxide dismutase. Striatal somatostatin mRNA is increased by stimulation of NMDA (*N*-methyl-D-aspartate) receptors. Ischaemia-induced cortical lesions also increase somatostatin gene expression in the striatum. In contrast, the levels of striatal somatostatin mRNA decrease after treatment with haloperidol, an antipsychotic agent that produces extrapyramidal symptoms, but not clozapine, which does not. Further evidence for a role for striatal somatostatin in extrapyramidal symptoms includes the observation that somatostatin mRNA levels decrease in the striatum after lesions are made in the dopaminergic pathway, a feature of Parkinson's disease. The largest change in somatostatin gene expression after dopaminergic lesions is the increase in somatostatin mRNA levels in neurons of the internal pallidum and lateral hypothalamus projecting to the lateral habenula. The results suggest that changes in brain somatostatin gene expression occur in pathological conditions and may be related to their symptoms.

1995 Somatostatin and its receptors. Wiley, Chichester (Ciba Foundation Symposium 190) p 51–64

The wide distribution of somatostatin in the brain suggests that it may function outside the hypothalamus, in regions such as the cerebral cortex, the striatum and the hippocampus (Johansson et al 1984, Kiyama & Emson 1990). Somatostatin depletion performed *in vivo* in adult rats by cysteamine or administration of somatostatin in discrete brain regions, suggests that the neuropeptide may play a role in motor control and cognitive processes (Fitzgerald & Dokla 1989, Raynor et al 1993). Furthermore, somatostatin depletion in the cerebral cortex has been correlated with the severity of dementia in patients with Alzheimer's disease, suggesting that somatostatin deficiency may also contribute to cognitive deficits in humans (Bissette & Myers 1992). The lack of a specific antagonist for somatostatin has greatly limited the ability to assess somatostatin function in the brain. Changes in somatostatinergic neurotransmission might be identified by measuring somatostatin gene expression in clinically relevant experimental models. Our studies have focused on the regulation of somatostatin mRNA levels *in vivo* in the the basal ganglia in animal models of Parkinson's disease, Huntington's disease and cortical lesions, as well as after treatments with typical and atypical neuroleptics.

Methods

Immunohistochemistry was performed as described in Chesselet & Graybiel (1986). *In situ* hybridization histochemistry with radiolabelled RNA probes was performed as described by Soghomonian & Chesselet (1991). Analysis of the intensity of labelling over single neurons in emulsion-coated sections was performed with a Morphon Image Analysis system. Detailed discussion of the quantification procedure can be found in Chesselet & Weiss-Wunder (1994).

Distribution of somatostatinergic neurons in the neostriatum

The striatum (caudate nucleus and putamen) contains the largest number of somatostatinergic neurons in the basal ganglia (Johansson et al 1984). These neurons are medium-sized aspiny interneurons (DiFiglia & Aronin 1982, Chesselet & Graybiel 1986). They also contain neuropeptide Y and express NADPH diaphorase activity, indicating the presence of nitric oxide synthase (Sandell et al 1986, Vincent & Johansson 1983, Dawson et al 1991). Although somatostatin and GABA (γ-aminobutyric acid) are often co-localized in the brain, this is not the case in the striatum, where somatostatinergic and GABAergic interneurons form two distinct neuronal populations (Chesselet & Robbins 1989). Somatostatinergic interneurons are preferentially located in the extrastriosomal matrix (Chesselet & Graybiel 1986, Sandell et al 1986), a striatal compartment connected preferentially to the sensorimotor system (Flaherty & Graybiel 1994). Somatostatinergic neurons are often found aligned along the border of striosomes, striatal compartments associated with the limbic system (Chesselet et al 1991). Some of their processes may cross the striosome/matrix border, suggesting that neurons may contribute to the transfer of information between striatal compartments (Chesselet & Graybiel 1986). These points of communication, however, are rare and their functional significance remains unclear.

Somatostatin mRNA in individual mouse neurons is more abundant in the lateral than in the medial striatum (Weiss & Chesselet 1989). This difference is not observed in rat neurons (Salin et al 1990a). Somatostatinergic interneurons are present in both species in the nucleus accumbens, a region rostral and ventral to the striatum which is part of the limbic system. Compared with the striatum, the nucleus accumbens contains more neurons, but the level of expression of somatostatin mRNA per neuron is lower in this region than in the striatum. This indicates that regional differences occur in the regulation of somatostatin gene expression in neighbouring brain areas (Salin et al 1990a). Heterogeneity in the level of binding sites for somatostatin was also noticed in the striatum and nucleus accumbens, with higher levels of the sstr2 agonist MK678 binding in the nucleus accumbens and ventromedial striatum than in the dorsolateral part of this region (Martin et al 1991).

Dopaminergic regulation of somatostatin gene expression

The striatum, nucleus accumbens and prefrontal cortex receive dopaminergic inputs from the mesencephalon and contain dopaminergic receptors that are blocked by antipsychotic agents. Although typical antipsychotic agents, such as haloperidol, induce extrapyramidal side effects, atypical antipsychotic agents, such as clozapine, do not (Baldessarini & Frankenburg 1991). We have examined the effects of repeated administration of haloperidol and clozapine on somatostatin mRNA levels in the striatum, nucleus accumbens and prefrontal cortex of adult rats (Salin et al 1990a). Chronic administration of haloperidol for 28 days at a dose which selectively blocks D2 dopamine receptors *in vivo* (1 mg/kg i.p.) decreased the levels of somatostatin mRNA in the nucleus accumbens, the lateral agranular region of the frontal cortex and the striatum of adult rats. In contrast, injection of clozapine (20 mg/kg i.p. for 28 days) increased the levels of somatostatin mRNA in the nucleus accumbens, but had no effect on somatostatin mRNA levels in the striatum or frontal cortex.

The results suggest that haloperidol-induced blockade of dopamine D2 receptors decreases somatostatin gene expression in the striatum. This is supported by studies in the mouse showing that fluphenazine-*N*-mustard, an irreversible antagonist of D2 dopamine receptors, also decreases somatostatin mRNA levels in the striatum (Weiss & Chesselet 1989). Furthermore, Augood et al (1991) have shown that acute exposure to either D2 or D1 antagonists decreases somatostatin mRNA levels in rat striatum. The decrease in somatostatin mRNA in the striatum after haloperidol treatment probably corresponds to a decrease in the synthesis of the peptide because haloperidol treatment also decreases the amount of somatostatin in this region (Beal & Martin 1984).

The mechanism underlying the differences in the effects of haloperidol and clozapine in the striatum, and the unique effect of clozapine in the nucleus accumbens, is unclear. In contrast to haloperidol, clozapine has a low affinity for dopamine D2 receptors and a broad pharmacological spectrum, including antagonistic effects at the D1 and D4 dopaminergic receptors, serotonin 2A and 2C, adrenergic, histaminergic and cholinergic receptors (Baldessarini & Frankenburg 1991). The action of clozapine on serotonergic receptors might contribute to its lack of effect on somatostatin gene expression in the striatum because lesions of serotonergic inputs to the striatum increase somatostatin mRNA levels there (Bendotti et al 1993). Antagonism of serotonergic transmission could therefore counteract the effects of blockade of dopaminergic receptors. Alternatively, blockade of muscarinic receptors by clozapine could affect somatostatin gene expression, because somatostatinergic interneurons in the striatum express mRNA encoding m1 muscarinic receptors (Bernard et al 1992). We are currently examining the effects of muscarinic antagonists on somatostatin mRNA levels in the striatum to test this hypothesis.

It is of particular interest that drugs which produce extrapyramidal side effects, such as haloperidol, have a different effect on somatostatin mRNA levels than drugs that do not produce side effects, such as clozapine. Although the effects of both drugs on a variety of neurotransmitters differ in the striatum, nucleus accumbens and frontal cortex, the results suggest that their distinct effects on somatostatin might contribute to the differences in their clinical side effects. In particular, the decrease in somatostatin mRNA in the striatum after haloperidol, but not clozapine treatment, may be related to the ability of haloperidol to elicit symptoms similar to those in Parkinson's disease (Baldessarini & Frankenburg 1991). This hypothesis is of particular interest in view of the preferential location of somatostatinergic neurons in the motor-related extrastriosomal matrix (Chesselet & Graybiel 1986). The decreased levels of striatal somatostatin after haloperidol treatment or dopamine depletion may down-regulate dopamine release, because dopamine release is increased by exogenous somatostatin *in vivo* and *in vitro* (Chesselet & Reisine 1983). Furthermore, somatostatinergic neurons form synapses with striatal efferent neurons (DiFiglia & Aronin 1982) and may therefore influence output pathways from the striatum which contribute to the regulation of movement (Flaherty & Graybiel 1994).

To test the hypothesis that somatostatinergic neurons are affected by interruption of dopaminergic transmission in the striatum, we examined somatostatin mRNA expression in the striatum after producing lesions in the nigrostriatal dopaminergic pathway, which are a feature of Parkinson's disease. As with haloperidol treatment, levels of somatostatin mRNA were decreased in the striatum two and three weeks after extensive lesions of the nigrostriatal dopaminergic pathway were made (Soghomonian & Chesselet 1991). As for many effects of dopamine depletion in the striatum, this effect was observed only after almost complete loss of dopamine in this region, which may explain why previous studies have not reported changes in somatostatin peptide levels after producing nigrostriatal lesions (Salin et al 1990b). Alternatively, a constant level of peptide may be associated with a parallel decrease in synthesis and release of somatostatin after the lesion. A correlation between decreased somatostatin levels in the cerebrospinal fluid and the severity of motor symptoms has been observed in patients with Parkinson's disease, suggesting that dopamine depletion may also lead to a decrease in somatostatin expression in humans (Strittmatter & Cramer 1992).

A surprisingly large increase in somatostatin mRNA levels was observed in neurons of the entopeduncular nucleus and adjacent lateral hypothalamus following the production of nigrostriatal lesions (Soghomonian & Chesselet 1991). Some of these neurons also contain glutamic acid decarboxylase (Y. Qin & M. F. Chesselet, unpublished observations), indicating that they are GABAergic, but levels of mRNA for the decarboxylase in these neurons were not as dramatically affected by the dopaminergic lesion (Soghomonian & Chesselet 1992). The mechanism responsible for this large increase in gene expression is unclear. We have recently found that this effect is prevented by lesions in the subthalamic nucleus which sends excitatory projections to the entopeduncular nucleus. This suggests that the entopeduncular nucleus may be directly or indirectly involved in the increased expression of somatostatin mRNA observed after the production of lesions using 6-hydroxydopamine (J. M. Delfs, V. M. Ciaramitaro, T. J. Parry & M.-F. Chesselet, unpublished work). As discussed in Soghomonian & Chesselet (1991), somatostatinergic neurons of the entopeduncular nucleus probably project to the lateral habenula. These neurons form a continuum with somatostatinergic neurons of the lateral hypothalamus, which also show a marked increase in somatostatin mRNA after dopaminergic lesions. This suggests that somatostatinergic neurons of the entopeduncular nucleus may have a similar function to those of the lateral hypothalamus. The lateral hypothalamus is involved in the regulation of food and water intake, physiological functions that are altered in Parkinson's disease. The recent development of antisense strategies for use *in vivo* may help to decipher the contribution of these changes in somatostatin gene expression to the multiple consequences of lesions in the dopaminergic nigrostriatal pathway.

Striatal somatostatinergic neurons and excitotoxicity

Beal et al (1986) have shown that somatostatinergic interneurons are relatively spared after striatal injections of quinolinic acid, an agonist of *N*-methyl-D-aspartate (NMDA) receptors which produces a massive loss of striatal efferent neurons. This observation is of particular interest because somatostatinergic neurons remain relatively preserved in the brains of patients with Huntington's disease, an autosomal dominant neurodegenerative illness characterized by the progressive loss of striatal efferent neurons, abnormal movements and cognitive dysfunction (Ferrante et al 1985). Some researchers have not been able to reproduce the observations of Beal and co-workers, possibly because of differences in experimental conditions and data analysis (see discussion in Roberts et al 1993).

We have confirmed that somatostatinergic neurons in rat striatum are relatively spared by quinolinic acid. We found that two weeks after a slow infusion of a low dose (60 nmol), the level of somatostatin mRNA increased in the surviving neurons (Qin et al 1992). This observation may be related to a direct stimulatory effect of the NMDA agonist on somatostatin gene expression which has been observed in cultures of cortical neurons (Patel et al 1991). Alternatively, the effect may be indirect, via the loss of inhibitory striatal efferent neurons that emit axon collaterals within the striatum and could be involved in the regulation of striatal somatostatinergic neurons.

After local injections of a larger dose of quinolinic acid (120 nmol), however, somatostatin mRNA levels decreased in the striatum. The relative preservation of the somatostatinergic cells could be demonstrated only indirectly, by measuring NADPH diaphorase activity, an enzymatic activity associated selectively with somatostatinergic interneurons in the striatum (Qin et al 1992). The relative preservation of NADPH diaphorase-positive neurons after lesions are made does not appear to be associated artifactually with the expression of the enzyme in other cell types, such as reactive astrocytes or microglia, because cells expressing NADPH diaphorase have the same neuronal morphology as in the absence of lesions. Therefore, the results suggest that quinolinic acid-induced changes in somatostatin expression in striatal interneurons may affect the detection of these cells based on their somatostatin content.

To test the hypothesis that striatal somatostatinergic neurons are relatively preserved by excitotoxic mechanisms, we examined the effects of transient ischaemia on striatal somatostatinergic neurons in the gerbil (Chesselet et al 1990). Interruption of carotid blood flow for five minutes in female Mongolian gerbils produced, in most animals, a massive loss of striatal neurons. We confirmed loss of striatal efferent neurons by immunostaining for the neuropeptides enkephalin and substance P in the striatum and in their respective target areas, the globus pallidus and entopeduncular nucleus. In contrast, somatostatinergic neurons were relatively spared in the lesioned area.

The relative preservation of somatostatinergic neurons in the striatum after injections of quinolinic acid and ischaemia is similar to that observed in Huntington's disease (Ferrante et al 1985). Resistance of the somatostatinergic neurons in the rat may be related to strong expression of mitochondrial super-oxide dismutase (Inagaki et al 1991), an observation we have recently confirmed in the human (Zhang & Chesselet, unpublished work). Superoxide dismutase may contribute to the resistance of somatostatinergic interneurons to free radical generation, a consequence of NMDA receptor stimulation (Lafon-Cazal et al 1993).

Corticostriatal inputs and somatostatinergic neurons

The striatum receives a massive glutamatergic projection from the cerebral cortex (Flaherty & Graybiel 1994). Although medium-sized spiny neurons are a preferential target of the corticostriatal input, evidence suggests that cortical projections also contribute to the regulation of somatostatinergic neurons in the striatum. Lesions of the frontoparietal cortex following thermocoagulation of pial blood vessels increase the number of immunoreactive somatostatinergic neurons detected in the dorsolateral striatum ipsilateral to the lesion (Salin et al 1990b). This effect is probably related to an increased synthesis of the neurotransmitter because these lesions induce a marked increase in the level of somatostatin mRNA in the dorsolateral striatum (P. Salin & M.-F. Chesselet 1994, unpublished observations). This effect may also be related to a stimulatory effect of glutamate on somatostatin gene expression. After thermocoagulation of pial blood vessels, which interrupts blood flow to the cerebral cortex, cortical neurons die progressively over a period of one week (Salin et al 1990b), during which they are likely to release their neurotransmitter, glutamate. Because the increases in levels of glutamate released under these conditions are not sufficient to be detected with current dialysis methods (M.-F. Chesselet, T. J. Parry, F. Szele, N. Clavel, S. Djali & M. Robinson, unpublished observations) pharmacological experiments to block glutamate receptors in the striatum will be necessary to test this hypothesis.

In conclusion, although the function of somatostatinergic neurons in the basal ganglia remains obscure, analysis of somatostatin gene expression in these brain regions suggests that synthesis of this peptide is altered in numerous experimental conditions related to neurodegenerative diseases or pharmacological treatments resulting in motor side effects. Two patterns of effects emerge: stimulation of glutamatergic NMDA receptors (as may occur after ischaemia or in degenerative diseases) increases somatostatin gene expression, whereas dopamine depletion or blocking dopamine receptors, e.g. in Parkinson's disease or after neuroleptic treatment, results in decreased somatostatin expression in the striatum. The results suggest that somatostatinergic agonists acting in the striatum may be beneficial in akinetic movement disorders. However, increases in somatostatin mRNA levels following dopamine depletion in another region of the basal ganglia, the entopeduncular nucleus, suggest that region-specific agents will be necessary to counteract the opposite changes occurring in various somatostatinergic systems.

Acknowledgements

We are grateful to Dr R. Goodman for the gift of preprosomatostatin cDNA. Supported by PHS grants NS-29230 and MH-44894, the Pharmaceutical Manufacturers Association and the Tourette Syndrome Association.

References

Augood SJ, Kiyama H, Faull RL, Emson PC 1991 Dopaminergic D1 and D2 receptor antagonists decrease prosomatostatin mRNA expression in rat striatum. Neuroscience 44:35–44

Baldessarini RJ, Frankenburg FR 1991 Clozapine-a novel antipsychotic agent. N Engl J Med 324:746–754

Beal MF, Martin JB 1984 Effects of neuroleptic drugs on brain somatostatin-like immunoreactivity. Neurosci Lett 47:125–130

Beal MF, Kowall NW, Ellison DW, Mazurek MF, Swartz KJ, Martin JB 1986 Replications of the neurological characteristics of Huntington's disease by quinolinic acid. Nature 321:168–171

Bendotti C, Tarizzo G, Fumagalli F, Baldessari S, Samanin R 1993 Increased expression of preproneuropeptide Y and preprosomatostatin mRNA in striatum after selective serotonergic lesions in rats. Neurosci Lett 160:197–200

Bernard U, Normand E, Bloch B 1992 Phenotypical characterization of rat striatal neurons expressing muscarinic receptor genes. J Neurosci 12:3591–3600

Bissette G, Myers B 1992 Somatostatin in Alzheimers disease and depression. Life Sci 51:1389–1410

Chesselet M-F, Graybiel AM 1986 Striatal neurons expressing somatostatin-like immunoreactivity: evidence for a peptidergic interneuronal system in the cat. Neuroscience 17:547–571

Chesselet M-F, Reisine TD 1983 Somatostatin regulates dopamine release in rat striatal slices and cat caudate nuclei. J Neurosci 3:232–236

Chesselet M-F, Robbins E 1989 Characterization of striatal neurons expressing high levels of glutamic acid decarboxylase messenger RNA. Brain Res 492:237–244

Chesselet M-F, Gonzales C, Lin C-S, Polsky K, Jin B-K 1990 Ischemic damage in the gerbil striatum: relative sparing of somatostatinergic and cholinergic interneurons contrasts with loss of efferent neurons. Exp Neurol 110:209–218

Chesselet M-F, Gonzales C, Levitt P 1991 Heterogeneous distribution of the limbic system-associated membrane protein in the caudate nucleus and substantia nigra of the cat. Neuroscience 40:725–733

Chesselet M-F, Weiss-Wunder L-T 1994 Quantification of in situ hybridization histochemistry. In: Eberwine J, Valentino KL, Barchas JD (eds) In situ hybridization in neurobiology. Oxford University Press, New York p114–123

Dawson TM, Bredt DS, Fotuhi M, Hwang PM, Snyder S 1991 Nitric oxide synthase and neuronal NADPH diaphorase are identical in brain and peripheral tissues. Proc Natl Acad Sci USA 88:7797–7801

DiFiglia M, Aronin AN 1982 Ultrastructural features of immunoreactive somatostatin neurons in the rat caudate nucleus. J Neurosci 2:1267–1274

Ferrante RJ, Kowell NW, Beal MF, Richardson EP, Martin JB 1985 Selective sparing of a class of striatal neurons in Huntington's disease. Science 230:561–563

Fitzgerald LW, Dokla CP 1989 Morris water task impairment and hypoactivity following cysteamine-induced reductions of somatostatin-like immunoreactivity. 505:246–250

Flaherty AW, Graybiel AM 1994 Input-output organization of the sensorimotor striatum in the squirrel monkey. J Neurosci 14:599–610

Inagaki S, Suzuki K, Taniguchi N, Takagi H 1991 Localization of Mn-superoxide dismutase (Mn-SOD) in cholinergic and somatostatinergic-containing neurons in the rat neostriatum. Brain Res 549:174–177

Johansson O, Hökfelt T, Elde RP 1984 Immunohistochemical distribution of somatostatin-like immunoreactivity in the central nervous system of the adult rat. Neuroscience 13:265–339

Kiyama H, Emson PC 1990 Distribution of somatostatin mRNA in the rat nervous system as visualized by a novel non-radioactive in situ hybridization histochemistry method. Neuroscience 38:223–244

Lafon-Cazal M, Pieti S, Culcasi M, Bockaert J 1993 NMDA-dependent superoxide production and neurotoxicity. Nature 364:535–537

Martin J-L, Chesselet M-F, Raynor K, Gonzales C, Reisine TD 1991 Differential distribution of somatostatin receptor subtypes in rat brain revealed by newly developed somatostatin analogs. Neuroscience 41:581–593

Patel SC, Papachristou DN, Patel YC 1991 Quinolinic acid stimulates somatostatin gene expression in cultured rat cortical neurons. J Neurochem 56:1286–1291

Qin Y, Soghomonian J-J, and Chesselet M-F 1992 Effects of quinolinic acid on messenger RNAs encoding somatostatin and glutamic acid decarboxylase in the striatum of adult rats. Exp Neurol 115:200–211

Raynor K, Lucki I, Reisine T 1993 Somatostatin 1 receptors in the nucleus accumbens selectively mediate the stimulatory effect of somatostatin on locomotor activity in rats. J Pharm Exp Ther 265:67–73

Roberts RC, Ahn A, Swartz KJ, Beal MF, DiFiglia M, 1993 Intrastriatal injections of quinolinic or kainic acid: differential patterns of cell survival and the effects of data analysis on the outcome. Exp Neurol 124:274–282

Salin P, Mercugliano M, Chesselet M-F 1990a Differential effects of chronic haloperidol and clozapine on preprosomatostatin mRNA in the striatum, nucleus accumbens and frontal cortex of the rat. Cell Molec Neurobiol 10:127–140

Salin P, Kerkerian-Le Goff L, Heidet V, Epelbaum J, Nieoullon A 1990b Somatostatin-immunoreactive neurons in the rat striatum: effects of corticostriatal and nigrostriatal dopaminergic lesions. Brain Res 521:23–32

Sandell JH, Graybiel AM, Chesselet M-F 1986 A new enzyme marker for striatal compartmentalization: NADPH diaphorase activity in the caudate nucleus and putamen of the cat. J Comp Neurol 243:325–334

Soghomonian J-J, Chesselet M-F 1991 Lesions of the nigro-striatal pathway alter preposomatostatin mRNA levels in the striatum, entopeduncular nucleus and lateral hypothalamus of the rat. Neuroscience 42:49–59

Soghomonian J-J, Chesselet M-F, 1992 Effects of nigrostriatal lesions on the levels of messenger RNAs encoding two isoforms of glutamate decarboxylase in the globus pallidus and entopeduncular nucleus of the rat. Synapse 11:124–133

Strittmatter MM, Cramer H 1992 Parkinson's disease and dementia:clinical and neurochemical correlations. Neuroreport 3:413–416

Vincent SR, Johansson O 1983 Striatal neurons containing both somatostatin and avian pancreatic polypeptide (APP)-like immunoreactivities and NADPH diaphorase activity: a light and electron microscopic study. J Comp Neurol 217:264–270

Weiss LT, Chesselet M-F, 1989 Regional distribution and regulation of preprosomatostatin mRNA in the striatum as revealed by in situ hybridization histochemistry. Mol Brain Res 5:121–130

DISCUSSION

Robbins: Many of the somatostatinergic neurons in regions of the brain such as the cortex and the hippocampus are a subpopulation of the GABAergic (γ-aminobutyric acid) system. Is this also true for the caudate putamen?

Chesselet: No. This has been shown by either the use of NADPH diaphorase as a marker or that somatostatin does not co-localize with parvalbumin which is present in the GABAergic neurons. All the evidence suggests that somatostatinergic and GABAergic neurons form separate populations in the caudate putamen.

Robbins: Beal's report (Beal et al 1988) on the toxicity of quinolinic acid is controversial. We removed neurons from the cortex and the hippocampus, cultured them and then did dose–response toxicity curves with glutamate and NMDA (*N*–methyl–D–aspartate) analogues. At low doses, most NMDA agonists stimulated somatostatin production then, after a certain developmental point, they became toxic to somatostatin interneurons. Does this system operate differently?

Chesselet: No, not in the striatum. 60 nmoles of quinolinic acid kills many striatal efferent neurons, but preserves somatostatinergic neurons and results in the stimulation of somatostatin mRNA expression. When we double this dose, we see a decrease in somatostatin mRNA expression, but not in the number of NADPH diaphorase-positive cells which suggests that down-regulation of the message occurs first, then the neurons die. The discrepancy could be explained by the use of different markers for the neurons.

Robbins: So that the interneurons in the striatum have somatostatin, are not part of the GABAergic subset and are therefore a unique somatostatin-secreting subset.

Patel: The controversy surrounding Beal's results on the selective preservation of striatal somatostatin neurons relates to comparisons with data from Davies & Roberts and has now been resolved. Davies & Roberts (1987) injected quinolinic acid directly into the striatum where it produced non-specific neuronal loss at the immediate peri-injection site. This led to their report describing the absence of preservation of somatostatin-containing neurons after intrastriatal injections of quinolinic acid. However, sparing of somatostatin neurons is observed at the interface between the site of acute damage and normal brain as subsequently pointed out by Beal et al (1989).

When we add quinolinic acid to primary cultures of rat striatal neurons, we find not only selective preservation of neurons that are positive for somatostatin, neuropeptide Y and NADPH diaphorase, but also up-regulation of the function of these neurons. Somatostatin secretion and mRNA levels are increased in these neurons (Patel et al 1991). In dose–response studies *in vitro* with quinolinic acid and NMDA, we have also shown dose-dependent stimulation of somatostatin synthesis and secretion with concentrations of up to 10 mM. These neurons survive such concentrations of the excitotoxins which kills all other striatal neurons.

Robbins: It seems that somatostatin neurons may be protected in some areas of the brain, such as the striatum; however, in the hippocampus of patients with epilepsy, somatostatin interneurons are exposed to bursts of excess glutamate. The somatostatinergic cells then disappear as the disease progresses, followed by the principal neurons. Are the different somatostatin receptor types expressed differentially in specific brain regions? If so, could this be correlated with neurological diseases that are associated with different regions of the brain; for instance in the caudate of patients with Parkinson's disease or in the cortex of those with Alzheimer's disease, where there are abnormalities in the population of somatostatinergic target cells?

Chesselet: We certainly hope that the receptors are expressed differentially. Does anyone know which receptor type is expressed in the lateral habenula?

Bell: Bito et al (1994) have shown by *in situ* hybridization that *sstr4* mRNA is present in the lateral habenula.

Reisine: Most of these studies have looked only at receptor mRNA. It has not been established whether the corresponding receptor itself is in the same location because, with the exception of sstr2, radioligands that can distinguish between different receptors are not available. There are differences in the pattern of *sstr1–3* mRNA expression in the hippocampus, but antibodies against the different receptors are required to localize them to particular neuronal populations.

Robbins: An additional problem is getting the ligand to the site in the brain where the receptor is located.

Reisine: Yes, new compounds that can be used in the CNS need to be developed.

Patel: Somatostatin neurons are preserved selectively in your ischaemia model. Do you also see an up-regulation of somatostatin mRNA in this system, perhaps following chronic ischaemia?

Chesselet: We have not measured somatostatin mRNA levels in our ischaemia model. The somatostatin neurons are the only neurons in the striatum that express a high level of superoxide dismutase (SOD). This may be the reason why they are preserved, because studies with transgenic mice overexpressing SOD and experiments using antisense oligonucleotides against SOD suggest that this enzyme protects neurons against the effects of ischaemia and glutamate induced-toxicity (Chan et al 1990, Kinouchi et al 1991).

Patel: Is there something special about somatostatinergic neurons in patients with Huntington's disease which causes the up-regulation of somatostatin mRNA and their survival?

Chesselet: Maybe, but they are not the only neurons which survive in Huntington's disease; for example, the GABAergic interneurons survive and express a very high level of parvalbumin. Landwehrmeyer et al (1993) have shown that somatostatinergic neurons may have different NMDA receptor subunits which may explain their resistance. These receptors may be involved in the regulation of somatostatin mRNA expression in somatostatinergic striatal interneurons.

Epelbaum: The resistance of somatostatinergic neurons is not specific to the striatum. We studied the effects of the toxin ibotenic acid in nucleus basalis-lesioned rats in collaboration with Alain Beaudet (Moyse et al 1993). We observed an increase in somatostatin levels in the frontal cortex at the site of neuronal projection from the nucleus basalis. *In situ* hybridization of somatostatin message in brains of Alzheimer's patients also reveals that there is no change in the number of somatostatin-expressing neurons in the hippocampus and frontal cortex (Dournaud et al 1994).

Patel: Has anyone looked at the diseased neurons in the cortex of Huntington's patients?

Chesselet: There is a loss of neurons in the cortex, but it's very variable from one patient to another.

Robbins: Beal et al (1988) measured somatostatin levels by radioimmunoassay and found that they were not reduced in the cortex of Huntington's patients.

Reichlin: Somatostatin levels are low in the temporal lobe of epileptic patients. The interpretation of Richard Robbin's group was that low concentrations of somatostatin contribute to the epileptogenic effects (DeLanerolle et al 1989). My interpretation was that somatostatin levels were low simply because all the cells had been damaged. I also speculated that the low levels of somatostatin in brains of Alzheimer's patients was an unspecific consequence of brain damage. Your results suggest that my interpretations may be wrong because somatostatinergic neurons were resistant when you induced unilateral ischaemia.

Chesselet: Yes, somatostatinergic neurons are resistant in the striatum, but I don't know whether this is also true for the somatostatinergic neurons in the cortex. Is SOD present in these cells?

Reichlin: Some neurons in the temporal lobe are SOD positive.

There are several classes of somatostatinergic neurons. Richard Robbins, which of these were destroyed in the epileptic brains that you looked at?

Robbins: We studied two types of epilepsy: idiopathic epilepsy and epilepsy in people who have had tumours in the cortex. These are two very different groups, because if you have a tumour that causes your epilepsy, you don't lose somatostatin-expressing interneurons in the dentate hilus. If you have the idiopathic kind, virtually all somatostatin-expressing interneurons are missing; however, somatostatin receptor levels are increased in the granular cells of the dentate, which are the next neurons downstream. These neurons therefore show evidence of denervation and the receptors on these cells are up-regulated tremendously. We've always thought this might be very interesting therapeutically in temporal lobe epilepsy. You could deliver a receptor-specific somatostatin analogue to the dentate granular cells to shut them down. The blood–brain barrier in the dentate hilus is very leaky in epileptic patients, so it might not be too difficult to deliver somatostatin analogues to these cells.

Beaudet: Given that you find an increase in somatostatin mRNA levels in the entopeduncular nucleus following dopamine depletion, is there an increase in the expression of somatostatin in the internal segment of the globus pallidus in patients with Parkinson's disease?

Chesselet: I don't know. The patients whose brains we study have had the disease, and have been treated for the disease, for many years. It is difficult to compare this situation with the short-term (two to three weeks) effect of lesions in experimental animals.

Reichlin: Do dopaminergic neurons inhibit somatostatin secretion?

Chesselet: I believe the inhibition is indirect, through the subthalamic nucleus.

Reichlin: What is the neurotransmitter in the subthalamic nucleus?

Chesselet: Glutamate. Increases in glutamate may up-regulate somatostatin mRNA expression.

Reichlin: If you destroy the dopaminergic input to the subthalamic nucleus, do you increase the activity of the GABAergic neurons?

Chesselet: We believe that when lesions are made in the nigrostriatal dopaminergic pathway, the dopaminergic input to the subthalamic nucleus is also decreased. It was previously thought that the whole system was regulated by dopamine in the striatum. We have evidence, however, that dopamine can act directly in the subthalamic nucleus (Parry et al 1994). That explains why, after lesions are made in the dopaminergic pathway, the subthalamic nucleus becomes overactive. This then plays a major role in behaviour and is probably more important than the activity of the striatum. Lesions of the subthalamic nucleus completely abolish this effect.

Robbins: If you culture those regions of the cortex, dopamine has no major effect on somatostatin secretion, whereas in the hypothalamus, dopamine is stimulatory.

Chesselet: We also observe weak changes in the striatum. Dopamine may play a direct or indirect role in this regulation.

References

Beal M, Mazurek M, Ellison D et al 1988 Somatostatin and NPY concentrations in pathologically graded cases of Huntington's disease. Ann Neurol 23:562–569

Beal MF, Kowall NW, Schwartz KJ, Ferranti RJ, Martin JB 1989 Differential sparing of somatostatin–neuropeptide Y and cholinergic neurons following striatal excitotoxin lesions. Synapse 3:38–47

Bito H, Mori M, Sakanaka C et al 1994 Functional coupling of SSTR4, a major hippocampal somatostatin receptor, to adenylate cyclase inhibition, arachidonate release, and activation of the mitogen-activated protein kinase cascade. J Biol Chem 269:12722–12730

Chan PH, Chu L, Chen SF, Carlson EJ, Epstein CJ 1990 Attenuation of glutamate-induced neuronal swelling and toxicity in transgenic mice overexpressing human CuZn–superoxide dismutase. Acta Neurochir Suppl 51:245–247

Davies SW, Roberts PJ 1987 No evidence for preservation of somatostatin-containing neurons after intrastriatal injections of quinolinic acid. Nature 327:326–329

de Lanerolle NC, Kim JH, Robbins RJ 1989 Hippocampal interneuron loss and plasticity in human temporal lobe epilepsy. Brain Res 495:387–395

Dournaud P, Cervera-Pierrot P, Hirsch E et al 1994 Somatostatin mRNA-containing neurons in Alzheimer's disease: an *in situ* hybridization study in hippocampus, parahippocampal cortex and frontal cortex. Neuroscience 61:755–764

Kinouchi H, Epstein CJ, Mizui T, Carlson E, Chen SF, Chan PH 1991 Attenuation of focal cerebral ischemia injury in transgenic mice overexpressing CuZn superoxide dismutase. Proc Natl Acad Sci USA 88:11158–11162

Landwehrmeyer GB, Standaert DG, Testa CM et al 1993 Expression of NMDA and metabotropic glutamate receptors in somatostatin positive striatal interneurons. Soc Neurosci Abstr 19:129

Moyse E, Szigethy E, Danger JM et al 1993 Short-term and long-term effects of nucleus basalis magnocellularis lesions on cortical levels of somatostatin and its receptors in the rat. Brain Res 607:154–160

Parry TJ, Eberle-Wang K, Lucki I, Chesselet M-F 1994 Dopaminergic stimulation of subthalamic nucleus elicits oral dyskinesia in rats. Exp Neurol 128:181–190

Patel SC, Papachristou DN, Patel YC 1991 Quinolinic acid stimulates somatostatin gene expression in cultured rat cortical neurons. J Neurochem 56:1286–1291

Molecular biology of somatostatin receptors

Graeme I. Bell*, Kazuki Yasuda*, Haeyoung Kong†, Susan F. Law†, Karen Raynor† and Terry Reisine†

*Howard Hughes Medical Institute and Departments of Biochemistry and Molecular Biology, and Medicine, The University of Chicago, Chicago, IL 60637; and †Department of Pharmacology, University of Pennsylvania School of Medicine, Philadelphia, PA 19104, USA

Abstract. The diverse physiological effects of somatostatin are mediated by a family of cell surface receptors that bind somatostatin selectively and with high affinity. The somatostatin receptors are members of the seven transmembrane segment receptor superfamily and molecular cloning studies have identified five types, designated sstr1–5. The human somatostatin receptors vary in size from 364 (sstr5) to 418 (sstr3) amino acids with 46–61% amino acid identity between receptors, and 105 amino acids are invariant. The sequences of the seven putative α-helical membrane-spanning domains are more highly conserved than those of the extracellular N- and intracellular C-terminal domains. Two forms of sstr2 have been identified in the mouse, sstr2A and sstr2B, which differ in size and sequence of the intracellular C-terminal domain. These two forms of sstr2 are products of a common gene and are generated by alternative splicing with sstr2A and sstr2B being the products of the unspliced and spliced forms, respectively, of *sstr2* mRNA. Thus, functional diversity within the somatostatin receptor family may result from the expression of multiple types as well as from alternative splicing. The five somatostatin receptors have distinct patterns of expression in the central nervous system and peripheral tissues. They have also been expressesd *in vitro* and shown to have different pharmacological properties. Somatostatin analogues selective for sstr2, sstr3 and sstr5 have been identified which will facilitate *in vivo* studies of the functions of these somatostatin receptors. Such studies to date suggest that sstr2 mediates inhibition of growth hormone secretion and sstr5 mediates inhibition of insulin secretion. The molecular cloning and functional characterization of the somatostatin receptor family is a first step in elucidating the diverse effects of somatostatin on cellular functions.

1995 Somatostatin and its receptors. Wiley, Chichester (Ciba Foundation Symposium 1995) p 65–88

Somatostatin is a 14 amino acid polypeptide with diverse physiological properties that is distributed widely throughout the central nervous system (CNS) and peripheral tissues (reviewed in Weil et al 1992). It potently inhibits basal and stimulated secretion from a wide variety of endocrine and exocrine cells and functions as a neurotransmitter/neuromodulator in the CNS with effects on locomotor activity and cognitive function. Somatostatin also has antiproliferative effects and may be an important hormonal regulator of cell proliferation and differentiation. Somatostatin is derived by specific proteolytic processing of a larger precursor, prosomatostatin, of 92 amino acids in mammalian cells, with somatostatin comprising the C-terminal 14 residues. The primary translation product of somatostatin mRNA is a 116 amino acid protein, preprosomatostatin. Proteolytic processing of prosomatostatin at dibasic and monobasic residues generates somatostatin and somatostatin-28, a 28 amino acid polypeptide with similar but not identical physiological properties to somatostatin. Its sequence includes the somatostatin sequence with an N-terminal extension of 14 amino acids.

The physiological actions of somatostatin are mediated by high affinity receptors on the surface of responsive cells (Schonbrunn & Tashjian 1978) that are coupled by G proteins (Reisine et al 1995, this volume) to multiple effector systems including adenylyl cyclase, ion channels and tyrosine phosphatases (this volume: Bruns et al 1995, Kleuss 1995, Delesque et al 1995).

Cloning a family of somatostatin receptors

Pharmacological and biochemical studies suggested that there might be multiple somatostatin receptors (Brown et al 1981, Tran et al 1985, Raynor & Reisine 1989, Rens-Domiano & Reisine 1992). In fact, biochemical studies identified a large number of proteins varying in size from 21–228 kDa that could be labelled with radioactive somatostatin agonists. However, recent studies indicate that not all these proteins are bona fide somatostatin receptors, and molecular cloning studies have shown that one of these 'putative' somatostatin receptors, a protein of 90 kDa identified in human HGT1 gastric tumour cells, is the p86 subunit of the nuclear autoantigen Ku which is present in patients with systemic lupus erythematosus (LeRomancer et al 1993).

The recent cloning of five somatostatin receptor subtypes that bind somatostatin and somatostatin-28 selectively and with high affinity established the functional identity of the somatostatin receptor and showed that the physiological actions of somatostatin were mediated by a family of structurally related proteins with different pharmacological properties and distinct patterns of expression in the CNS and peripheral tissues.

Two different experimental approaches were used to clone the somatostatin receptor. The approach that we employed (Yamada et al 1992a) took advantage of the fact that somatostatin receptors were expressed in pancreatic islets. Using a polymerase chain reaction (PCR)-based strategy for cloning new G protein-coupled receptors and human pancreatic islet RNA as the substrate for the PCR, we identified two clones encoding novel G protein-coupled receptors. Binding studies using ligands for receptors known to be present on insulin-secreting β-cells showed that one of these receptors bound ([^{125}I] Tyr^{11}) somatostatin. We called this receptor sstr1, the IUPHAR (International Union of Pharmacology) committee on somatostatin receptors has recently proposed that lower case letters be used to designate the recombinant somatostatin receptors. The identification of the other somatostatin receptors was facilitated by the fact that the sequences of the somatostatin receptor genes were sufficiently similar and lacked introns, at least in their protein coding regions, that related genes could be isolated readily from cDNA or genomic libraries by low stringency cross-hybridization using the cloned somatostatin receptor genes as hybridization probes, or by PCR-based strategies using oligonucleotide primers that would amplify somatostatin receptor-like sequences: *sstr2* (Yamada et al 1992a), *sstr3* (Yasuda et al 1992, Yamada et al 1992b) and *sstr4* (Bruno et al 1992) were first identified in this manner. With the publication of the sequences of human and mouse *sstr1* and *sstr2*, searches of databases of 'orphan' G protein-coupled receptors, i.e. receptors for which the ligands were unknown, for sequences related to *sstr1* and *sstr2* identified rat *sstr2* (Meyerhof et al 1991) and *sstr5* (O'Carroll et al 1992). Kluxen et al (1992) employed an alternative approach to isolate somatostatin receptor cDNA clones. They used ([^{125}I] Tyr^{11}) somatostatin and ([^{125}I] Tyr^{3}) octreotide to screen a rat brain cDNA library expressed in mammalian cells and identified rat *sstr2* in this manner.

The sequences of five different somatostatin receptors have been reported (Table 1, Fig.1) designated sstr1–5, based on the order in which they were reported. The sequences of a rat brain somatostatin receptor (Bruno et al 1992) and rat pituitary somatostatin receptor (O'Carroll et al 1992) were reported simultaneously and both were named sstr4. The IUPHAR committee on somatostatin receptors proposed that sstr4 denotes the rat brain somatostatin receptor and sstr5 describes the rat pituitary somatostatin-28-preferring receptor.

The amino acid sequences of human (Yamada et al 1992a,b, Rohrer et al 1993, Yamada et al 1993a) and rat sstr1–5 (Meyerhof et al 1991, Kluxen et al 1992, Meyerhof et al 1992, Bruno et al 1992, O'Carroll et al 1992), mouse sstr1–3 (Yamada et al 1992a,b, Yasuda et al 1992), and bovine (Xin et al 1992) and porcine sstr2 (Matsumoto et al 1994) have been reported from the analysis of cDNA and/or genomic sequences. The amino acid sequences of sstr1, sstr2, sstr3 and sstr5 have been obtained from both cDNA and genomic clones whereas the amino acid sequence of sstr4 has been deduced only from the sequence of the gene. Comparisons of cDNA and genomic sequences are consistent with the absence of introns in the somatostatin receptor genes.

Although the somatostatin receptor genes in general lack introns, there is a cryptic intron in the mouse *sstr2* gene, and two forms of *sstr2* mRNA, *sstr2A* and *sstr2B*, have been identified (Vanetti et al 1992). They arise by alternative splicing with *sstr2A* and *sstr2B* respresenting the unspliced and spliced forms, respectively, of this subtype. Mouse *sstr2A* and *sstr2B* mRNAs encode proteins of 369 and 346 amino acids, respectively. The sequences of residues 1–331 are identical but the length and sequence of the intracellular C-termini differ (the site at which the sequences of mouse sstr2A and sstr2B diverge is noted by the arrow in Fig. 2). Transcripts of *sstr2B* have only been reported in mouse (Vanetti et al 1992) and rat tissues (Y Tokuyama & GI Bell, unpublished work) and there is presently no evidence for two forms of *sstr2* mRNA in human tissues.

The somatostatin receptors comprise a distinct subgroup of seven transmembrane domain receptors (Bell & Reisine 1993) that is most closely

```
                                                      <------ TM1 ------>
hsstr1   MFPNGTASSPSSSPSPSPGSCGEGGGSRGPGAGAADGMEEPGRNASQNGTLSEGQGSAILISFIYSVVCLVGLCG   75
hsstr2   M---DMADEPL-----------NGSHTWLSIPFDLNGSVVSTNTSNQTEPYYDLTSNAVL-TFIYFVVCIIGLCG   60
hsstr3   M---DMLHPSS-----------VSTTSEPENASSAWPPDATLGNVSAGPSPAGLAVSGVLIPLVYLVVCVVGLLG   61
hsstr4   MSAPSTLPPGGEEGL---------GTAWPSAANASSAPAEAEEAV--AGPGDARAAGMVAIQCIYALVCLVGLVG   64
hsstr5   M-EP--LFPAS-------------TPSWNASSPGAASGGGDNRTL--VGPAPSAGARAVLVPVLYLLVCAAGLGG   57
Consensus M-....L.P.S-----------....SW...A..A.............GP.......AVLI..IY.VVC.VGL.G

         ------>          <-------- TM2 --------->          <------ TM3 ------->
hsstr1   NSMVIYVILRYAKMKTATNIYILNLAIADELLMLSVPFLVTSTLLRHWPFGALLCRLVLSVDAVNMFTSIYCLTV   150
hsstr2   NTLVIYVILRYAKMKTITNIYILNLAIADELFMLGLPFLAMQVALVHWPFGKAICRVVMTVDGINQFTSIFCLTV   135
hsstr3   NSLVIYVVLRHTASPSVTNVYILNLALADELFMLGLPFLAAQNALSYWPFGSLMCRLVMAVDGINQFTSIFCLTV   136
hsstr4   NALVIFVILRYAKMKTATNIYILNLAVADELFMLSVPFVASSAALRHWPFGSVLCRAVLSVDGLNMFTSVFCLTV   139
hsstr5   NTLVIYVVLRFAKMKTVTNIYILNLAVADVLYMLGLPFLATQNAASFWPFGPVLCRLVMTLDGVNQFTSVFCLTV   132
Consensus N.LVIYVILRYAKMKT.TNIYILNLA.ADELFMLGLPFLA.Q.AL.HWPFG..LCRLVM.VDG.NQFTSIFCLTV

         ---->           <--------- TM4 --------->
hsstr1   LSVDRYVAVVHPIKAARYRRPTVAKVVNLGVWVLSLLVILPIVVFS-RTAANSDGT----VACNMLMPE-PAQRW   219
hsstr2   MSIDRYLAVVHPIKSAKWRRPRTAKMITMAVWGVSLLVILPIMIYA-GLRSNQWGR----SSCTINWPG-ESGAW   204
hsstr3   MSVDRYLAVVHPVRVARTAPVARTVSAAVWVASAVVVT--PRGM----STCHMQWPE-PAAAW   202
hsstr4   LSVDRYVAVVHPLRAATYRRPSVAKL-------INLGVWLASLLVTLPIAIFADTRPARGGQAVACNLQWPHPAW   207
hsstr5   MSVDRYLAVVHPLSSARWRRPRVAKLASAAAWVLSLCMSLPLLVF-------ADVQ--EGGTCNASWPE-PVGLW   197
Consensus MSVDRYLAVVHP..SARWRRP.VAK....AVWV.SL.V.LP..VF.-.....DG.-----.C...WPE-P..AW

         <--------- TM5 --------->                     <------ TM6 ------->
hsstr1   LVGFVLYTFLMGFLLPVGAICLCYVLIIAKMRMVALKA---GWQQRKRSERKITLMVMMVVMVFVICWMPFYVVQ   291
hsstr2   YTGFIIYTFILGFLVPLTIICLCYLFIIIKVKSSGIRV---GSSKRKKSEKKVTRMVSIVVAVFIFCWLPFYIFN   276
hsstr3   RAGFIIYTAALGFFGPLLVICLCYLLIVVKVRSAGRRVWAPSCQRRRRSERRVTRMVVAVVALFVLCWMPFYVLN   277
hsstr4   SAVFVVYTFLLGFLLPVLAIGLCYLLIVGKMRAVALRA---GWQQRRSEKKITRLVLMVVVVFVLCWMPFYVVQ   279
hsstr5   GAVFIIYTAVLGFFAPLLVICLCYLLIVVKVRAAGVRV---GCV-RRRSERKVTRMVLVVVLVFAGCWLPFFTVN   268
Consensus .AGFIIYTF.LGFL.PLL.ICLCYLLIV.KVR...G.RV---G.Q.RRRSERKVTRMV..VV.VFV.CWMPFYVVN

         ----->          <--------- TM7 --------->
hsstr1   LVNV-FAEQDDATVS---QLSVILGYANSCANPILYGFLSDNFKRSFQRIL--CLSWM---------------D   344
hsstr2   VSSVSMAISPTPALKGMFDFVVVLTYANSCANPILYAFLSDNFKKSFQNVL--CLVKV----------------   332
hsstr3   IVNVVCPLPEEPAFFGLYFLVVALPYANSCANPILYGFLSYRFKQGFRRILLRPSRRVRSQEPTVGPPEKTEEED   352
hsstr4   LLNLVVT----SLDATVNHVSLILSYANSCANPILYAFLSDNFRRSFQRVL--CLR-----------CCLLEGA   336
hsstr5   IVNLAVALPQEPASAGLYFFVVILSYANSCANPVLYGFLSDNFRQSFQKVL--CLR-----------KGS   325
Consensus .VNV..A....PA..G....VVIL.YANSCANPILYGFLSDNFK.SFQRVL--CLR..--------------...

hsstr1   NAAEEPVDY----YATALKSRAYSVE-------DFQPENLES--GGVFRNGTCTS------RITTL   391
hsstr2   SGTDDGERS----DSKQDKSRLNETT-------ETQR-TLLN--GDL--------------QTSI   369
hsstr3   EEEEDGEESREGGKGKEMNGRVSQITQPGTSGQERPPSRVASKEQQLLPQEASTGEKSSTMRISYL   418
hsstr4   GGAAEEEPLDYYAT---ALKSK---GGAGCMCPPLPCQQEALQ--PEPGRKRIPLT------RTTTF   388
hsstr5   GAKDADATE----------PR---PD------RIRQQQEATP--PAHRAAANGLM------QTSKL   364
Consensus ...E.....---....KSR..........----...Q......--...........------R.T.L
```

FIG. 1.   Comparison of the amino acid sequences of human (h) somatostatin receptors. Dashes indicate gaps inserted to produce this alignment. Invariant residues in the five human somatostatin receptors are shaded. The consensus indicates residues that are conserved in the sequences of three or more receptors. This alignment was generated using GeneWorks® (IntelliGenetics, Mountain View, California). The membrane-spanning seven α-helical domains (TM1–TM7) are noted.

TABLE 1. Somatostatin receptors

	Size (amino acids)		Chromosomal
Receptor	Human	Rat	location (human)
sstr1	391	391	14q13
sstr2	369	369	17q24
sstr3	418	428	22q13.1
sstr4	388	384	20p11.2
sstr5	364	363	16p13.3

As noted in the text, there are two forms of sstr2, termed sstr2A and sstr2B, in the mouse and rat that arise by alternative splicing. The sizes of mouse sstr2A and sstr2B are 369 and 346 amino acids, respectively. The chromosomal locations of the human somatostatin receptor genes are from Yamada et al (1993b), Yasuda et al (1993b), Panetta et al (1994) and J. Takeda, A. A. Fernald, K. Yamagata, M. M. Le Beau & G. I. Bell, unpublished observations. The human somatostatin gene is located in chromosome band 3q28 (Naylor et al 1983).

related to the opioid receptor family (Yasuda et al 1993a). There is approximately 30% identity between the sequences of somatostatin and opioid receptors. The human somatostatin receptors vary in size from 364 (sstr5) to 418 amino acids (sstr3) (Fig. 1, Table 1) with 46–61% identity between the sequences of subtypes (Table 2), and 105 residues (about 25%) are invariant (Fig. 2). The sequences of the seven α-helical transmembrane regions are most similar and those of the N- and C-termini are the most divergent.

Comparisons of the sequences of the receptors from different species indicate the sequence of sstr1 is the most highly conserved with 97% identity between the human and rat proteins, and the sequence of sstr5 is the most divergent with only 81% identity between the human and rat sequences. There is 92, 86 and 89% identity between human and rat sstr2, sstr3 and sstr4, respectively. The amino acid differences within a receptor subtype are located primarily in the regions of the extracellular N-domain and intracellular C-termini with the sequences of the membrane-spanning α-helical domains and intracellular connecting loops being more highly conserved.

The five human somatostatin receptor genes are located on different chromosomes (Table 1). DNA polymorphisms have been identified in the *sstr1* and *sstr2* genes that will facilitate genetic studies of these loci (Yamada et al 1993b).

Model for the membrane topology of the somatostatin receptors

The predicted orientation of the somatostatin receptors in the plasma membrane is similar to that proposed for other members of the seven transmembrane domain receptors. A model for the orientation of human sstr2 in the plasma membrane is shown in Fig. 2 and the probable arrangement of seven membrane-spanning α-helical domains is shown in the inset.

The somatostatin receptors are glycoproteins. The carbohydrate component may be involved in promoting high affinity ligand binding (Rens-Domiano & Reisine 1991). There are one or more consensus sequences (Asn-X-Ser/Thr; N-X-S/T) for Asn-linked glycosylation in the extracellular N-terminal domain of each subtype consistent with a role for oligosaccharide addition in somatostatin receptor function.

Several conserved sequence patterns that characterize many seven transmembrane domain receptors are also present in the somatostatin receptors (Probst et al 1992, Baldwin 1994). These sequence motifs may be required to generate the appropriate tertiary structure necessary for functional activity and include the sequences GN..V, NLA.AD, S...L...S.DRY, W..S.....P, F..P, F..CW.P and NS..NP..Y in transmembrane domains 1–7, respectively. The somatostatin receptors also have conserved Cys residues in extracellular loops 1 and 2 that may form a disulphide bond and stabilize the receptor structure. The intracellular C-terminal domains of sstr1, sstr2, sstr4 and sstr5 contain a sequence, LCL, which is palmitoylated in the β_2-adrenergic receptor (O'Dowd et al 1989) and anchors this region of the β_2-adrenergic receptor to the membrane. These somatostatin receptor subtypes may be similarly modified; however, this motif is not present in human, mouse or rat sstr3 implying that palmitoylation may not play a critical role in somatostatin receptor function.

In addition to structural motifs shared with other G protein-coupled seven transmembrane domain receptors, Rohrer et al. (1993) examined the somatostatin receptor sequences for structural features that may uniquely characterize the somatostatin receptor family. They suggested that the motif N.FTS located in transmembrane domain 3 may be specific for the somatostatin receptors but recent studies indicate that this motif is also present in the related opioid receptors. A putative phosphorylation site R..SE in the third intracellular loop and the motif L.YANSCANP.LY.FLS in transmembrane domain 7 appear to be specifically conserved in the somatostatin receptors. In this regard, a somatostatin receptor-like sequence has been reported in the nematode

TABLE 2. Amino acid identity between human somatostatin receptors

| | Receptor | | | | |
	sstr1	sstr2	sstr3	sstr4	sstr5
sstr1	–	47%	46%	61%	49%
sstr2		–	51%	47%	52%
sstr3			–	46%	59%
sstr4				–	51%
sstr5					–

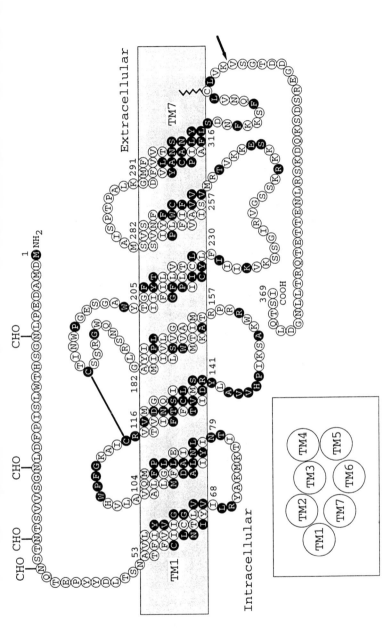

FIG. 2. Model for the orientation of the somatostatin receptor in the plasma membrane. The sequence of human sstr2 is shown. The seven membrane-spanning α-helical domains are shown. The invariant amino acid residues among the human somatostatin receptors are shown as filled circles. The four potential sites for Asn-linked glycosylation, indicated by 'CHO', in the extracellular N-terminal domain are noted. Cys-115 and Cys-193 are predicted to be linked by a disulphide bond. Cys-328 may be palmitoylated and anchor this region of the intracellular C-terminal domain in the plasma membrane. The inset shows the possible arrangement of the seven membrane-spanning α-helical domains (TM1–TM7) based on the structure of rhodopsin, a structurally related protein (Schertler et al 1993). The single letter abbreviations for the amino acids are used. The arrow indicates the corresponding location in the sequence mouse sstr2 where the sequences of sstr2A and sstr2B diverge because of alternative splicing.

Caenorhabditis elegans (Wilson et al 1994). However, this protein lacks the somatostatin receptor signature motif in transmembrane domain 7 and we predict is likely to be a receptor for a ligand other than somatostatin.

Tissue distribution of somatostatin receptor subtypes

The tissue distribution of transcripts encoding the five cloned somatostatin receptors has been examined using several different procedures including Northern blotting, RNase protection, reverse transcriptase-polymerase chain reaction amplification of cellular RNA and *in situ* hybridization histochemistry. The mRNAs for the five somatostatin receptors are expressed widely in human and rodent tissues and have distinct but overlapping patterns of expression (Yamada et al 1992a,b, Yasuda et al 1992, Breder et al 1992, Meyerhof et al 1991, 1992, Bruno et al 1992, 1993, O'Carroll et al 1992). All receptors are expressed in the CNS and *sstr1*, *sstr2*, *sstr3* and *sstr4* mRNAs can be readily detected in adult rat brain by RNA blotting. The levels of *sstr5* mRNA are much lower and can only be detected using a more sensitive RNase protection assay.

In situ hybridization studies (Breder et al 1992, Yasuda et al 1992, Kaupmann et al 1993, Kong et al 1994, Pérez et al 1994, Bito et al 1994) have shown that *sstr1*, *sstr2*, *sstr3* and *sstr4* mRNAs are present throughout the neocortex, the hippocampal formation and amygdala, consistent with a role for somatostatin in the regulation of complex integrative activities such as locomotor activity, learning and memory, and suggesting that these effects of somatostatin are mediated by multiple receptors. In addition, the expression of *sstr1*, *sstr2*, *sstr3* and *sstr4* mRNAs in the piriform cortex, the primary olfactory cortex in the rodent, suggests that somatostatin plays an important role in the processing and modulation of primary sensory information and that this effect of somatostatin is also mediated by multiple receptors. There are also high levels of *sstr3* mRNA in the olfactory tract, again suggesting a specific role of somatostatin in sensory processing. *sstr2* and *sstr4* mRNAs are specifically expressed at high levels in the habenula (medial habenula, *sstr2* mRNA; lateral habenula, *sstr4* mRNA). The localization of *sstr2* and *sstr4* in this region suggests that somatostatin may regulate communication between the basal ganglia and mesolimbic structures and the raphe nuclei. *sstr1*, *sstr2*, *sstr3* and *sstr5* mRNAs are expressed in the hypothalamus suggesting that these receptors may be involved in the regulation of autonomic and neuroendocrine function.

Ligand binding studies have shown that there are few somatostatin receptors in the adult cerebellum (Leroux et al 1995, this volume). The RNA blotting, RNase protection and *in situ* hybridization studies showed no significant expression of *sstr1*, *sstr2*, *sstr4* or *sstr5* mRNA in this brain region. However, there were very high levels of *sstr3* mRNA in the granular cell layer of cerebellum. In fact, the level of *sstr3* mRNA in this region was higher than in other regions of the brain. The granular cell layer contains a large number of densely packed small neurons that do not project from the cerebellar cortex but function as inhibitory interneurons to modulate the activity of the Purkinje neurons in the Purkinje cell layer. These neurons also express somatostatin (Shiraishi et al 1993). We do not have an explanation for the apparently paradoxical presence of *sstr3* mRNA but absence of somatostatin binding sites in adult cerebellum. We do not believe that the neurons in the granular layer express high levels of *sstr3* mRNA but do not translate it. Rather, sstr3 in the cerebellum may be in a low affinity state and thus unable to be labelled specifically by somatostatin. In this regard, the sequence of sstr3 is unique in that the intracellular C-terminal domain contains a region that is rich in Glu residues, e.g. 13 of 16 Glu residues in mouse sstr3 and 9 and 12 amino acids in the human sequence. This Glu-rich sequence may confer novel properties to this receptor in the interneurons of the granular layer.

Somatostatin receptor mRNA has also been identified in peripheral tissues. Several tissues such as pituitary and spleen expressed all five receptors with sstr2 expressed most strongly in the pituitary, followed by sstr1 and sstr3, then sstr5, then sstr4; and sstr3 expressed most strongly in the spleen, followed by sstr1, sstr4 and sstr5, then sstr2 (Bruno et al 1993). Adrenal glands express high levels of *sstr2* mRNA (Kong et al 1994) and heart expresses *sstr4* mRNA (Bruno et al 1993). Somatostatin receptor mRNAs are also expressed in the gastrointestinal tract (*sstr1*), kidney (*sstr3* and *sstr4*) and lung (*sstr4*). Somatostatin receptor mRNA has also been identified in tumours (Kubota et al 1994, Greenman & Melmed 1994, Reubi et al 1994) and the efficacy of the somatostatin analogue, octreotide (SMS, Sandostatin™), in inhibiting tumour proliferation and hormone secretion may depend upon the expression of sstr2 by the tumour.

Comparison of the pattern of somatostatin receptor expression in human and rodent tissues suggests that there may be species differences in sites and/or levels of expression of the different receptors. Somatostatin receptor mRNA is expressed generally at low levels in most tissues which may either be due to differences in the sensitivity of the methods used to detect different transcripts or to factors such as age or overall metabolic status. For example, 'normal' human tissues are usually obtained from trauma victims who have been maintained on life-support systems or from patients undergoing surgery for various reasons and thus may not represent truly normal tissue.

Pharmacological properties of the cloned receptors

The cloned somatostatin receptors have been expressed in a variety of cultured cells having few endogenous somatostatin receptors and these cell lines have facilitated pharmacological studies of the individual receptors (Table 3) (Rens-Domiano et al 1992, Raynor et al 1993a,b, O'Carroll et al 1994). These studies have shown that sstr1 binds somatostatin and somatostatin-28 with high and similar affinities. However, few analogues of somatostatin bind to sstr1 with high affinity and none bind selectively to this receptor.

TABLE 3. Ligand binding properties of the cloned somatostatin receptors

Receptor	IC_{50} (nM)							
	Somatostatin	Somatostatin-28	Octreotide	MK 678	NC4 28B	BIM 23052	BIM 23056	L-362,855
human sstr1	0.1	0.1	>1000	>1000	>1000	23	>1000	>1000
mouse sstr2	0.3	0.4	2.4	0.2	0.002	32	>1000	29
mouse sstr3	0.1	0.1	150	268	100	0.4	0.02	30
human sstr4	1.2	0.3	>1000	>1000	>1000	18	160	63
rat sstr5	0.3	0.05	0.2	1.3	1	0.004	43	0.3
human sstr5	0.2	0.05	32	23	–	4	–	0.016

These data are from Raynor et al (1993a,b) and O'Carroll et al (1994).
The structures of the somatostatin analogues are:

octreotide   D-Phe-c[Cys-Phe-D-Trp-Lys-Thr-Cys]-Thr-ol
MK 678   c[N-Me-Ala-Tyr-D-Trp-Lys-Val-Phe]
NC4-28B   D-Phe-c[Cys-Tyr-D-Trp-Lys-Ser-Cys]-Nal-NH$_2$
BIM-23052   D-Phe-Phe-Phe-D-Trp-Lys-Thr-Phe-Thr-NH$_2$
BIM-23056   D-Phe-Phe-Tyr-D-Trp-Lys-Val-Phe-Nal-NH$_2$
L-362,855   c[Aha-Phe-p-Cl-Phe-D-Trp-Lys-Thr-Phe]
Aha, 7-aminoheptanoic acid; Nal, β-(2-naphthyl) alanine.

sstr2 exhibits high and similar affinity for somatostatin and somatostatin-28 and can be specifically labelled with [^{125}I] MK 678, a hexapeptide analogue of somatostatin. There are no apparent differences in the ligand-binding properties of mouse sstr2A and sstr2B (Reisine et al 1993). sstr2 has the highest affinity of the cloned somatostatin receptors for octreotide.

sstr3 has a high affinity for somatostatin and somatostatin-28 but low affinity for MK 678 and octreotide. The linear somatostatin analogue BIM 23056 appears to bind selectively to sstr3 and may be useful for *in vivo* studies of the functional properties of this receptor subtype.

There is 61% identity between the sequences of sstr1 and sstr4, and like sstr1, sstr4 has a high affinity for somatostatin and somatostatin-28 and a low affinity for synthetic somatostatin analogues. No sstr4-selective agonists have been identified.

The ligand-binding properties of sstr5 are unique and this receptor has a four to sixfold greater affinity for somatostatin-28 than somatostatin. Several stable, rigid analogues of somatostatin have been identified that specifically bind to sstr5. One of these, L-362,855, shows a 1000-fold higher affinity for human sstr5 than for other somatostatin receptor subtypes.

The human and rodent forms of sstr1, sstr2, sstr3 and sstr4 appear to have similar ligand-binding properties. However, there are significant differences in the affinity of human and rat sstr5 for octapeptide analogues of somatostatin (O'Carroll et al 1994). Human sstr5 has a 160-fold lower affinity than rat sstr5 for octreotide. This result implies that the ligand-binding domains of human and rat sstr5 differ subtly, since they show similar high affinities for somatostatin-28 and somatostatin. It may be possible to use this difference in affinity to identify key residues in rat and human sstr5 responsible for high affinity binding to octreotide.

Future directions

Somatostatin has attracted attention because of its diverse physiological actions. Some of these effects, such as the inhibition of growth hormone secretion from both normal pituitaries and growth hormone secreting tumours, the basal and stimulated secretion by other endocrine and exocrine cells and inhibition of cell proliferation, are targets for specific therapeutic agents. However, studies of somatostatin action have been hampered by lack of a detailed understanding of its receptors, including numbers, sequences, tissue distribution and biochemical and pharmacological properties. The recent cloning and characterization of the somatostatin receptor gene family means that we now have a better understanding of how somatostatin exerts its physiological effects.

The cloning and functional characterization of the somatostatin receptors have shown that the diverse physiological effects of somatostatin are mediated by five structurally related proteins with distinct ligand binding properties and patterns of expression in the CNS and periphery. It is now possible to begin to examine critically the role of each subtype in mediating specific physiological effects of somatostatin. The availability of agonists selective for sstr2, sstr3 and sstr5 will facilitate studies of the specific physiological functions of these receptors. Although selective agonists have not yet been identified for either sstr1 or sstr4, the availability of cell lines expressing high levels of these proteins will aid in the search for agonists that bind with high affinity.

Information gained from the cloning and pharmacological studies provides a better understanding of somatostatin receptor function. Thus, sstr2 is likely to be the major receptor subtype that mediates the inhibition of growth hormone release by somatostatin (Raynor et al 1993a). The predominant role of sstr2 in mediating the inhibition of growth hormone release is consistent with the high potency of sstr2 agonists such as octreotide in inhibiting growth hormone secretion in patients with pituitary adenomas, and the high levels of *sstr2* mRNA in pituitary compared to other somatostatin receptors.

Somatostatin-28 is more potent than somatostatin in reducing insulin release from pancreatic islets (Brown et al 1981). The higher affinity of sstr5 for somatostatin-28 compared to somatostatin suggests that this receptor mediates this effect and recent studies of Rossowski and Coy (1993), showing that sstr5-selective agonists were as potent as somatostatin in reducing insulin secretion *in vivo* in rats, have confirmed this expectation.

Somatostatin analogues such as octreotide are used clinically in patients with acromegaly, metastatic carcinoid and vasoactive intestinal peptide-secreting tumours to inhibit hormone secretion (Lamberts et al 1995, this volume). Somatostatin and several of its analogues are potentially useful antiproliferative agents for treating certain types of tumours and in many cases may even cause tumour shrinkage. Somatostatin analogues have also been used as imaging agents to detect certain tumours that express somatostatin receptors. The diversity of somatostatin actions suggests that it will have other potential therapeutic uses. The localization of specific somatostatin receptors in regions of the brain implicated in locomotor activity, sensory perception, learning and memory suggest that non-peptide somatostatin analogues that can cross the blood–brain barrier may have potential therapeutic applications. The elucidation of the three-dimensional structures of the somatostatin analogues MK 678 (Huang et al 1992) and RC 160 (Varnum et al 1994) will aid the design of new selective agonists including non-peptide compounds. The cloning of the somatostatin receptor family represents an important beginning in unravelling the myriad physiological functions of somatostatin.

Acknowledgements

These studies were supported by the Howard Hughes Medical Institute and U.S. Public Health Service grants MH 45533, MH 48518, DK 20595, DK 42086 and DK 44840. K.Y. was supported by a mentor-based fellowship from the American Diabetes Association, Inc.

References

Baldwin JM 1994 Structure and function of receptors coupled to G proteins. Curr Opin Cell Biol 6:180–190

Bell GI, Reisine T 1993 Molecular biology of somatostatin receptors. Trends Neurosci 16:34–38

Bito H, Mori M, Sakanaka C et al 1994 Functional coupling of sstr4, a major hippocampal somatostatin receptor, to adenylyl cyclase inhibition, arachidonate release, and activation of the mitogen-activated protein kinase cascade. J Biol Chem 269:12722–12730

Breder CD, Yamada Y, Yasuda K, Seino S, Saper CB, Bell GI 1992 Differential expression of somatostatin receptor subtypes in brain. J Neurosci 12:3920–3934

Brown M, Rivier J, Vale W 1981 Somatostatin-28: selective action on the pancreatic β-cell and brain. Endocrinology 108:2391–2396

Bruno JF, Xu Y, Song JF, Berelowitz M 1992 Molecular cloning and functional expression of a brain-specific somatostatin receptor. Proc Natl Acad Sci USA 89:11151–111555

Bruno JF, Xu Y, Song JF, Berelowitz M 1993 Tissue distribution of somatostatin receptor subtype messenger ribonucleic acid in the rat. Endocrinology 133:2561–2567

Bruns Ch, Weckbecker G, Raulf F et al 1995 Characterization of somatostatin receptor subtypes. In: Somatostatin and its receptors. Wiley, Chichester (Ciba Found Symp 190) p 89–110

Delesque N, Buscail L, Estève JP et al 1994 A tyrosine phosphatase is associated with the somatostatin receptor. In: Somatostatin and its receptors. Wiley, Chichester (Ciba Found Symp 190) p 187–203

Greenman Y, Melmed S 1994 Heterogeneous expression of two somatostatin receptor subtypes in pituitary tumors. J Clin Endocrinol & Metab 78:398–403

Huang Z, He Y-B, Raynor K, Tallent M, Reisine T, Goodman M 1992 Main chain and side chain chiral methylated somatostatin analogs: syntheses and conformational analyses. J Am Chem Soc 114:9390–9410

Kaupmann K, Bruns C, Hoyer D, Seuwen K, Lübbert H 1993 Distribution and second messenger coupling of four somatostatin receptor subtypes expressed in brain. FEBS Lett 331:53–59

Kleuss C 1995 Somatostatin modulates voltage-dependent calcium channels in GH_3 cells via a specific G_0 splice variant. In: Somatostatin and its receptors. Wiley, Chichester (Ciba Symp 190) p 171–186

Kluxen F-W, Bruns C, Lübbert H 1992 Expression cloning of a rat brain somatostatin receptor cDNA. Proc Natl Acad Sci USA 89:4618–4622

Kong H, DePaoli AM, Breder CD, Yasuda K, Bell GI, Reisine T 1994 Differential expression of messenger RNAs for somatostatin receptor subtypes SSTR1, SSTR2 and SSTR3 in adult rat brain: analysis by RNA blotting and *in situ* hybridization histochemistry. Neuroscience 59:175–184

Kubota A, Yamada Y, Kagimoto S et al 1994 Identification of somatostatin receptor subtypes and an implication for the efficacy of somatostatin analogue SMS 201–955 in treatment of human endocrine tumors. J Clin Invest 93:1321–1325

Lamberts SWJ, de Herder WW, van Koetsveld PM et al 1995 Somatostatin receptors: clinical implications for endocrinology and oncology. In: Somatostatin and its receptors. Wiley, Chichester (Ciba Found Symp) p 222–239

LeRomancer M, Cherifi Y, Reyl-Desmars F, Lewin MJM 1993 Somatostatin specifically binds p86 subunit of the autoantigen Ku. CR Acad Sci Ser III Sci Vie 316:1283–1285

Leroux P, Bodenant C, Bologna E, Gonzalez B, Vaudry H 1995 Transient expression of somatostatin receptors in the brain during development. In: Somatostatin and its receptors. Wiley, Chichester (Ciba Found Symp) p 127–141

Matsumoto K, Yokogoshi Y, Fujinaka Y, Zhang C, Saito S 1994 Molecular cloning and sequencing of porcine somatostatin receptor 2. Biochem Biophys Res Commun 199:298–305

Meyerhof W, Paust H-J, Schönrock C, Richter D 1991 Cloning of a cDNA encoding a novel putative G-protein-coupled receptor expressed in specific brain regions. DNA Cell Biol 10:689–694

Meyerhof W, Wulfsen I, Schönrock C, Fehr S, Richter D 1992 Molecular cloning of a somatostatin-28 receptor and comparison of its expression pattern with that of a somatostatin-14 receptor in rat brain. Proc Natl Acad Sci USA 89:10267–10271

Naylor SL, Sakaguchi AY, Shen L-P, Bell GI, Rutter WJ, Shows TB 1983 Polymorphic human somatostatin gene is located on chromosome 3. Proc Natl Acad Sci USA 80:2686–2689

O'Carroll A-M, Lolait SJ, König M, Mahan LC 1992 Molecular cloning and expression of a pituitary somatostatin receptor with preferential affinity for somatostatin-28. Mol Pharmacol 42:939–946

O'Carroll A-M, Raynor K, Lolait SJ, Reisine T 1994 Characterization of cloned human somatostatin receptor sstr5. Mol Pharmacol 46:291–298

O'Dowd BF, Hnatowich M, Caron MG, Lefkowitz RJ, Bouvier M 1989 Palmitoylation of the human β-adrenergic receptor: mutation of Cys[341] in the carboxyl tail leads to an uncoupled nonpalymitoylated form of the receptor. J Biol Chem 264:7564–7569

Panetta R, Greenwood MT, Warszynska A et al 1994 Molecular cloning, functional characterization, and chromosomal localization of a human somatostatin receptor (somatostatin receptor type 5) with preferential affinity for somatostatin-28. Mol Pharmacol 45:417–427

Pérez J, Rigo M, Kaupmann K et al 1994 Localization of somatostatin (SRIF) SSTR1, SSTR2 and SSTR3 receptor mRNA in rat brain by in situ hybridization. Naunyn-Schmiedebergs Arch Pharmakol 349:145–160

Probst WC, Snyder LA, Schuster DI, Brosius J, Sealfon SC 1992 Sequence alignment of the G-protein coupled receptor superfamily. DNA Cell Biol 11:1–20

Raynor K, Reisine T 1989 Analogues of somatostatin selectively label distinct subtypes of somatostatin receptors in rat brain. J Pharmacol Exp Ther 251:510–517

Raynor K, Murphy WA, Coy DH et al 1993a Cloned somatostatin receptors: identification of subtype-selective peptides and demonstration of high affinity binding of linear peptides. Mol Pharmacol 43:838–844

Raynor K, O'Carroll A-M, Kong HY et al 1993b Characterization of cloned somatostatin receptors SSTR4 and SSTR5. Mol Pharmacol 44:385–392

Reisine T, Kong H, Raynor K et al 1993 Splice variant of the somatostatin 2 subtype, somatostatin receptor 2B, couples to adenylyl cyclase. Mol Pharmacol 44:1008–1015

Reisine T, Woulfe D, Raynor K et al 1995 Interaction of somatostatin receptors with G proteins and cellular effector systems. In: Somatostatin and its receptors. Wiley, Chichester (Ciba Found Symp 190) p 160–170

Rens-Domiano S, Reisine T 1991 Structural analysis and functional role of the carbohydrate component of somatostatin receptors. J Biol Chem 226:20094–20102

Rens-Domiano S, Reisine T 1992 Biochemical and functional properties of somatostatin receptors. J Neurochem 58:1987–1996

Rens-Domiano S, Law SF, Yamada Y, Seino S, Bell GI, Reisine T 1992 Pharmacological properties of two cloned somatostatin receptors. Mol Pharmacol 42:28–34

Reubi JC, Schaer JC, Waser B, Mengold G 1994 Expression and localization of somatostatin receptor SSTR1, SSTR2 and SSTR3 messenger RNAs in primary human tumours using in situ hybridization. Cancer Res 54:3455–3459

Rohrer L, Raulf F, Bruns C, Buettner R, Hofstaedter F, Schüle R 1993 Cloning and characterization of a fourth human somatostatin receptor. Proc Natl Acad Sci USA 90:4196–4200

Rossowski WJ, Coy DH 1993 Potent inhibitory effects a type four receptor-selective somatostatin analog on rat insulin release. Biochem Biophys Res Commun 197:366–371

Schertler GFX, Villa C, Henderson R 1993 Projection structure of rhodopsin. Nature 362:770–772

Schonbrunn A, Tashijian AH Jr 1978 Characterization of functional receptors for somatostatin in rat pituitary cells in culture. J Biol Chem 253:6473–6483

Shiraishi S, Kuriyama K, Saika T et al 1993 Autoradiographic localization of somatostatin mRNA in the adult rat lower brainstem: observation of the double illumination technique. Neuropeptides 24:71–79

Tran VT, Beal MF, Martin JB 1985 Two types of somatostatin receptors differentiated by cyclic somatostatin analogs. Science 228:492–495

Vanetti M, Kouba M, Wang XM, Vogt G, Höllt V 1992 Cloning and expression of a novel mouse somatostatin receptor (sstr2B). FEBS Lett 311:290–294

Varnum JM, Thakur ML, Schally AV, Jansen S, Mayo KH 1994 Rhenium-labelled somatostatin analogue RC-160: ^{1}H NMR and computer modeling conformational analysis. J Biol Chem 269:12583–12588

Weil C, Müller EE, Thorner MO (eds) 1992 Somatostatin. Springer-Verlag, Berlin (Basic Clin Aspects Neurosci 4)

Wilson R, Ainscough R, Anderson K et al 1994 2.2 Mb of contiguous nucleotide sequence from chromosome III of C. elegans. Nature 368:32–38

Xin WW, Wong M-L, Rimland J, Nestler EJ, Duman RS 1992 Characterization and functional expression of somatostatin receptor isolated from locus coeruleus. GenBank, accession no L06613

Yamada Y, Post SR, Wang K, Tager HS, Bell GI, Seino S 1992a Cloning and functional characterization of a family of human and mouse somatostatin receptors expressed in brain, gastrointestinal tract, and kidney. Proc Natl Acad Sci USA 89:251–255

Yamada Y, Reisine T, Law SF et al 1992b Somatostatin receptors, an expanding gene family: cloning and functional characterization of human SSTR3, a protein coupled to adenylyl cyclase. Mol Endocrinol 6:2136–2142

Yamada Y, Kagimoto S, Kubota A et al 1993a Cloning, functional expression and pharmacological characterization of a fourth (hSSTR4) and a fifth (hSSTR5) human somatostatin receptor subtype. Biochem Biophys Res Commun 195:844–852

Yamada Y, Stoffel M, Espinosa R et al 1993b Human somatostatin receptor genes: localization to human chromosomes 14, 17 and 22 and identification of simple tandem repeat polymorphisms. Genomics 15:449–452

Yasuda K, Rens-Domiano S, Breder CD et al 1992 Cloning of a novel somatostatin receptor, sstr3, coupled to adenylyl cyclase. J Biol Chem 267:20422–20428

Yasuda K, Raynor K, Kong J et al 1993a Cloning and functional comparison of kappa and delta opioid receptors from mouse brain. Proc Natl Acad Sci USA 90:6736–6740

Yasuda K, Espinosa R, Davis EM, Le Beau MM, Bell GI 1993b Human somatostatin receptor genes: localization of sstr5 to human chromosome 20p 11.2 Genomics 17:785–786

DISCUSSION

Reichlin: Are the orphan receptors that you encountered in the search for glucagon receptors members of the glucagon receptor family?

Bell: No, the glucagon receptor is a member of a different gene family (Jelinek et al 1993) and primers that we selected would not have amplified the glucagon receptor sequence.

Robbins: You showed how the seven transmembrane structures might look (Fig. 2). Agi Schonbrunn showed that somatostatin seems to conduct K^+ very quickly after it is applied to the membrane (Koch et al 1988, White et al 1991). Is the somatostatin receptor an ionophore? Could we predict from the primary structure whether the seven transmembrane domains could form a circular structure that transports ions through the middle?

Bell: Yatani et al (1987) have shown that a pertussis toxin-sensitive G protein directly couples the somatostatin receptor to K^+ channels which rules out that the somatostatin receptor itself is a ligand-gated ion channel.

Schonbrunn: That's correct. Somatostatin's effect on K^+ channel activity requires coupling of the receptor to a pertussis toxin-sensitive G protein and involves protein dephosphorylation (Koch et al 1988, White et al 1991).

Robbins: Are there any regions in the intracellular domains that look like they have affinities for other enzymes, kinases for instance?

Bell: The intracellular regions of the somatostatin receptors contain putative phosphorylation sites for various protein kinases; however, there is no direct evidence showing phosphorylation of these sites and modulation of receptor function.

Reichlin: You showed that sequences of the putative transmembrane segments are, in some instances, highly conserved (Fig. 2). I thought that transmembrane regions were not very important functionally and largely provided topographic requirements for transmembrane localization. However, your results suggest that these regions could be functionally important. Could they represent sites that interact with G proteins?

Bell: I don't believe that G proteins interact with the transmembrane domains. The evidence suggests that G proteins interact primarily with the third intracellular loop and perhaps the intracellular C-terminal domain (Reisine et al 1995, this volume).

Reichlin: Does anyone have any concept of how that might work?

Reisine: The ligand binding domains of the adrenergic and muscarinic receptors are hydrophobic pockets formed by the transmembrane regions (Dixon et al 1987). So, for many classic G protein-coupled receptors, but not peptide receptors, the transmembrane regions may contribute to ligand binding. It is quite possible, therefore, that these regions may mediate somatostatin-14 or somatostatin-28 binding, but there is no direct evidence as yet.

Reichlin: But they are excluded physically from the external environment.

Reisine: They form a pocket. Catecholamines are small ligands that fit within this pocket and presumably interact with multiple sites within the transmembrane region. The problem with peptides, and in particular somatostatin, is that these ligands are very large. Possibly, one part of the ligand fits into the pocket and other parts interact with the extracellular loops.

Reichlin: Studies of the structure and function of somatostatin suggest that the middle sequence of the molecule is required for activity.

Reisine: That's not completely true because somatostatin-28 and somatostatin-14 have different receptor selectivities. Somatostatin-28 contains the somatostatin-14 sequence, so its unique 14 amino acid N-terminal must cause a conformational change of the C-terminal 14 residues, which is the somatostatin-14-containing region that differs from somatostatin-14 itself.

Stork: The work of Roger Cone suggests that transmembrane residues of peptide receptors also contribute to ligand binding (R. Cone, personal communication).

Reisine: One of the opiate receptors also has transmembrane regions and extracellular loops involved in ligand binding; but they are a different family of receptors and have different ligands, so we cannot generalize.

Montminy: You mentioned that sstr5 may mediate somatostatin inhibition of insulin secretion, but that its expression is not restricted to the insulin-secreting β-cells of the pancreas. Could you expand on this point?

Bell: sstr5 is the somatostatin-28-preferring receptor and is expressed at its highest levels in the pituitary (O'Carroll et al 1992). RNase protection studies have shown expression of rat *sstr5* mRNA in brain (hypothalamus and pre-optic area), small intestine, spleen and anterior pituitary (Bruno et al 1993). Reverse transcriptase polymerase chain reaction methodology (RT-PCR) shows that human *sstr5* mRNA is expressed selectively in small intestine, heart, adrenal gland, cerebellum, pituitary, placenta and skeletal muscle, but not in kidney, liver, pancreas, uterus, thymus, testis, spleen, lung, thyroid, ovary or mammary gland (O'Carroll et al 1994).

O'Carroll: We don't detect *sstr5* mRNA in the human pancreas using RT-PCR (O'Carroll et al 1994).

Bell: The islets constitute less than 5% of the volume of the pancreas so if you're using total pancreatic RNA you might not be able to detect *sstr5* mRNA.

Montminy: What receptors do you see in total pancreatic mRNA?

O'Carroll: We just looked for sstr5, we haven't looked for the other four receptors.

Patel: We detected all five *sstr* mRNAs in isolated rat islets using RT-PCR (S. Grigorakis, S. Panetta, Y. C. Patel, unpublished work). *sstr2* is the most abundant, followed by *sstr5*, *sstr1*, *sstr3* and *sstr4*.

Two mRNA transcripts of 8.5 and 2.5 kb were detected in your Northern blots of human tissues (Yamada et al 1992). Does the 8.5 kb transcript represent the spliced variant of the human sstr2 gene? We cannot detect *sstr2B* mRNA in any human tissue using exon-specific primers, although we have found it in the rat (Patel et al 1993).

Bell: I do not know.

Patel: You detect abundant expression of *sstr3* mRNA in the cerebellum, yet direct membrane binding studies and *in vitro* autoradiography have not detected membrane receptors in this tissue. It is therefore unlikely that receptor antibodies are going to show that the receptors are present. Can you explain this?

Bell: If you compare the sequences of all the somatostatin receptors, sstr3 is unique in that it lacks the cysteine residue that could anchor it in the lipid bilayer. The intracellular C-terminal segment is also highly charged—it has a glutamic acid-rich region which is unique among the somatostatin receptors. I predict that these two features affect the affinity of sstr3 for somatostatin. So, it is possible that it is present, but not detected by binding studies.

Eppler: Are you suggesting that the receptor may become detached from the membrane if it lacks this cysteine residue? Why would a seven transmembrane receptor need fatty acid attachment to be anchored in the lipid bilayer?

Bell: No, I am suggesting that it may alter its functional properties. O'Dowd et al (1989) have shown that a Cys[341] to Gly mutation of the palmitoylation site of the human β_2-adrenergic receptor results in a non-palmitoylated form of the receptor that exhibits a dramatically reduced ability to mediate agonist stimulation of adenylyl cyclase. This suggests that post-translational modification of the β_2-adrenergic receptor by palmitoyl groups may play a crucial role in the normal coupling of the receptor to the adenylyl cyclase signal transduction system. In contrast, mutations of the α_{2A}-adrenergic receptor that eliminate detectable palmitoylation do not perturb G protein coupling (Kennedy & Limbird 1993), suggesting that this modification may have different roles at different receptor–G protein interfaces.

Leroux: In humans, there are somatostatin receptors in the granule cell and molecular layers of the cerebellum (Laquerrière et al 1994). Have you looked at somatostatin receptor mRNA in the human cerebellum?

Bell: No, not yet.

Patel: Is sstr3 present in these layers?

Leroux: I don't know.

Reisine: Binding studies are dependent upon receptors having a high affinity for their radioligand. We all use extremely low concentrations of iodinated ligands. If the affinity of the receptor is significantly reduced, even by only 10-fold, you will probably not detect ligand binding. So, if there is a modification of the sstr3 expressed in the cerebellum which results in a slight decrease in affinity, you may not detect the receptor by binding assays, even if the protein is expressed. We need to confirm that sstr3 is present in the cerebellum using antibodies, then we can ask why it has a different affinity there than in other tissues and whether differential processing is involved.

Reichlin: The cerebellum has the lowest amount of somatostatin of any part of the brain. It's odd that the area that has the highest level of receptors is the area with the lowest levels of somatostatin.

Reisine: It contains the highest level of receptor mRNA, but not necessarily the highest level of receptor protein. There are not many examples of membrane-bound proteins where there is a high level of mRNA and a low level of protein, but it is possible that there is a translational problem.

Humphrey: Even if there are high numbers of receptors present, it doesn't necessarily mean that they are functionally relevant. For instance, 5-HT$_2$ receptors are ubiquitous in human and rat brains, yet 5-HT$_2$ antagonists are relatively ineffective, so there's no major central function mediated through that receptor.

There are other examples where the receptors are present at a high density, but they're not functional. The reverse is also true—situations where you cannot detect binding and yet the receptors are functional. For example, you can't detect 5-HT$_{1B}$ receptors transfected into the CHO (Chinese hamster ovary) cell line by binding assays and yet they're functionally active (Giles et al 1994). Also, the 5-HT$_3$ receptor does seem to have important pathophysiological role(s) in the brain and yet it is sparsely distributed (Humphrey 1992). This suggests that low concentrations of receptors are so efficiently coupled that they can mediate a response.

Patel: A similar situation exists in a number of other tissues. For instance, it is not clear if there are somatostatin receptors in the liver.

Reisine: I have not detected any.

Patel: I have not either. Yet Bruno et al (1993) have detected *sstr3* mRNA expression in the liver. All five receptor mRNAs are present in the spleen (Bruno et al 1993), yet no one has detected receptors there by binding studies. *sstr4* mRNA is also detected in the lung (Bruno et al 1993, Roher et al 1993), so it is possible that it has some physiological relevance there too.

Eppler: There's also a lot of β-adrenergic receptor message in the liver, but no one can find it pharmacologically (Arner et al 1990).

Reisine: A lot of people are using either RT-PCR or Northerns to look at mRNA expression in tumours, but they still have to demonstrate that these tumours are making the protein.

Berelowitz: I agree. We can't theorize from the message to protein until we have antibodies, but it is fascinating that the message is expressed in places that are unexpected.

Lamberts: The situation with regard to somatostatin receptors in the liver and spleen is complicated. Normal liver and spleen are always visualized with *in vivo* somatostatin receptor imaging; however, no specific somatostatin receptors have been demonstrated in intact liver (Sato et al 1988), although there may be specific somatostatin binding sites in the basolateral membranes of hepatocytes (Raper et al 1992). The *in vivo* visualization is probably related to a saturable, non-receptor-mediated uptake of somatostatin by the liver. The situation is different in the spleen. Reubi et al (1993a) demonstrated that the human spleen expresses high affinity somatostatin receptors that are located preferentially in the red pulpa. It is difficult to identify the cell type bearing these receptors, but they are probably lymphocytes. Consequently, an organ may seem receptor positive, but this positivity may be based on the presence of lymphocytes.

The local production of somatostatin itself may also interfere with the presence of somatostatin receptors on some organs. Phaeochromocytomas and medullary thyroid cancers often synthesize and release relatively high amounts of somatostatin resulting in an increase in somatostatin concentration in the peripheral circulation (Lundberg et al 1979, Roos et al 1981). Reubi et al (1993b) suggested that high levels of somatostatin mRNA, and consequently of radioimmunoassayable somatostatin, in many neuroendocrine tumours might explain why somatostatin receptors are not always found in all neuroendocrine tumours investigated by ligand-binding studies and/or receptor autoradiography. In these cases, the absence of somatostatin receptors may be due to down-regulation by excess locally produced somatostatin (Lamberts et al 1991). A false negative somatostatin receptor status obtained with *in vivo* scanning may therefore be caused by an excessive dilution of the intravenously injected tracer by the locally produced somatostatin (Reubi et al 1993b).

Berelowitz: I agree. There are many functional autocrine or localized systems where endogenously produced somatostatin binds to locally active receptors. This may be the explanation for many of the failures to detect binding. Unless locally produced somatostatin is eliminated from those receptors, it is not possible to say whether or not there are binding sites available.

Lamberts: Even if both monoclonal antibodies against the five different receptors and a good system to detect mRNA expression were available, these two techniques would not demonstrate whether a tissue expresses functional somatostatin receptors *in vivo*. You have to eliminate the high levels of endogenous somatostatin that are present in some tissues in order to demonstrate whether functional receptors are present. It is also possible that the locally produced somatostatin plays a role in the local control of tumour growth via an autocrine loop.

Patel: Somatostatin receptors can be identified by *in vitro* autoradiography in brain sections (Reubi & Maurer 1985). Endogenous somatostatin has not been a problem in these studies.

Lamberts: I agree, perhaps there are tissues in which this doesn't apply.

Schonbrunn: Or it may vary between different somatostatin receptor types— the ligand dissociating from one receptor more rapidly than another.

Eppler: Would it be too radical to suggest that prosomatostatin can be processed into something other than somatostatin-28 or somatostatin-14, which could be the ligand for the sstr3 that's expressed in the cerebellum? That would be consistent with Graeme Bell's suggestion that sstr3 is expressed in the cerebellum in a form that can't bind [^{125}I]somatostatin-14.

Chesselet: It is possible that we should consider the absence of somatostatin as an indication that the ligand may not be somatostatin-14 or somatostatin-28.

Robbins: Have you studied the regulation of the receptors? Do the genes have hormone response elements in their promoter regions or any other regulatory elements?

Bell: We haven't pursued studies of the structure of the promoter or the regulation of expression of the somatostatin receptor genes; however, Wolfgang Meyerhof has recently reported the sequence and characterization of the promoter of the rat *sstr1* gene (Hauser et al 1994).

Meyerhof: We have only looked at the promoter sequence of *sstr1*. We found that there is no TATA box or CCAAT box, but we did find GC factor-binding sequences, Ap-2 sites and two Pit-1 sites. This would explain the relatively high level of expression of the *sstr1* gene in rat pituitary tumour GH$_3$ cells because they also express prolactin which is regulated by Pit-1.

sstr3 mRNA is differentially expressed during development. Expression starts in the cerebellum very clearly at E7 (embryonic day 7) and reaches a maximum at E14. This level is constant throughout further development. The young rats are weaned around E14, so some social behaviour or external input may regulate gene expression. This could possibly also account for the differences in expression observed in different labs in different tissues, e.g. the pituitary, the striatum and the hippocampus.

Susini: We also observe differential regulation of receptor expression in AR42J pancreatic tumour cells. Epidermal growth factor, basic fibroblast growth factor and gastrin all increase the levels of *sstr2A* and *sstr3* mRNA, but not *sstr1*. This results in an increase in binding of somatostatin by about two to threefold in these cells (Vidal et al 1994).

Schonbrunn: Substantial literature already exists showing that many different hormones regulate somatostatin receptors in different target cells. Somatostatin receptors are modulated by glucocorticoids, oestradiol, thyroid hormones, heterologous peptide/protein hormones as well as by somatostatin itself (Schonbrunn & Tashjian 1980, Schonnbrunn 1982, Hinkle et al 1981, Osborne & Tashjian 1982, Sakamoto et al 1984, Heisler & Srikant 1985, Katakami et al 1985, Viguerie et al 1987, Presky & Schonnbrunn 1988, Mahy et al 1988, Kimura et al 1989, Slama et al 1992). Now that the molecular tools are available, the time is ripe to go back to this literature and define the receptor types involved in each instance. Even more exciting is the prospect of determining the molecular mechanisms by which somatostatin receptors are regulated, at both the mRNA and protein levels.

References

Arner P, Engfeldt P, Lonnqvist F et al 1991 β-adrenoceptor subtype expression in human liver. J Clin Endocrinol & Metab 71:1119–1126

Bruno JF, Xu Y, Song J, Berelowitz M 1993 Tissue distribution of somatostatin receptor subtype messenger ribonucleic acid in the rat. Endocrinology 133: 2561–2567

Dixon RAF, Sigal IS, Rands E et al 1987 Ligand binding to the β-adrenergic receptor involves its rhodopsin-like core. Nature 326:73–77

Giles H, Lansdell SJ, Fox P, Lockyer M, Hall V, Martin GR 1994 Characterisation of a 5-HT$_{1B}$ receptor on CHO cells: functional responses in the absence of radioligand binding. Br J Pharmacol (suppl) 112:317P (abstr)

Hauser F, Meyerhof W, Wulfsen I, Schönrock C, Richter D 1994 Sequence analysis of the promotor region of the rat somatostatin receptor subtype 1 gene. FEBS Lett 345:225–228

Heisler S, Srikant CB 1985 Somatostatin-14 and somatostatin-28 pretreatment down-regulate somatostatin-14 receptors and have biphasic effects on forskolin-stimulated cyclic adenosine, 3′5′-monophosphate synthesis and adrenocorticotropin secretion in mouse anterior pituitary tumor cells. Endocrinology 117: 217–225

Hinkle PM, Perrone MH, Schonnbrunn A 1981 Mechanism of thyroid hormone inhibition of thyrotropin-releasing hormone action. Endocrinology 108:199–205

Humphrey P 1992 5-hydroxytryptamine receptors and drug discovery. Int Acad Biomed Drug Res 1:129–139

Jelinek LJ, Lok S, Rosenberg GB et al 1993 Expression, cloning and signal properties of the rat glucagon receptor. Science 259:1614–1616

Katakami H, Berelowitz M, Marbach M, Frohman LA 1985 Modulation of somatostatin binding to rat pituitary membranes by exogenously administered growth hormone. Endocrinology 117:557–560

Kennedy ME, Limbird LE 1993 Mutations of the α_{2A}-adrenergic receptor that eliminate detectable palmitoylation do not perturb receptor-G-protein coupling. J Biol Chem 268:8003–8011

Kimura N, Hayafuji C, Kimura NO 1989 Characterization of 17-beta-estradiol-dependent and -independent somatostatin receptor subtypes in rat anterior pituitary. J Biol Chem 264:7033–7044

Koch BD, Blalock JB, Schonbrunn A 1988 Characterization of cyclic AMP-independent actions of somatostatin in GH cells. An increase in potassium conductance is responsible for both the hyperpolarization and the decrease in intracellular free calcium produced by somatostatin. J Biol Chem 263:216–225

Lamberts SWJ, Krenning EP, Reubi J-C 1991 The role of somatostatin and its analogs in the diagnosis and treatment of cancer. Endocr Rev 12:450–482

Laquerrière A, Leroux P, Bodenant C, Gonzalez B, Tayot J, Vaudry H 1994 Quantitative autoradiographic study of somatostatin receptors in the adult human cerebellum. Neuroscience 62:1147–1154

Lundberg JM, Hamberger B, Schultzberg M et al 1979 Enkephalin- and somatostatin-like immunoreactivities in human adrenal medulla and pheochromocytoma. Proc Natl Acad Sci USA 76:4079–4083

Mahy N, Woolkalis M, Manning D, Reisine T 1988 Characteristics of somatostatin desensitization in the pituitary tumor cell line AtT-20. J Pharmacol Exp Ther 247:390–396

O'Carroll A-M, Lolait SJ, König M, Mahan LC 1992 Molecular cloning and expression of a pituitary somatostatin receptor with preferential affinity for somatostatin-28. Mol Pharmacol 42:939–946

O'Carroll A-M, Raynor K, Lolait SJ, Reisine T 1994 Characterization of cloned human somatostatin receptor sstr5. Mol Pharmacol 46:291–298

O'Dowd BF, Hnatowich M, Caron MG, Lefkowitz RJ, Bouvier M 1989 Palmitoylation of the human β_2-adrenergic receptor: mutation of Cys[341] in the carboxyl tail leads to an uncoupled nonpalmitoylated form of the receptor. J Biol Chem 264: 7564–7569

Osborne R, Tashjian AHJr 1982 Modulation of peptide binding to specific receptors on rat pituitary cells by tumor-promoting phorbol esters: decreased binding of thyrotropin-releasing hormone and somatostatin as well as epidermal growth factor. Cancer Res 42:4375–4381

Patel YC, Greenwood M, Kent G, Panetta R, Srikant CB 1993 Multiple gene transcripts of the somatostatin receptor sstr2: tissue-selective distribution and cAMP regulation. Biochem Biophys Res Commun 192:288–294

Presky DH, Schonbrunn A 1988 Somatostatin pretreatment increases the number of somatostatin receptors in GH_4C_1 pituitary cells and does not reduce cellular responsiveness to somatostatin. J Biol Chem 263:714–721

Raper SE, Kothary PC, Delvalle J 1992 Identification and partial characterization of a somatostatin-14-binding protein on rat liver plasma membranes. Hepatology 16:433–439

Reisine T, Woulfe D, Raynor K et al 1995 Interaction of somatostatin receptors with G proteins and cellular effector systems. In: Somatostatin and its receptors. Wiley, Chichester (Ciba Found Symp 190) p 160–170

Reubi J-C, Maurer R 1985 Autoradiographic mapping of somatostatin receptors in the rat central nervous system and pituitary. Neuroscience 15:1183–1193

Reubi J-C, Waser B, Horisberger U et al 1993a In vitro autoradiographic and in vivo scintigraphic localization of somatostatin receptors in human lymphatic tissue. Blood 82:2143–2151

Reubi J-C, Waser B, Lamberts SWJ 1993b Somatostatin (SRIH) messenger ribonucleic acid expression in human neuroendocrine and brain tumors using in situ hybridization histochemistry: comparison with SRIH receptor content. J Clin Endocrinol & Metab 76:642–647

Rohrer L, Raulf F, Bruns C, Buettner R, Hofstaedter F, Schule R 1993 Cloning and characterization of a 4th human somatostatin receptor. Proc Natl Acad Sci USA 90:4196–4200

Roos BA, Lindall AW, Ells R, Lamberts BW, Birnbaum RS 1981 Increased plasma and tumor somatostatin-like immunoreactivity in medullary thyroid carcinoma and small cell lung cancer. J Clin Endocrinol Metab 52:187–191

Sakamoto C, Goldfine ID, Williams JA 1984 The somatostatin receptor on isolated pancreatic acinar cell plasma membranes: Identification of subunit structure and direct regulation by cholecystokinin. J Biol Chem 259:9623–9627

Sato H, Sugiyama Y, Sawada Y et al 1988 Dynamic determination of kinetic parameters for the interaction between polypeptide hormones and cell-surface receptors in the perfused rat liver by the multiple-indicator dilution method. Proc Natl Acad Sci USA 85:8355–8359

Schonbrunn A 1982 Glucocorticiods down-regulate somatostatin receptors on pituitary cells in culture. Endocrinology 110:1147–1154

Schonbrunn A, Tashjian AHJr 1980 Modulation of somatostatin receptors by thyrotropin-releasing hormone in a clonal pituitary cell strain. J Biol Chem 255:190–198

Slama A, Videau C, Kordon C, Epelbaum J 1992 Estradiol regulation of somatostatin receptors in the arcuate nucleus of the female rat. Neuroendocrinology 56:240–245

Vidal C, Ranly I, Zeggari M et al 1994 Up-regulation of somatostatin receptors by epidermal growth factor and gastrin in pancreatic cancer cells. Mol Pharmacol 46:97–104

Viguerie N, Esteve J-P, Susini C, Logsdon CD, Vaysse N, Ribet A 1987 Dexamethasone effects on somatostatin receptors in pancreatic acinar Ar4-2J cells. Biochem Biophys Res Commun 147:942–948

White RE, Schonbrunn A, Armstrong DL 1991 Somatostatin stimulates Ca^{2+}-activated K^+ channels through protein dephosphorylation. Nature 351:570–573

Yamada Y, Post SR, Wang K, Taga HS, Bell GI 1992 Cloning and functional characterization of a family of human and mouse somatostatin receptors expressed in brain, gastrointestinal tract and kidney. Proc Natl Acad Sci USA 89:251–255

Yatani A, Codina J, Sekura RD, Birnbaumer L, Brown AM 1987 Reconstitution of somatostatin and muscarinic receptor mediated stimulation of K^+ channels by isolated G_K protein in clonal rat anterior pituitary cell membranes. Mol Endocrinol 1:283–289

Characterization of somatostatin receptor subtypes

Ch. Bruns, G. Weckbecker, F. Raulf, H. Lübbert and D. Hoyer

Preclinical Research, SANDOZ Pharma AG, CH 4002 Basle, Switzerland

Abstract. Somatostatin regulates endocrine and exocrine secretion, possesses antiproliferative properties and acts as a neurotransmitter/neuromodulator in the central nervous system. These effects are mediated by G protein-coupled receptors, of which at least five types have been cloned (sstr1–5). In radioligand-binding studies we have compared the binding properties of sstr1–5 with their activities as somatostatin receptors. All receptors identified so far bind somatostatin-14 and somatostatin-28 with high affinity. The similarities in receptor sequence and in the binding profiles of short synthetic somatostatin analogues such as octreotide, MK 678 or RC 160 for sstr1–5 indicate the existence of two classes of receptors: sstr1/sstr4 with virtually no or very low affinity and sstr2/sstr3/sstr5 with intermediate to high affinity for the short somatostatin analogues. All five receptors mediate inhibition of adenylyl cyclase; this inhibition is sensitive to pertussis toxin. *In vitro* and *in vivo* studies suggest the importance of sstr2 and/or sstr5 in the inhibition of growth hormone release. The sstr2 receptor is apparently the predominant subtype expressed in somatostatin receptor-positive tumours. Evidence exists for the importance of sstr5 receptors in insulin secretion and sstr1 receptors in oncology. Somatostatin receptor-selective agonists and antagonists will help to explore new therapeutic opportunities in oncology as well as in endocrine and gastrointestinal disorders and those of the central nervous system.

1995 Somatostatin and its receptors. Wiley, Chichester (Ciba Foundation Symposium 190) p 89–110

Somatostatin is a cyclic peptide that is distributed widely throughout the body and regulates both endocrine and exocrine secretion. It inhibits the release of many hormones, such as growth hormone (GH), insulin, glucagon, gastrin, secretin and vasoactive intestinal peptide (Reichlin 1983). The *in vivo* effects of somatostatin are limited by its rapid proteolytic degradation (plasma half-life not more than three minutes); consequently, its therapeutic application would require continuous infusion regimens. For that reason, several short synthetic somatostatin analogues with increased metabolic stability, for example octreotide (SMS 201–995, Sandostatin™), have been synthesized (Bauer et al 1982, Veber et al 1979). The primary therapeutic indications for these analogues are the treatment of GH-secreting pituitary adenomas and the relief of symptoms associated with gastroenteropancreatic endocrine tumours because of the inhibition of excessive hormone secretion (Lamberts et al 1991). Furthermore, there is both preclinical and clinical evidence that somatostatin analogues are able to inhibit, under certain conditions, the proliferation of tumour cells (Weckbecker et al 1993, Arnold et al 1993). In the brain, somatostatin acts as a neurotransmitter and affects locomotor activity as well as cognitive and behavioural processes (Epelbaum & Bertherat 1993). All these regulatory effects are mediated by specific, high-affinity membrane receptors for somatostatin on the target tissues such as brain, pituitary, pancreas, gastrointestinal tract and on tumour tissues.

On the basis of radioligand-binding studies, it has been suggested for a long time that at least two somatostatin receptor subtypes exist (sstr1 and sstr2). This receptor heterogeneity was first demonstrated in rat cortex, where octreotide displaced specific $[^{125}I]$ somatostatin-14 binding biphasically (Reubi 1984); the sites with nanomolar affinity for octreotide were called SS1, those with micromolar affinity for octreotide were named SS2. Somatostatin-14, somatostatin-28 and short analogues such as octreotide displayed very high affinity for the SS1 sites, as assessed by displacement of $[^{125}I]204–090$ (Tyr^3-octreotide). Tran et al (1985) confirmed these observations using octreotide and other short-chain analogues such as RC 160 and MK 678; they named the high affinity sites SS_A and the low affinity sites as SS_B. Additional evidence for receptor heterogeneity was reported by Reubi & Maurer (1986): in brain membranes $[^{125}I]$ somatostatin-14 binding to SS2 sites (performed in the presence of an excess of octreotide to block SS1 sites) was significantly increased in the presence of 120 mM NaCl. Under these conditions, SS2 sites labelled with $[^{125}I]$ somatostatin-14 showed very low affinity for octreotide. Autoradiographic studies confirmed these findings (Reubi et al 1987). $[^{125}I]$ somatostatin-14 binding sites were shown to have a wider distribution than $[^{125}I]204–090$ binding sites, suggesting that SS1 sites (labelled by $[^{125}I]204–090$) represent only a subset of somatostatin receptors labelled with $[^{125}I]$ somatostatin-14. Martin et al (1991) subsequently reported similar findings using $[^{125}I]MK$ 678 and $[^{125}I]CGP$ 23996 as somatostatin receptor-specific ligands. Binding sites labelled by $[^{125}I]MK$ 678 were called SRIF1 and were different from $[^{125}I]CGP$ 23996 binding sites in pituitary membranes. The latter were shown to have very low affinity for MK 678 and octreotide and were called SRIF2.

Besides binding data, results from functional assays supported the concept of receptor heterogeneity. Various somatostatin analogues differ in their ability to inhibit the release of neurotransmitters or hormones, such as GH, glucagon and insulin. For instance, several halogenated Trp^8 analogues of somatostatin, while equipotent to somatostatin-14 at inhibiting gastric acid secretion, proved to be more potent than somatostatin as inhibitors of GH release (Meyers et al 1978). Furthermore, octreotide, MK 678 and RC 160 inhibit GH or glucagon release much more potently than they inhibit insulin release. In contrast, somatostatin-28 appears to be more potent than somatostatin-14 at inhibiting insulin release (Srikant & Patel 1987). The final proof for somatostatin receptor heterogeneity came from the cloning of at least five somatostatin receptor genes.

The somatostatin receptor gene family

Over the last three years, molecular cloning has demonstrated the existence of at least five somatostatin receptors, sstr1–5 (Fig. 1) (for review see: Bell & Reisine 1993, Patel et al 1994, Bruns et al 1994). They belong to the superfamily of G protein-coupled receptors with seven membrane-spanning domains. Two rat sstr4 clones that are structurally and pharmacologically different have been reported (Bruno et al 1992, O'Carroll et al 1992). On the basis of submission date, we and others use the nomenclature sstr4 for the receptor cloned by Bruno et al (1992) and sstr5 for the receptor cloned by O'Carroll et al (1992).

The diversity of the five somatostatin receptors identified so far is increased by alternative splicing of the sstr2 gene to give two splice variants, sstr2A and sstr2B, which differ in the length and sequence of their C-terminal tails (Vanetti et al 1992). The overall amino acid identity among somatostatin receptors is around 48% (about 70% similarity), with the highest levels of identical sequences found in the transmembrane regions. sstr1 and sstr4 form a closely related subgroup within the somatostatin receptor family, showing about 61% overall amino acid sequence identity. This close relationship is reflected by their similar pharmacological properties which will be discussed below.

The binding properties of sstr1–5

A variety of radiolabelled somatostatin analogues have been used to study the binding properties of the five cloned somatostatin receptor subtypes. Most commonly used are $([^{125}I]Tyr^{11})$somatostatin-14, $([^{125}I]Tyr^{25})$somatostatin-28 and $([^{125}I]Tyr^3)$octreotide. In general, somatostatin-14 or somatostatin-28 show little differences in binding affinities to sstr1–5 (Raynor et al 1993a,b, Bruns et al 1994, Patel et al 1994). It was demonstrated by several groups that the natural somatostatins bind with high affinity ($pK_i \geqslant 9$) to all somatostatin receptors. The short synthetic analogues, however, such as MK 678, RC 160, BIM 23014 and octreotide, display a different binding profile. High affinity binding is observed only for types 2 and 5 ($pK_i \geqslant 9$), whereas low affinity binding was demonstrated for types 1 and 4 ($pK_i < 7$). The type 3 receptor is somewhat different in that it binds the short-chain analogues with an intermediate affinity ($pK_i \approx 8$). The binding profiles of MK 678, RC 160, BIM 23014 and octreotide for sstr1–5 are nearly identical (Raynor et al 1993a,b, Patel et al 1994, Bruns et al 1994). These analogues, which are used in clinical trials, are selective, high affinity ligands for sstr2 and sstr5, although they may show

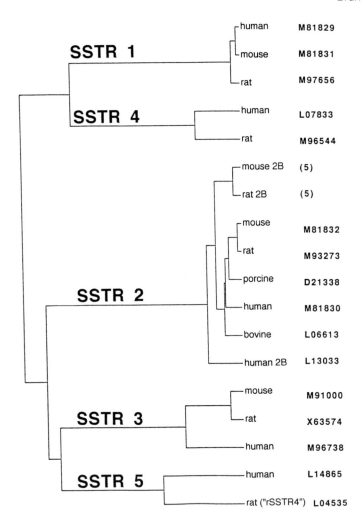

FIG. 1. Molecular cloning of five somatostatin receptor types (sstrs). Similarity plot of the aligned amino acid sequences of the five known types from five species. The Gen Bank accession numbers are given on the right, if available. Note the clustering into two subgroups comprising sstr1 and sstr4 versus sstr2, sstr3 and sstr5. (5), Vanetti et al (1992).

lower affinity for human sstr5 than rat sstr5 (O'Carroll et al 1992, Patel et al 1994). None of the cloned somatostatin receptors shows a clear preference for somatostatin-14 or somatostatin-28, which have nanomolar or higher affinities, although there is a tendency for somatostatin-28 selectivity of the type 5 receptor (O'Carroll et al 1992, Patel et al 1994).

In summary, as suggested by structural similarities of the five cloned somatostatin receptor types, their binding properties also indicate two subfamilies of receptors. Therefore, the historical division of somatostatin receptors into two main classes is still valid (Table 1). Our binding data obtained with a series of somatostatin analogues reveal a significant correlation between ($[^{125}I]$Tyr^3)octreotide (SS1) binding, $[^{125}I]$MK 678 binding in rat brain and sstr2 binding in transfected cells (Fig. 2). Furthermore, we have demonstrated

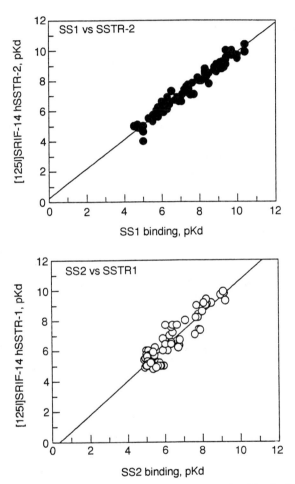

FIG. 2. Correlation between SS1 and SS2 binding, as defined in the brain by Reubi & Maurer (1986), and sstr2 and sstr1 binding (CHO cells) respectively. SS1 binding was performed in rat cortex using ($[^{125}I]$Tyr^3)octreotide, SS2 binding was performed using $[^{125}I]$SRIF-14 ($[^{125}I]$somatostatin-14) in the presence of 120 mM NaCl. Binding to CHO cells expressing recombinant hsstr-1 and hsstr-2 receptors (human sstr1 and sstr2 respectively) was performed using $[^{125}I]$SRIF-14 as the specific ligand.

by autoradiography that the distributions of sites labelled with ($[^{125}I]$ Tyr^3)-octreotide and $[^{125}I]$ MK 678 are superimposable and correspond to the expression pattern of *sstr2* mRNA in rat brain (Hoyer et al 1994). The binding results obtained in rat brain using $[^{125}I]$ somatostatin-14 in the presence of 120 mM NaCl (SS2 sites) correlate very significantly with sstr1 binding (Fig. 2).

It can be concluded that the SS1 sites correspond to the sstr2 receptor type. The SS2 sites described previously are equivalent to the sstr1 receptor type (Table 1).

Somatostatin receptor function

There is now ample evidence for somatostatin receptor heterogeneity as confirmed by the cloning of five somatostatin receptors. Whether these mediate all effects produced by somatostatin and its stable analogues remains open. In the brain, somatostatin shows a variety of effects and acts as a neuromodulator/neurotransmitter (Epelbaum & Bertherat 1993). It has been reported to modulate the release of 5-hydroxytryptamine in the hypothalamus (Tanaka & Tsujimoto 1981) and was shown to facilitate dopamine release in the nucleus accumbens (Chesselet & Reisine 1983). Although many results regarding the distribution of somatostatin and its receptors in the brain have been reported, only a few functional responses can be assigned to a specific receptor type. This is largely because there are few, if any, type-selective ligands (especially antagonists) available. Effects produced by analogues such as octreotide, MK 678 or RC 160 were assumed to be mediated by the same receptor (sstr2) but the similar pharmacological profiles of sstr2 and sstr5 do not allow this assumption to be made. Although some ligands selective for sstr2, sstr3 and sstr5 have been described recently (Raynor et al 1993a,b), the lack of somatostatin receptor antagonists is a major problem when investigating the importance of a certain somatostatin receptor type. Thus, hyperlocomotor activity observed in the rat after local application of MK 678 (Raynor et al 1993c) may be mediated by sstr2 and/or sstr5. However, CGP 23996, which shows high affinity for all somatostatin receptors, has no effect.

In the pituitary, mRNAs for most somatostatin receptors are expressed (i.e. sstr1, sstr2, sstr3 and sstr5); the pharmacology of the effects of somatostatin and its analogues on the release of GH, thyroid-stimulating hormone and prolactin release may be more complex than expected. The inhibition of GH release, from its pharmacological characteristics, is mediated by sstr2 and/or sstr5 (Hoyer et al 1994).

The expression patterns of somatostatin receptors in various organs are different, although overlapping. For example, in the pancreas where somatostatin inhibits hormone release (insulin, glucagon, somatostatin) effectively and interferes with exocrine secretion, sstr2, sstr3 and sstr5 are present. Furthermore, the pharmacology of the somatostatin receptor that mediates the inhibition of insulin release from pancreatic β cells suggests that sstr3 and sstr5 receptors are involved (Rossowski & Coy 1993).

TABLE 1 Nomenclature of somatostatin receptors

Receptor	Gene cloning		Ligand binding (human; pIC_{50})			Coupling	Other nomenclatures		
	Human	Rat	Somatostatin-28	Somatostatin-14	Octreotide		Reisine/ O'Carroll	Reubi et al	IUPHAR/ Hoyer et al
sstr1	Yamada et al 1992	Li et al 1992	9.1	9.4	<6.0	Adenylyl cyclase inhibition	SSTR1	SS2	$SRIF_{2A}$
sstr2A[a]	Yamada et al 1992	Kluxen et al 1992	9.4	10.0	9.5	Adenylyl cyclase inhibition K^+/Ca^{2+} channels?	SSTR2A	SS1	$SRIF_{1A}$
sstr2B[a]	Patel et al 1993	Vanetti et al 1992				Adenylyl cyclase inhibition	SSTR2B	—	—
sstr3	Yamada et al 1992	Meyerhof et al 1992	9.2	9.5	7.5	Adenylyl cyclase inhibition	SSTR3	—	$SRIF_{1C}$
sstr4	Roher et al 1993	Bruno et al 1992	9.0	9.1	<6.0	Adenylyl cyclase inhibition	SSTR4	—	$SRIF_{2B}$
sstr5	Panetta et al 1994	O'Carroll et al 1992	9.5	9.2	8.1	Adenylyl cyclase inhibition	SSTR5	—	$SRIF_{1B}$

[a]sstr2A and sstr2B are splice variants.

In the periphery, several other regulatory effects have been reported for somatostatin and its analogues. In the atria, somatostatin exerts potent negative inotropic effects (Dimech et al 1993). The very low potency and absence of efficacy of MK 678 in the atria rules out effects mediated by either sstr2 or sstr5 and hints at the function of sstr1, sstr3, or sstr4. Somatostatin analogues inhibit neurogenically induced contractions in guinea-pig ileum and vas deferens (Feniuk et al 1993): in the ileum, an sstr2-like receptor profile has been reported, whereas in the vas deferens, the sstr3 receptor is a candidate because of the intermediate potency of octreotide.

It has become evident that not only classical somatostatin target tissues such as pituitary, pancreas or the gastrointestinal tract express somatostatin receptors, but also a variety of tumour cells express high affinity somatostatin receptors. Various human tumours, such as gastroenteropancreatic endocrine tumours, have been demonstrated to be sensitive to octreotide therapy (Lamberts et al 1991). By inhibiting the secretory activity of these tumours, their symptoms can be controlled effectively. More importantly, anti-proliferative properties of somatostatin and its short-chain analogues have been demonstrated in a number of *in vitro* and *in vivo* tumour models (Weckbecker et al 1993).

Evidence as to which somatostatin receptor type is involved in the inhibition of cancer cell proliferation has been obtained from cell growth experiments using various somatostatin analogues that bind selectively and with high affinity to sstr2 and sstr5. Octreotide and related analogues effectively inhibit proliferation of various kinds of tumour cells cultured *in vitro*, such as the human breast adenocarcinoma cell line MCF7, the human breast carcinoma ZR-75-1 and the rat pancreatic tumour cell line AR42J. These analogues have also shown antiproliferative properties in appropriate animal tumour models (Weckbecker et al 1993). The effectiveness of octreotide and related somatostatin analogues in a wide range of preclinical cancer models suggest that (i) the receptor types expressed in somatostatin-susceptible cancers are sstr2 and/or sstr5, because they both bind octreotide and related analogues specifically and with high affinity and (ii) there is a potential clinical role for somatostatin analogues in the treatment of tumour cells that express somatostatin receptors. Recent studies by Buscail et al (1994) using cell lines transfected with cDNAs for sstr2 confirmed the central role of sstr2 in mediating cell growth inhibitory effects. In monkey kidney COS-7 cells and mouse NIH/3T3 fibroblasts expressing sstr2, the short-chain somatostatin analogues octreotide and RC 160 inhibit serum-driven cell proliferation potently, with IC_{50} values in the picomolar range. In contrast, the analogues failed to inhibit the growth of control cells not expressing sstr2. Expression of sstr1 has also been demonstrated to be involved in somatostatin analogue-mediated inhibition of cell growth (Buscail et al 1994). Whether other somatostatin receptors, such as sstr5, are involved in mediating antiproliferative effects remains to be elucidated.

Reverse transcriptase polymerase chain reaction (RT-PCR) methodology was used to investigate the expression of somatostatin receptors in well established tumour cell lines. Using this sensitive technique, we were able to demonstrate the expression of *sstr2* mRNA in rat AR42J cells and human ZR-75-1 cells. (Fig. 3). The growth of these *sstr2*-expressing tumour cells is potently inhibited by somatostatin and the short-chain analogues (Fig. 4), again demonstrating the relevance of sstr2 in oncology (Weckbecker et al 1993). Eden & Taylor (1993) used RT-PCR to determine the expression pattern of somatostatin receptor types in a panel of human and rodent solid tumours and tumour cell lines. *sstr2* was clearly the dominant somatostatin receptor: it was detectable in all breast, prostate, pancreatic and small-cell lung cancers, as well as in melanomas, hepatomas and all tumour cell lines analysed. *sstr1* mRNA was found in small-cell lung cancer cell lines and a hepatoma cell line. *sstr3* mRNA was detectable in a melanoma and a hepatoma cell line. These results from RT-PCR studies, while still limited, suggest that sstr2 is the most important somatostatin receptor type in tumour cells.

The oncological relevance of *sstr2* expression is supported by autoradiographic studies which demonstrate the expression of high affinity binding sites for short-chain analogues, such as octreotide, in a wide range of human tumours (Lamberts et al 1991). In a small percentage of pituitary tumours, insulinomas and meningiomas (up to 15%), sstr1 receptors with low affinity for short-chain somatostatin analogues have been identified. These receptors are functional because somatostatin-14 or somatostatin-28 but not octreotide effectively inhibits hormone release of such tumour cells cultured *in vitro*. Now that five somatostatin receptors have been cloned, the current emphasis is on the analysis of the tissue distribution of sstr1–5 in human tumours.

Recently, a new aspect regarding the importance of somatostatin receptor expression on tumour growth regulation has emerged. Since tumour growth is

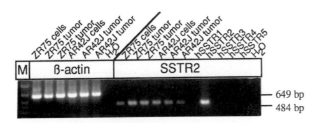

FIG. 3.   *sstr2* mRNA expression in human breast carcinoma ZR-75–1 and rat pancreatic cancer AR42J cells and the corresponding tumours. RNA was detected using the reverse transcriptase polymerase chain reaction (RT-PCR). Lanes 2–8, β-actin control with the expected product length of 649 bp. Lanes 9–20, *sstr2* with the expected product length of 484 bp. Lanes 8 and 20, H_2O as negative control; lane 15, hSSTR1; lane 16, hSSTR2; lane 17, hSSTR3; lane 18, hSSTR4; lane 19, hSSTR5 plasmid controls. hSSTR1–5 represent human sstr1–5 respectively. M, molecular weight marker (700, 500, 400, 300 bp).

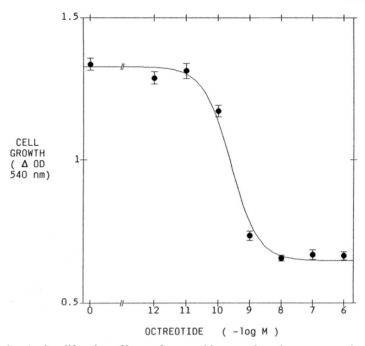

FIG. 4. Antiproliferative effects of octreotide on cultured rat pancreatic tumour cells (AR42J). The cells were exposed to octreotide continuously for five days in 96-well plates without medium change. The change in cell number was determined using the sulphorhodamine B colorimetric assay. The IC_{50} for a five-day exposure was 2.5×10^{-10} M.

a result of the interaction between tumour cells with normal surrounding tissue, including blood vessels and infiltrating immune cells, the understanding of the role of somatostatin receptors in tumorigenesis requires investigation of their expression in these tumour-associated cells. It was demonstrated that veins surrounding human cancer tissue express high levels of somatostatin receptors (Reubi et al 1994). Furthermore, octreotide and RC 160 potently inhibit new vessel formation (angiogenesis), an important process in tumorigenesis (Barrie et al 1993). Activated lymphocytes were demonstrated to bind octreotide with high affinity (Van Hagen et al 1994). From these findings, it can be assumed that sstr2 and/or sstr5 expression in tumour-surrounding tissue may be relevant for tumour growth.

Conclusions

The availability of the cloned somatostatin receptors will be very useful for the identification of receptor type-selective agonists and antagonists to elucidate the physiological role of these receptors in the brain and in the periphery. For example, site-directed mutagenesis studies will allow the identification of ligand-binding domains of the receptors which could help to develop new structures that bind specifically to distinct somatostatin receptor types. Therefore, somatostatin receptors may provide new therapeutic opportunities for the treatment of endocrine and oncological disorders and those of the central nervous system.

Acknowledgements

The technical support of Ch. Bourquin, D. Eckert and B. Urban is gratefully acknowledged. We thank Drs I. Lewis, P. Smith-Jones and J. Schloos for critical reading of the manuscript.

References

Arnold R, Neuhaus C, Benning R et al 1993 Somatostatin analogue Sandostatin and inhibition of tumour growth in patients with metastatic endocrine gastroenteropancreatic tumours. World J Surg 17:511–519

Barrie R, Woltering EA, Hajarizadeh H, Mueller C, Ure T, Fletcher WS 1993 Inhibition of angiogenesis by somatostatin and somatostatin-like compounds is structurally dependent. J Surg Res 55:446–450

Bauer W, Briner U, Doepfner W et al 1982 SMS 201–995: a very potent and selective octapeptide analogue of somatostatin with prolonged action. Life Sci 31:1133–1140

Bell GI, Reisine T 1993 Molecular biology of somatostatin receptors. Trends Neurosci 16:34–38

Bruno JF, Xu Y, Song JF, Berelowitz M 1992 Molecular cloning and functional expression of a brain-specific somatostatin receptor. Proc Natl Acad Sci USA 89:11151–11155

Bruns C, Weckbecker G, Raulf F et al 1994 Molecular pharmacology of somatostatin receptor subtypes. In: Wiedenmann B, Kvols L, Arnold R, Riecken E (eds) Molecular and cell biological aspects of gastroenteropancreatic tumour disease. New York Academy of Sciences, New York 733:138–147

Buscail L, Delesque N, Estève JP et al 1994 Stimulation of tyrosine phosphatase and inhibition of cell proliferation by somatostatin analogues: mediation by human somatostatin receptor subtypes SSTR1 and SSTR2. Proc Natl Acad Sci USA 91:2315–2319

Chesselet M-F, Reisine TD 1983 Somatostatin regulates dopamine release in rat striatal slices and cat caudate nuclei. J Neurosci 3:232–236

Dimech J, Feniuk W, Humphrey PPA 1993 Antagonist effects of seglitide (MK 678) at somatostatin receptors in guinea-pig isolated rat atria. Br J Pharmacol 109:898–899

Eden PA, Taylor JE 1993 Somatostatin receptor subtype gene expression in human and rodent tumours. Life Sci 53:85–90

Epelbaum J, Bertherat J 1993 Localization and quantification of somatostatin receptors by light microscopic autoradiography. Methods Neurosci 12:279–292

Feniuk W, Dimech J, Humphrey PPA 1993 Characterization of somatostatin receptors in guinea-pig isolated ileum, vas deferens and right atrium. Br J Pharmacol 110:1156–1164

Hoyer D, Lübbert H, Bruns C 1994 Molecular pharmacology of somatostatin receptor subtypes. Naunyn-Schmiedeberg's Arch Pharmacol 350:441–453

Kluxen FW, Bruns C, Lübbert H 1992 Expression cloning of a rat brain somatostatin receptor cDNA. Proc Natl Acad Sci USA 89:4618–4622

Lamberts SWJ, Krenning EP, Reubi J-C, 1991 The role of somatostatin and its analogs in the diagnosis and treatment of tumors. Endocr Rev 12:450–482

Li XJ, Forte M, North RA, Ross CA, Snyder SH 1992 Cloning and expression of a rat somatostatin receptor enriched in brain. J Biol Chem 267:21307–21312

Martin J-L, Chesselet M-F, Raynor K, Gonzales C, Reisine T 1991 Differential distribution of somatostatin receptor subtypes in rat brain revealed by newly developed somatostatin analogs. Neuroscience 41:581–593

Meyerhof W, Wulfsen I, Schönrock C, Fehr S, Richter D 1992 Molecular cloning of a somatostatin-28 receptor and comparison of its expression pattern with that of a somatostatin-14 receptor in rat brain. Proc Natl Acad Sci USA 89: 10267–10271

Meyers CA, Coy DH, Huang WY et al 1978 Highly active position eight analogues of somatostatin and separation of peptide diastereomers by partition chromatography. Biochemistry 17:2326–2331

O'Carroll A-M, Lolait SJ, König M, Mahan LC 1992 Molecular cloning and expression of a pituitary somatostatin receptor with preferential affinity for somatostatin-28. Mol Pharmacol 42:939–946

Panetta R, Greenwood MT, Warszynska A et al 1994 Molecular cloning, functional characterization and chromosomal localization of a human somatostatin receptor (somatostatin receptor type 5) with preferential affinity for somatostatin-28. Mol Pharmacol 45:417–427

Patel YC, Greenwood M, Kent G, Panetta R, Srikant CB 1993 Multiple gene transcripts of the somatostatin receptor SSTR2: tissue selective distribution and cAMP regulation. Biochem Biophys Res Commun 192:288–294

Patel YC, Greenwood MT, Warszynska A, Panetta R, Srikant CB 1994 All five cloned human somatostatin receptors (hSSTR1–5) are functionally coupled to adenylyl cyclase. Biochem Biophys Res Commun 198:605–612

Raynor K, Murphy WA, Coy DH et al 1993a Cloned somatostatin receptors: identification of subtype-selective peptides and demonstration of high-affinity binding of linear peptides. Mol Pharmacol 43:838–844

Raynor K, O'Carroll A-M, Kong HY et al 1993b Characterization of cloned somatostatin receptors SSTR4 and SSTR5. Mol Pharmacol 44:385–392

Raynor K, Lucki I, Reisine T 1993c Somatostatin receptors in the nucleus accumbens selectively mediate the stimulatory effect of somatostatin on locomotor activity in rats. J Pharmacol Exp Ther 265:67–73

Reichlin S 1983 Somatostatin. N Engl J Med 309:1495–1501, 1556–1563

Reubi JC 1984 Evidence for two somatostatin-14 receptor types in rat brain cortex. Neurosci Lett 49:259–263

Reubi JC, Maurer R 1986 Different ionic requirements for somatostatin receptor subpopulations in the brain. Regul Pept 14:301–311

Reubi JC, Probst A, Cortes R, Palacios JM 1987 Distinct topographical localization of two somatostatin receptor subpopulations in the human cortex. Brain Res 406:391–396

Reubi JC, Horisberger U, Laissue J 1994 High density of somatostatin receptors in veins surrounding human cancer tissue: role in tumour–host interaction? Int J Cancer 56:681–688

Rohrer L, Raulf R, Bruns C, Buettner R, Hofstaedter F, Schuele R 1993 Cloning and characterization of a fourth human somatostatin receptor. Proc Natl Acad Sci USA 90:4196–4200

Rossowski WJ, Coy DH 1993 Potent inhibitory effects of a type four receptor-selective somatostatin analogue on rat insulin release. Biochem Biophys Res Commun 197:366–371

Srikant CB, Patel YC 1987 Somatostatin receptor evidence for structural and functional heterogeneity. In: Reichlin S (ed) Somatostatin, basic and clinical aspects. Proceedings of the international conference on somatostatin. Plenum, New York, p 89–102

Tanaka S, Tsujimoto A 1981 Somatostatin facilitates the serotonin release from rat cerebral cortex, hippocampus and hypothalamus slices. Brain Res 208:219–222

Tran VT, Beal MF, Martin JB 1985 Two types of somatostatin receptors differentiated by cyclic somatostatin analogs. Science 228:492–495

Vanetti M, Kouba M, Wang XM, Vogt G, Höllt V 1992 Cloning and expression of a novel mouse somatostatin receptor (SSTR2B). FEBS Lett 311:290–294

Van Hagen PM, Krenning EP, Kwekkeboom DJ et al 1994 Somatostatin and the immune and the haematopoetic system: a review. Eur J Clin Invest 24:91–99

Veber DF, Holly FW, Nutt RF et al 1979 Highly active cyclic and bicyclic somatostatin analogues of reduced ring size. Nature 280:512–514

Weckbecker G, Raulf F, Stolz B, Bruns C 1993 Somatostatin analogues for diagnosis and treatment of cancer. Pharmacol & Ther 60:245–264

Yamada Y, Post SR, Wang K, Tager HS, Bell GI, Seino S 1992 Cloning and functional characterization of a family of human and mouse somatostatin receptors expressed in brain, gastrointestinal tract and kidney. Proc Natl Acad Sci USA 89:251–255

DISCUSSION

Reisine: There is a major difference between rat and human sstr5 pharmacology. We find that BIM 23052 is at least 100-fold less potent on human sstr5 than on rat sstr5 (O'Carroll et al 1994). This is comparable to the potency that you observed. We also have a Merck compound (L362-855) which has a high selectivity towards human and rat sstr5.

Octreotide (SMS 201-995, Sandostatin™) binds to rat sstr5 as strongly as it does to sstr2; however, octreotide is at least 100-fold less potent on human sstr5 than sstr2 (Raynor et al 1993a). This difference creates a problem if the rat is used for preclinical studies. I will also show (Reisine et al 1995, this volume) that BIM 23052 acts on sstr5 but not sstr2, and that other compounds act on sstr2 but not sstr5. I do not know why our results differ for sstr3—we observe a high affinity. It is possible that there are structural variations between the analogues that David Coy synthesized and those of Christian Bruns' group.

Bruns: In contrast to your group (Raynor et al 1993a,b), we were unable to reproduce the marked selectivities for sstr3 and/or sstr5 using BIM 23056 and BIM 23052; however, we studied human somatostatin receptors whereas you studied rat so, as you say, we may not be able to compare results between species.

Also, we showed that both somatostatin-14 and octreotide bind to human sstr5 with high affinities in the nanomolar range. We observed a difference in binding of only one order of magnitude between the two peptides, in contrast to the two orders of magnitude that you have reported (Raynor et al 1993a).

Patel: I have compared our binding results (Patel & Srikant 1994) for the five human somatostatin receptors with Christian Bruns. Except for CGP 23996, we agreed on the binding affinities of all the analogues that Terry Reisine tested. However, these binding affinities are different from those that Terry Reisine has described.

Let us take sstr2 and sstr3 as examples. Terry Reisine's report of IC_{50} values in the picomolar range for NC4-28B, BIM 23023, BIM 23034, BIM 23059 and BIM 23056 against sstr2, and for BIM 23056 and BIM 23058 against sstr3, differ by 100–1700-fold from our findings (Patel & Srikant 1994). The degree of selectivity of these analogues for sstr2 and sstr3 in Terry Reisine's studies is several orders of magnitude greater than what we have found.

Do you really think that these differences in potency ratios can be explained simply by species differences, because the differences are very large? Alternatively, could they result from methodological differences or differences in the peptides?

Reisine: The peptides we tested (Raynor et al 1993a,b) were either provided by Biomeasure (the BIM peptides) or David Coy's group.

Patel: We tested peptides supplied by David Coy's group. They were the same ones that Terry Reisine used.

Coy: But Bruns et al (1995, this volume) make their own peptides.

Reisine: With regards to species variation, we have only ever found this with human and rat sstr5. We have never seen clear species differences with sstr1, sstr2, sstr3 or sstr4. Other explanations could be the use of different tissues or cells.

Reichlin: Terry Reisine, what are the different expression systems that workers in the field use?

Reisine: People transiently express receptors in monkey COS-1 or COS-7 kidney cells. However, this system often results in the expression of variable levels of receptors. The level of receptors is obviously important with regard to binding studies.

Receptors can also be expressed stably in CHO (Chinese hamster ovary) cells. We used a cell line that Graeme Bell originally transfected called CHO-DG44. These cells have different G proteins than CHO-K1 cells, which are the CHO cells that are more frequently used. A human kidney fibroblast cell line called HEK293 has also been used. These cells have different G protein subunits, different receptor levels and different cellular environments. However, it is unlikely that these differences explain the binding differences. There may simply be differences in the stability of the peptides or how they are made. For instance, one dogma of the somatostatin field for many years has been that you need a cyclic structure to bind to a somatostatin receptor. Many of the peptides we are using, however, are linear peptides. They may form a cyclic structure in solution because of their hydrophobic tails, but they are essentially linear and this could explain the binding differences.

Reichlin: Perhaps these results could only be compared if you all used the same expression systems?

Reisine: About 90% of the compounds would probably give similar results. For example, Christian Bruns mentioned in his talk the correlation between peptide binding to sstr2 and the inhibition of growth hormone (GH) release (Bruns et al 1994). We reported this also (Raynor et al 1993a). Disparities occur with the receptor-selective compounds.

Schonbrunn: Everyone has focused on the receptor-selective analogues because it is critical to determine whether they do or do not exist.

Reisine: There are receptor-selective agonists for sstr2 and probably also for sstr5, but not for sstr3.

Coy: We have tested various preparations of peptides using the same assay system and have never observed batch to batch variation. BIM 23056 is inactive with respect to sstr3 in all systems studied so far except a gastric smooth muscle cell system which we looked at with Bob Jensen (Gu et al 1994). BIM 23056 has a high affinity for these cells, presumably this reflects a high affinity for sstr3.

Taylor: I would like to talk about sstr2. We've made hundreds of somatostatin peptides and have screened them *in vitro* with several different assay systems: CHO sstr2 cells that we have from Graeme Bell and Terry Reisine; AR42J rat pancreatic tumours cells; and human small-cell lung carcinoma cell lines H69 and H345. We found that the rank order of 200 peptides in terms of binding to sstr2 is similar in all these assay systems and that the most potent compound was the BIM 23027, a cyclic hexapeptide. The potency of the peptides in sstr2-transfected CHO cells therefore reflects their potency in tumour cell lines and suggests that there are no discrepancies between transfected and untransfected cells.

Humphrey: That's exactly what you would hope to see in all the systems, but occasionally you do observe a variation in pharmacology. For instance Mary Richards (1991) reviewed the estimated equilibrium dissociation constants of various ligands for different muscarinic receptors and found that there were up to 100-fold differences for the same ligand between different cell lines. The particular cell line one uses is therefore absolutely critical. The G protein concentration, relative to that of the transfected receptor, and differences in the G protein types may explain these differences in pharmacology. Different cell lines may also metabolize peptides differently—perhaps linear peptides are more prone to metabolism?

Coy: But these linear peptides are quite stable compared to somatostatin.

Humphrey: We've shown that in the guinea-pig vas deferens, somatostatin-14 is metabolized whereas somatostatin-28 is not (Feniuk et al 1993), so there are differential metabolic characteristics that need to be studied. My preferred explanation is the one involving G protein concentrations and other aspects of the intracellular milieu of the cells used.

Epelbaum: Christian Bruns, in your *in vivo* experiment studying tumour growth in nude mice, you mention that the cells you use express sstr2 and respond to octreotide. Adenomatous cells also express sstr2, yet octreotide does not elicit a reduction in the growth of this tumour. Can you explain this?

Bruns: I did not make a general statement that the growth of every tumour cell that expresses sstr2 is inhibited by octreotide. I just gave one example (AR42J cells) where octreotide binds with a high affinity to sstr2. In parallel, octreotide interferes with tumour cell proliferation *in vitro* (Fig. 4) as well as tumour growth *in vivo* (nude rats). In contrast, we have reported examples of human tumour cells where *sstr2* is not expressed, but inhibition of tumour growth *in vivo* occurs when octreotide is administered (Weckbecker et al 1993). It is likely that, in this case, octreotide interferes with the release of growth factors such as IGF1 (insulin-like growth factor 1) and/or IGF binding proteins (IGF BP). It was demonstrated by Ren et al (1992) that expression of IGF BP1 is stimulated by octreotide independently of GH regulation; therefore, mechanisms that interfere with tumour growth might be completely independent of the expression of high affinity somatostatin receptors in the tumour cell.

Epelbaum: Our study on rat GC pituitary tumour cells shows a moderate effect in response to octreotide. There is also a poor antiproliferative response in acromegalic patients.

Lamberts: It is difficult to prove that an antimitotic effect occurs in acromegaly because pituitary tumours can take 15–25 years to grow to a diameter of 1–2 cm. In nude mice, however, transplanted cancer cells often have doubling times of days, so it would be easier to demonstrate the potential antimitotic effects of somatostatin analogues in these mice.

Christian Bruns presented a growth curve showing very rapid cell division of control cells, but no division in somatostatin-treated cells (G. Weckbecker, L. Tolcsvai & C. Bruns, unpublished observations). However, if they had prolonged that experiment for another four to five weeks, then the somatostatin-treated tumour would have started to grow at the same rate as the controls. Do you agree Christian?

Bruns: Yes. After a treatment period of five to six weeks, tumour growth starts again; however, tumour growth is delayed significantly in octreotide-treated animals when compared to the control group. This inhibitory response is also dependent on the treatment regime. We have studied nude mice where octreotide was administered by minipumps (10μg/kg per h) or by a new slow release octreotide formulation, Sandostatin-LAR™ (long-acting repeatable), where a single injection maintains high plasma levels over a period of four to five weeks. Both administration methods produce the same inhibitory effect on tumour growth. If octreotide is injected twice a day, however, the inhibitory effect on tumour growth is less pronounced (Weckbecker et al 1992). Therefore, we have evidence that the inhibitory effects depend on the route of administration and the maintenance of high plasma levels of octreotide. The question arises whether this is true for all tumour types, because somatostatin receptor desensitization may occur in some tumours if they are exposed to high levels of somatostatin for a long period of time. For example, if patients who have gastroenteropancreatic tumours are treated with somatostatin analogues, after six to nine months of treatment, some tumours begin to escape from the inhibitory effect on hormone release, indicating that somatostatin receptor desensitization has occurred (Arnold et al 1994).

Somatostatin receptor internalization may or may not occur, since conflicting results have been reported for different types of cells. Agi Schonbrunn does not observe internalization in rat pituitary tumour GH_4C_1 or GH_3 cells, whereas Christiane Susini does in pancreatic tumour cells. Our lab has no evidence for internalization.

Epelbaum: VIPomas and glucagonomas have also been reported to escape suddenly from octreotide treatment (Kvols et al 1986, Vinik et al 1986). Is there a similar desensitization process in acromegalic patients?

Lamberts: No desensitization has been reported so far during long-term treatment of acromegaly. We treated our first group of acromegalic patients with three 100 μg subcutaneous injections of octreotide in the spring of 1984. Now, after more than 10 000 injections per patient, these tumours still demonstrate a high, unchanged sensitivity to octreotide with respect to its inhibitory effect on hormonal hypersecretion. None of these tumours have escaped from the inhibitory effect of octreotide therapy; if the tumour responds to the first injection, it will continue to respond for at least 10 years.

Humphrey: In many situations you get tachyphylaxis quite markedly to somatostatin, i.e. a rapid loss of effect caused by the continued presence of somatostatin. The extent to which this occurs depends on which receptor or which response you're looking at; for example, in the guinea-pig ileum, somatostatin's ability to inhibit neurogenic contraction is very tachyphylactic whereas its inhibition of gastrin-induced acid secretion in rat gastric mucosa is maintained.

We've recently reported that synthetic agonists like octreotide can behave as partial agonists, although admittedly this appears not be the case clinically in the treatment of acromegaly. Nevertheless, we've been able to modify the experimental conditions to turn octreotide from an agonist into an antagonist (Dimech et al 1993). So, if an agonist that acts on somatostatin receptors is developed, it must be determined whether or not it is a full agonist and whether it produces tachyphylaxis when it is administered chronically.

Lamberts: Sometimes, within a few days or weeks of the initiation of octreotide therapy, you observe desensitization of its inhibitory effects on gall bladder or gastrointestinal tract contractility. Relatively benign pituitary tumours, such as GH-secreting or thyrotropin-secreting tumours, do not desensitize *in vivo* or *in vitro*. In contrast, many somatostatin receptor-bearing malignant tumours, such as islet cell cancers and carcinoids, often escape from the inhibitory effect on hormone secretion during continuous exposure to the analogue and can sometimes also start to grow. This occurs both *in vivo* and *in vitro* and can be explained by two simultaneously occurring phenomena: dedifferentiation of the tumour, in which the tumour becomes more aggressive and loses its somatostatin receptors; or real desensitization of the receptors. Desensitization should be defined by the reappearance of these receptors after stopping somatostatin analogue treatment.

Taylor: When cell growth is inhibited with somatostatin or its analogues, other cells lacking a somatostatin receptor may increase their rate of proliferation.

Lightman: Although initially a high proportion of the cells in the tumour may be sensitive to treatment, there will be a subpopulation of cells that are insensitive. These will be selected for during treatment and eventually new clones of insensitive dedifferentiated cells take over.

Schonbrunn: Identifying receptor types is going to be important in somatostatin receptor desensitization studies.

Reisine: We've looked at the molecular mechanisms of desensitization of sstr2 and found that an enzyme called Bark (β-adrenergic receptor kinase) is involved. We have a *bark* dominant negative mutant that prevents the phosphorylation and desensitization of sstr2. The problem is that *bark* is not expressed in all cells—it may not be expressed in the pituitary where there is very little evidence of somatostatin receptor desensitization. In other tissues, e.g. in brain neurons, sstr2-selective agonists desensitize sstr2 receptors very rapidly. The same receptor may therefore be present in different tissues, but the expression of different kinases may cause heterogeneous desensitization.

Schonbrunn: So it is important to look at each receptor in the cell type of interest. You cannot necessarily extrapolate what happens in a particular expression system to the normal response in the physiological target.

Taylor: We've looked at *sstr2* mRNA and receptor binding in about 40–50 tumour clonal cell lines including small-cell lung carcinoma, breast cancer and prostate cancer, but we haven't evaluated functional activity.

Reichlin: Tumours seem to respond differently to long-term administration of octreotide—one third of acromegalic tumours show regression of size whereas growth sometimes arrests in vipomas and carcinoids, but continued proliferation and metastases are more common. Can you ever expect to find a specific analogue that turns off growth?

Taylor: I don't think that you can expect to turn off growth entirely, only to slow the proliferation rate. We find that small-cell lung carcinoma cells and prostate tumours *in vivo* continue to grow, although the rate is retarded by somatostatin peptides (Taylor et al 1988, Bodgen et al 1990a,b).

Lightman: There are several differences between the different acromegalic tumours. We have studied the regulation of intracellular Ca^{2+} concentrations in acromegalic tumours. Usually, we get a fall in intracellular Ca^{2+} levels after the addition of somatostatin or its analogues, but in a few tumours we see a rise in intracellular Ca^{2+} (Z. Chen-Ping, A. Levy & S. L. Lightman, unpublished results). Some patients also respond to treatment with somatostatin analogues by an increase in GH secretion, so there are large differences either in the receptors or the receptor coupling.

Robbins: There is also a subset of acromegalic patients who have *gsp* mutations and high levels of intracellular cAMP (Vallar et al 1987). We've looked at the effect of somatostatin on cAMP levels in a number of different cells. cAMP levels don't decrease significantly in normal cells, but if you add an analogue that raises the level of cAMP, somatostatin prevents that rise. Is it possible that the acromegalic patients, in whom you see a reduction in tumour size over a relatively short time, are the same subset as those who have the *gsp* mutation? Steven Lamberts, have you looked at cAMP levels in acromegalic patients?

Lamberts: No, we have not. Spada et al (1990) demonstrated that the presence of *gsp* mutations in GH-secreting pituitary adenomas does not result in differences in the clinical features of acromegaly, the duration of the disease at diagnosis or the cure rate after transplantation. Most patients with *gsp* mutations develop the full clinical manifestations of the disease when their tumours are still small, suggesting that these adenomas have a high rate of secretory activity. In general, tumours with *gsp* mutations are also sensitive to the inhibitory effect of somatostatin on cAMP production resulting in good GH secretion control (Spada et al 1990).

It is impossible to compare the results from animal tumour models, especially monoclonal tumour cell lines, with observations in patients with different somatostatin receptor-positive tumours. These human tumours are often heterogeneous—some tumour cells have few or no somatostatin receptors whereas others have many somatostatin receptors. Treatment with somatostatin analogues often results in controlled hormonal hypersecretion with an instant beneficial effect on the quality of life of these patients. However, the effect on different cell populations that express variable numbers of somatostatin receptors is unknown. It is possible that the somatostatin receptor-negative tumour cells, which do not respond to long-term somatostatin analogue therapy, are responsible for further growth, despite beneficial effects of the analogue on somatostatin receptor-positive cells within the same tumour.

Susini: I agree with Steven Lamberts that there are different types of cells in tumours; however, an additional problem is that the antiproliferative effect of somatostatin is biphasic. When a high concentration (up to 10 nM) of somatostatin analogue is used, the inhibitory effect is decreased.

Stork: Octreotide therapy sometimes inhibits GH secretion but not tumour growth and vice versa. It is not known what effector pathways inhibit growth but they are probably not the same pathways that inhibit secretion. A lot can be learned by comparing somatostatin with dopamine, in particular with the effect of bromocriptine in pituitary GH-secreting adenomas and prolactinomas where hormonal secretion and growth effects are not always coupled (Molitch et al 1985).

Bruns: We are investigating human breast cancer cells *in vitro* and *in vivo* to find out the mechanism of action of hormones such as somatostatin, but it is difficult to make assumptions that the same situation occurs in human breast cancer. The studies by Christiane Susini and Philip Stork on phosphatase activity (Florio et al 1994, Delesque et al 1995, this volume) are interesting since the somatostatin receptor-dependent stimulation of phosphatase activity might be one important mechanism to explain the antiproliferative response to octreotide.

Has anyone examined effector pathways in AR42J cells? In particular, is the inhibition of growth by somatostatin sensitive to vanadate?

Susini: We observe an inhibition of the antiproliferative effect of somatostatin by vanadate.

Reichlin: Philip Stork, you suggested that the activation of the receptor could influence secretion and growth separately via parallel pathways. Could you elaborate on this?

Stork: We need to find out whether, in tumours, one receptor can simultaneously couple to inhibition of growth and inhibition of secretion, or whether these responses are mediated by separate receptors.

Reichlin: Do you think that growth and secretion could be separated if you only had one activated receptor?

Stork: Certainly, a single receptor may couple to multiple G proteins, each of which couples to distinct effector pathways of secretion and growth inhibition.

Reisine: Let us suppose that sstr2 couples to a Ca^{2+} channel via $G_o\alpha$ and to a tyrosine phosphatase, lets say via G_x. In this case, somatostatin would inhibit GH release and you would also get antiproliferative effects. You may have cells that don't have G_x, in which case somatostatin would inhibit GH release because you could still block Ca^{2+} influx, but it might not couple with the tyrosine phosphatase. The reverse may also be true—many peripheral cells don't have G_o, so GH secretion would be blocked, but you would still have antiproliferative effects.

Schonbrunn: Dr Susini showed that the antiproliferative effects of somatostatin in AR42J cells are pertussis toxin insensitive (Viguerie et al 1989), yet when you look at secretion it's always pertussin toxin sensitive, indicating that two different mechanisms are involved, presumably via different G proteins.

Eppler: Liu et al (1994) and Moxham et al (1993) have also knocked out $G_i\alpha2$ in two different systems by expression of stably transfected antisense constructs and shown that you lose the ability of somatostatin to lower cAMP levels.

Reichlin: In order to define the molecular/structural component required for an antiproliferative effect, could you use secretion as a marker to screen your tumour lines or must you look at proliferation?

Reisine: It depends on what you want to look at. Screening compounds for the inhibition of GH release may not be the best approach if your intention is to make a compound that could be used for treating tumours. In this case, you should look at inhibition of tumour growth.

Stork: I agree with Terry Reisine. At one time, GH was thought to be the mitogenic agent for somatostatin-responsive tumours and that somatostatin's inhibition of GH release accounted for its antiproliferative effects. Now people think that this may not be the case.

References

Arnold R, Neuhaus C, Benning R et al 1994 Somatostatin analog Sandostatin and inhibition of tumor growth in patients with metastatic endocrine gastroenteropancreatic tumors. World J Surg 17, in press

Bodgen AE, Taylor JE, Moreau JP, Coy DH 1990a Treatment of R-3327 prostate tumors with a somatostatin analogue (somatuline) as adjuvant therapy following surgical castration. Cancer Res 50:2646–2650

Bodgen AE, Taylor JE, Moreau JP, LePage D, Coy DH 1990b Response of human lung tumor xenographs to treatment with a somatostatin analogue (somatuline). Cancer Res 50:4360–4365

Bruns C, Weckbecker G, Raulf F et al 1994 Molecular pharmacology of somatostatin receptor subtypes. In: Wiedenmann B, Kvols L, Arnold R, Riecken E (eds) Molecular and cell biological aspects of gastroenteropancreatic tumour disease. New York Academy of Sciences, New York 733:138–147

Delesque N, Buscail L, Estève JP 1995 A tyrosine phosphatase is associated with the somatostatin receptor. In: Somatostatin and its receptors. Wiley, Chichester (Ciba Found Symp 190) p 187–203

Dimech J, Feniuk W, Humphrey PPA 1993 Antagonist effects of seglitide (MK 678) at somatostatin receptors in guinea-pig isolated right atria. Br J Pharmacol 109:898–899

Feniuk W, Dimech J, Humphrey PPA 1993 Characterisation of somatostatin receptors in guinea-pig isolated ileum, vas deferens and right atrium. Br J Pharmacol 110:1156–1164

Florio T, Rim C, Pan MG, Stork PJS 1994 The somatostatin receptor SSTR1 is coupled to phosphotyrosine phosphatase activity in CHO-K1 cells. Mol Endocrinol 8:1289–1297

Gu Z-F, Mantey SA, Coy DH, Maton PN, Jensen RT (1994) sstr3 subtype of somatostatin receptor mediates the inhibitory action of somatostatin on smooth muscle cells. Am J Physiol, in press

Kvols LK, Moertel CG, O'Connell MJ, Schutt AJ, Rubin J, Hahn RG 1986 Treatment of malignant carcinoid syndrome: evaluation of a long-acting somatostatin analogue. N Engl J Med 315:663–666

Liu YF, Jakobs KH, Rasenick MM, Alberts PR 1994 G protein specificity in receptor-effector coupling. Analysis of the roles of G_o and G_{i2} in GH_4C_1 pituitary cells. J Biol Chem 269:13880–13886

Molitch ME, Elton RL, Blackwell RE et al 1985 Bromocriptine as primary therapy for prolactin-secreting macroadenomas: results of a prospective multicenter study. J Clin Endocrinol Metab 60:698–707

Moxham CM, Hod Y, Malbon CC 1993 Induction of $G_{i\alpha2}$-specific antisense RNA *in vivo* inhibits neonatal growth. Science 260:991–995

O'Carroll A-M, Raynor K, Lolait S, Reisine T 1994 Characterization of clonal human somatostatin receptor sstr5. Mol Pharmacol 46:291–298

Patel YC, Srikant CB 1994 Subtype selectivity of peptide analogs for all five cloned human somatostatin receptors (hsstr1-5). Endocrinology 135:2814–2817

Raynor K, O'Carroll A-M, Kong HY et al 1993a Characterization of cloned somatostatin receptors sstr4 and sstr5. Mol Pharmacol 44:385–392

Raynor K, Murphy WA, Coy DH et al 1993b Cloned somatostatin receptors, identification of subtype selective peptides and demonstration of high affinity binding of linear peptides. Mol Pharmacol 43:838–844

Reisine T, Woulfe D, Raynor K et al 1995 Interaction of somatostatin receptors with G proteins and cellular effector systems. In: Somatostatin and its receptors. Wiley, Chichester (Ciba Found Symp 190) p 160–170

Ren SG, Ezzat S, Melmed S, Braunstein GD 1992 Somatostatin analog induces insulin-like growth factor binding protein-1 (IGFBP-1) expression in human hepatoma cells. Endocrinology 131:2470–2481

Richards MH 1991 Pharmacology and second messenger interaction of cloned muscarinic receptors. Biochem Pharmacol 42:1645–1653

Spada A, Arosio M, Bochicchio D et al 1990 Clinical, biochemical and morphological correlates in patients bearing growth hormone secreting tumours with or without constitutively active adenylyl cyclase. J Clin Endocrinol Metab 71:1421–1426

Taylor JE, Bogden AE, Moreau JP, Coy DH 1988 *In vitro* and *in vivo* inhibition of human small-cell lung carcinoma (NCI-H69) growth by a somatostatin analogue. Biochem Biophys Res Comm 153:81–86.

Vallar L, Spada A, Giannattasiv G 1987 Altered G_s and adenylate cyclase activity in human GH-secreting pituitary adenomas. Nature 330:566–568

Viguerie N, Tahiri-Jouti N, Ayral AM et al 1989 Direct inhibitory effects of a somatostatin analog, SMS201-995, on AR4-2J cell proliferation via pertussis toxin-sensitive guanosine triphosphate-binding protein-independent mechanisms. Endocrinology 124:1017–1025

Vinik AI, Tsai ST, Moattri AR, Cheung P, Eckhauser FE, Cho K 1986 Somatostatin analog (SMS 201-995) in the management of gastroenteropancreatic tumors and diarrhea syndromes. Am J Med 81:23–40

Weckbecker G, Tolcsvai L, Liu R, Bruns C 1992 Preclinical studies on the anticancer activity of the somatostatin analogue octreotide (SMS 201-995). Metabolism 41:99–103

Weckbecker G, Raulf F, Stolz B, Bruns C 1993 Somatostatin analogs for diagnosis and treatment of cancer. Pharmacol & Ther 60:245–264

Regulation of somatostatin receptor mRNA expression

Michael Berelowitz, Yun Xu, Jinfen Song and John F. Bruno

Division of Endocrinology and Metabolism, Health Sciences Center 15-060, SUNY at Stony Brook, Stony Brook, New York 11794, USA

Abstract. The five somatostatin receptor mRNAs are expressed with distinct though overlapping patterns of distribution in the CNS and peripheral tissues. All receptor types are expressed in the anterior pituitary and hypothalamus and could therefore be modulated in states of growth hormone (GH) dysregulation. Metabolic perturbations such as food deprivation and diabetes mellitus lead to suppression of GH levels in the rat, in part due to increased somatostatin tone. In rats deprived of food, pituitary *sstr1, 2* and *3* mRNAs were reduced by 80% compared to fed controls; *sstr4* and *sstr5* mRNAs were unchanged. Hypothalamic *sstr* mRNA expression was unaltered. In diabetic rats pituitary *sstr1, 2* and *3* mRNAs were reduced by 50–80% with *sstr1* mRNA restored in part by insulin therapy. Pituitary *sstr4* mRNA and hypothalamic expression of these four types was unaffected. *sstr5* mRNA is reduced by 70% in the pituitary and by 30% in the hypothalamus with restoration of both by insulin treatment. Altered pituitary sstr expression in food deprivation and diabetes could result from chronic exposure to increased plasma somatostatin. In rat GH₃ pituitary tumour cells exposed to 1 μM somatostatin for up to 48 h, *sstr1, 3, 4* and *5* mRNA increased dramatically while *sstr2* mRNA exhibited a biphasic response. We observed a net increase in receptor binding associated with increased *sstr* mRNA. Somatostatin receptor expression is regulated in a tissue- and type-specific manner, adding further complexity to the action of the multifaceted peptide somatostatin.

1995 Somatostatin and its receptors. Wiley, Chichester (Ciba Foundation Symposium 190) p 111–126

Somatostatin exerts a variety of biological effects, predominantly inhibitory, on endocrine and exocrine secretion and neural function. Biologically active somatostatin peptides known to occur *in vivo* (somatostatin-14 and somatostatin-28) act by binding to five receptor types, coupled via G proteins to several signal transduction pathways. Such complexity may focus a particular somatostatin action to a particular target tissue through specificity of tissue distribution, signal transduction pathway or homeostatic regulation.

Hypothalamic somatostatin, synthesized in the periventricular nucleus and released into the hypophyseal–portal system from nerve endings in the median eminence, acts on the anterior pituitary to inhibit growth hormone (GH) and thyroid-stimulating hormone (TSH) secretion. Ligand-binding studies demonstrated that high affinity somatostatin receptors are present on pituitary plasma membranes. These receptors exhibit different ligand specificities compared with neural receptors. Somatostatin-28 competes more potently for ($[^{125}I]$Tyr^{11})somatostatin binding in the pituitary than in brain plasma membranes (Skikant & Patel 1981) and somatostatin binding is more completely displaced by the analogue octreotide (SMS 201–995, Sandostatin™) in the pituitary than in the cortex (Tran et al 1985). Biochemical evidence of heterogeneity among pituitary plasma membrane somatostatin receptors is provided by affinity cross-linking studies. Our group used a panel of bifunctional cross-linking agents to demonstrate that two protein species exist with molecular masses of 66 and 69 kDa (Bruno & Berelowitz 1989).

With the cloning of five somatostatin receptors (sstr1–5) and the demonstration that they have different analogue specificities and, perhaps, different transmembrane signalling systems, it will now be possible to define the role that each receptor plays in mediating the physiological effects of somatostatin. These studies focus on regulation of *sstr* expression in the hypothalamus and pituitary.

Methods

Animals and tissue handling

Male Sprague–Dawley rats (Taconic Farms, Germantown, NY) weighing 175–225 g were housed in pairs under constant temperature (22°C) and a 12 h light, 12 h dark cycle (lights on at 0700 h) and with free access to rat chow (Ralston Purina, St. Louis, MO) and tap water. Animals were kept under these conditions for 3–7 days prior to all experimental procedures.

Three experimental groups of animals were used in these studies.

(a) Normal animals: for studies of *sstr* distribution.

(b) Animals deprived of food for 72 h: pituitary and hypothalamic *sstr* mRNA levels were determined in groups of rats deprived of food for 72 h or allowed free access to food; both groups had access to drinking water. Animals were weighed immediately before food deprivation then killed 72 h later.

(c) Diabetic animals: rats received an intraperitoneal injection of 100 mg/kg streptozocin freshly prepared in 0.05 M citric acid buffer (pH 4.5). Control animals received an injection of buffer alone. Streptozocin-induced diabetes was defined by the appearance of gross glycosuria within 1 day using Diastix (Eli Lilly, Indianapolis, IN) which corresponded in all cases to plasma glucose concentrations in fed rats of more than 4 g/l). Insulin-treated rats received NPH human insulin (Novolin-N; Novo-Nordisk, Princeton, NJ) by subcutaneous injection (5–7 U) twice a day between 0800–0900 h and 1600–1700 h from Day 1 after streptozocin treatment. Control and untreated diabetic rats received parallel injections with equal volumes of saline. Animals were killed after 72 h of insulin treatment.

Rats were killed by decapitation and tissues were rapidly dissected then immediately frozen on dry ice for later RNA extraction. Freshly dissected anterior pituitaries were prepared as previously described (Bruno & Berelowitz 1989).

Cell cultures

Rat GH_3 pituitary tumour cells were maintained in Ham's F-10 medium supplemented with 15% horse serum and 2.5% fetal calf serum under 5% CO_2/95% air at 37 °C. Cells at a density of $1-2 \times 10^6$ cells/35 mm dish (for binding) or $1-2 \times 10^7$ cells/100 mm dish (for RNA) were exposed to somatostatin (1 μM) for various times up to 48 h.

RNA extraction

Total RNA was isolated from dissected tissue blocks or cells using guanidine isothiocyanate/phenol chloroform extraction. RNA concentrations were estimated by measuring absorbance at 260 nm and identical concentrations of RNA from each sample were used for nuclease protection assays. Aliquots (3-4 μg) of total RNA were subjected to gel electrophoresis using a 1% agarose gel, stained with ethidium bromide, then examined visually to confirm quality and integrity and to estimate RNA concentration.

Preparation of probes

sstr cDNA constructs in pBluescript II KS + (pBS), used to generate ^{32}P-labelled antisense RNA for nuclease protection assays, were made as previously described (Bruno et al 1993).

Nuclease protection assay

Solution hybridization/nuclease protection assays were performed, as described by White et al (1986), on 5-20 μg total cellular RNA. After separation of stable hybrids on 8% polyacrylamide 8 M urea gels, the dried gels were exposed to Kodak (Rochester, NY) X-Omat X-ray film. Autoradiographic densities were quantitated using an LKB (Rockville, MD) laser densitometer.

Somatostatin receptor binding studies

Equilibrium binding of somatostatin to pituitary plasma membranes was performed as previously described by Bruno & Berelowitz (1989). After exposure to somatostatin, GH_3 cells were washed in acidic medium to dissociate surface-bound ligand, then intact cells were exposed to ($[^{125}I]$ Tyr)somatostatin with (non-specific binding) or without (total binding) 1 μM labelled somatostatin. Cells were then solubilized and membrane-associated counts were determined.

Analysis of data

Where appropriate, results are expressed as the mean $\pm$ SEM. Densitometric values were normalized to controls, which were arbitrarily set to equal one. Comparisons of data between experimental groups were performed using one way analysis of variance, followed by Fisher's least significant difference test.

Rat tissue *sstr* mRNA distribution

The hypothalamus, like most areas in the CNS, expresses multiple *sstr* mRNAs. *sstr1* and *sstr2* mRNAs are expressed most strongly, followed by *sstr3* then *sstr5* then *sstr4*. The hypothalamus represents the primary region where *sstr5* expression is observed while the other receptors are more widely distributed.

All five *sstr* mRNAs are also expressed in the anterior pituitary. *sstr2* is expressed most strongly, followed by *sstr1* and *sstr3* then *sstr5* then *sstr4* (Bruno et al 1993).

Although *sstr* mRNAs in the hypothalamus and pituitary cannot be correlated with the levels of sstr protein expression until subtype-specific ligands are available, some conclusions can be drawn. Somatostatin binding sites in the hypothalamus are not displaced by octreotide. These receptors are therefore probably sstr1 because sstr1 has a low affinity for octreotide in the CNS. (Rens-Domiano et al 1992). In contrast, since somatostatin binding in the pituitary is fully displaceable by octreotide, this implies the functional presence of sstr2, 3 and/or 5, which have high affinity for this analogue (Rens-Domiano et al 1992, Yasuda et al 1992, O'Carroll et al 1992). The higher affinity of somatostatin-28 in pituitary membranes (compared to the CNS) supports the presence of sstr5, which is selective for this form of the peptide (O'Carroll et al 1992).

Metabolic regulation of *sstr* mRNA regulation

Prolonged food deprivation and insulin deficient diabetes mellitus in the rat result in a loss of GH secretory episodes. We have previously demonstrated a marked decrease in GH-releasing hormone (GHRH) mRNA expression in the hypothalamus of food-deprived (Bruno et al 1990) and streptozocin-induced diabetic rats (Olchovsky et al 1990). GH secretion can be restored, in part, by *in vivo* immunoneutralization in both states using somatostatin antiserum (Tannenbaum et al 1978). In addition, our group has documented that decreased pituitary plasma membrane somatostatin receptor binding is due to a decrease in receptor concentration and not a decrease in affinity in both food-deprived rats (Bruno et al 1994) and streptozocin-induced diabetic rats (Olchovsky et al 1990). This suggests *in vivo* somatostatin receptor down-regulation. Pituitary cells from these animals incubated *in vitro* demonstrated resistance to the GH-suppressive effects of somatostatin reflecting somatostatin receptor down-regulation (Walsh & Szabo 1988, Olchovsky et al 1990). To evaluate sstr specificity (if any) of the *in vivo* decrease in total sstr concentration, we examined expression of *sstr* mRNA in the hypothalamus and pituitary of food-deprived and streptozocin-induced diabetic rats.

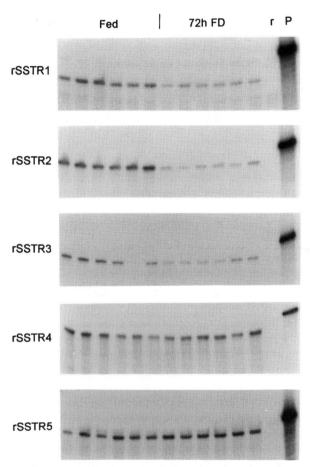

FIG. 1. *sstr* mRNA expression in pituitaries of individual rats allowed free access to food (Fed) or food deprived for 72 h (72 h FD). Autoradiograph of a representative nuclease protection assay is shown. The lane marked r represents a negative control (absence of protected bands in *Escherichia coli*-derived ribosomal RNA). The lane marked P represents the mobility of undigested cRNA probe. rSSTR1–5 represent rat sstr1–5.

The expression of *sstr1*, *sstr2* and *sstr3* mRNA was reduced by about 80% in the pituitary of food-deprived rats compared to fed controls, while *sstr4* and *sstr5* mRNA levels were unaffected (Fig. 1). In contrast, hypothalamic *sstr* mRNA expression was unchanged (Fig. 2). The situation in streptozocin-induced diabetic rats resembled that seen in food-deprived rats in that pituitary *sstr1*, *sstr2* and *sstr3* mRNA expression was reduced by 50–80% and *sstr4* mRNA expression was unchanged (Fig. 3). Hypothalamic expression of these receptors was also unchanged (Fig. 4). In contrast to food-deprived rats, *sstr5* mRNA expression was reduced by 70% in the pituitary and by 30% in hypothalamus of streptozocin-induced diabetic rats, with complete restoration by insulin treatment (Bruno et al 1994).

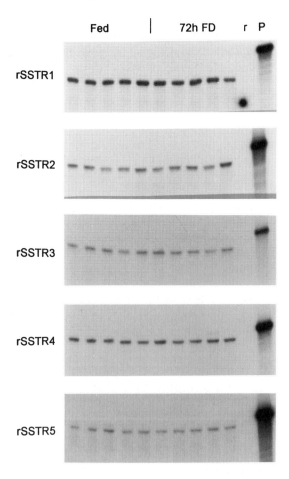

FIG. 2.   *sstr* subtype mRNA expression in hypothalamus of Fed and 72 h FD rats as outlined for Fig. 1.

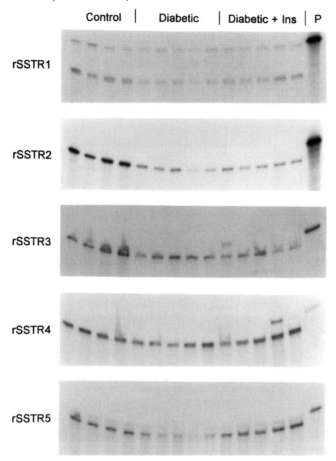

FIG. 3.   Effect of streptozocin-induced diabetes on expression of pituitary *sstr* mRNA. Representative autoradiograph of nuclease protection analysis for *sstr* mRNA expression in control, diabetic and insulin-treated diabetic (Diabetic + Ins) rats. The lane marked P represents the mobility of undigested cRNA probe. rSSTR1–5 represent rat sstr1–5.

In the rat pituitary, *sstr1*, *sstr2* and *sstr3* mRNAs are the major types represented, thus, it is not surprising that a decline in expression of all three *sstr*s is associated with the decrease in total pituitary somatostatin receptor binding. In addition, it appears from studies with somatostatin analogues that *sstr2* has characteristics of the pharmacologically defined pituitary somatostatin receptor (Raynor et al 1993), thus, a decrease in *sstr2* expression alone could explain much of the decrease in pituitary somatostatin receptor binding.

For many G protein-coupled receptors, chronic exposure to saturating concentrations of agonist desensitizes target tissues to further stimulation

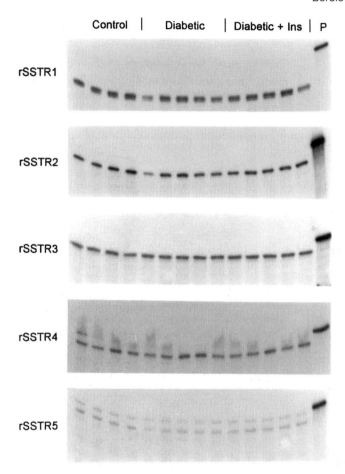

FIG. 4.   Effect of streptozocin-induced diabetes on hypothalamic *sstr* mRNA expression as outlined for Fig. 3.

(homologous desensitization). This desensitization involves rapid uncoupling of the receptor–G protein complex, followed by a reduction in receptor numbers. This is accompanied, in most cases, by a decrease in receptor gene expression (Collins et al 1992).

Increased exposure to somatostatin *in vivo* leading to homologous down-regulation may explain the observed decrease in receptor concentration and expression of *sstr* mRNAs. *In vitro* transfection of *sstr2* and *sstr3*, but not *sstr4*, into transformed African green monkey COS kidney cells or CHO (Chinese hamster ovary) cells results in homologous desensitization (Rens-Domiano et al 1992, Yasuda et al 1992, Xu et al 1993). Previous evidence for *in vivo* homologous desensitization of pituitary somatostatin receptors was provided by studies where a transient increase in endogenous somatostatin (caused by GH injection) was followed by a transient decrease in pituitary somatostatin receptor binding (Katakami et al 1985).

Decreased expression of *sstr1*, *2*, *3* and *4* in the pituitary but not the hypothalamus of food-deprived and streptozocin-induced diabetic rats may be explained by exposure of the pituitary to high somatostatin levels in hypophyseal–portal blood or a limited access of circulating somatostatin to the hypothalamus.

Decreased expression of *sstr5* mRNA in the hypothalamus and pituitary of streptozocin-induced diabetic but not food-deprived rats is an interesting finding. *sstr5* differs from the other receptor types since its distribution in the CNS is limited to the hypothalamus and pre-optic area, it is highly expressed in the pituitary and shows preferential affinity for somatostatin-28. Exposure of the hypothalamus and the pituitary to higher concentrations of somatostatin-28 relative to somatostatin-14 could explain these findings. Alternatively, it is possible that the metabolic consequences of diabetes are distinct from those of food deprivation, such as hyperglycaemia or absolute insulin deficiency. Normalization of *sstr* mRNA by insulin therapy supports this. Further studies are required to distinguish between these possibilities.

Homologous regulation of *sstr* mRNA expression

It is possible that the process of homologous desensitization explains the decrease in pituitary somatostatin receptor binding and *sstr* mRNA expression observed in food-deprived and diabetic rats. To investigate this process, our group studied the effect of somatostatin on somatostatin receptor binding and *sstr* mRNA expression in cells of rat pituitary origin that express all five *sstrs*.

Several pituitary tumour cell lines have been used in studies of somatostatin receptor desensitization. For initial studies of the effect of somatostatin on somatostatin receptors, we chose to use the rat GH_3 pituitary tumour cell line. Somatostatin has previously been shown to up-regulate rather than down-regulate somatostatin binding in related GH_4C_1 cells (Presky & Schonbrunn 1988); however, the molecular mechanisms involved in this process can be identified more easily in GH_3 cells and the principles later applied to normal rat pituitary cells.

Prolonged exposure of GH_3 cells to 1 μM somatostatin resulted in a time-dependent increase in specific ($[^{125}I]Tyr^{11}$)somatostatin binding which reached 280% of control by 24 h and maximal effects (350% of control) by 48 h. This effect was specific in that it was not observed with carbamylcholine chloride (carbachol) or phenylisopropyl adenosine—agents that similarly decrease intracellular cAMP and Ca^{2+} in GH_3 cells.

Increases in somatostatin binding to GH_4C_1 cells following somatostatin treatment result from an increase in receptor number with no change in affinity (Presky & Schonbrunn 1988). Up-regulation could involve an increase in receptor biosynthesis controlled at the level of transcription, RNA processing and/or translation. To evaluate the role of altered mRNA expression, we measured *sstr* mRNA expression in GH$_3$ cells following exposure to somatostatin. *sstr1* mRNA levels increased with duration of exposure, reaching maximum levels of more than 400% from 4–24 h. Levels of *sstr3*, *sstr4* and *sstr5* mRNA (which varied in untreated GH$_3$ cells) were dramatically increased at 24 h and 48 h. In contrast, *sstr2* mRNA levels exhibited a biphasic response, showing an early

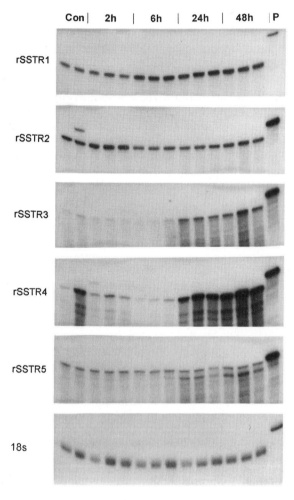

FIG. 5   *sstr* mRNA expression in GH$_3$ cells following culture in medium alone (Con) or in the presence of somatostatin 1 μM for the times indicated. The lane marked P represents the mobility of undigested cDNA probe. rSSTR1–5 represent rat sstr1–5. 18s mRNA was used as an internal control for sample loading.

(2 h) increase to 150% then decreasing to 50% of control levels by 6 h with restoration to control levels by 48 h (Fig. 5).

These results provide a model on which to base future studies of homologous regulation, though they clearly do not provide a mechanism by which pituitary *sstr* mRNA is regulated in food-deprived or diabetic rats—states of *in vivo* increases in somatostatin tone. It is interesting to note, however, that the changes in *sstr2* at 6 h and the net change in *sstr* mRNA expression at 24–48 h are consistent with the decrease in pituitary somatostatin receptor binding following the injection of a single dose of GH into rats (Katakami et al 1985).

Summary

These studies indicate an additional level of regulation (at the level of specific *sstr* mRNA expression) in the already complex cascade of events involved in GH homeostasis. Food-deprivation and streptozocin-induced diabetes in the rat are metabolic states that involve abnormalities at all levels of the GH regulatory axis; thus, it is difficult to identify the mechanism(s) leading to altered pituitary *sstr* mRNA expression outlined above. In contrast, GH$_3$ cells, which express *sstr* mRNAs in a regulated manner, provide an excellent model in which to test hypotheses regarding somatostatin receptor regulation. These experiments can then be repeated in normal pituitary cells *in vitro* and whole animals *in vivo*. Taken together with an emerging understanding of the relationships between somatostatin receptor structure, function and signal transduction pathways, such studies will provide greater insight into the multiple roles of somatostatin in physiology and pathophysiology.

References

Bruno JF, Berelowitz M 1989 Covalent labelling of the somatostatin receptor in rat anterior pituitary membranes. Endocrinology 124:831–837

Bruno JF, Olchovsky D, White JD, Leidy JW, Song J, Berelowitz M 1990 Influence of food deprivation in the rat on hypothalamic expression of growth hormone-releasing factor and somatostatin. Endocrinology 127:2111–2116

Bruno JF, Xu Y, Song JF, Berelowitz M 1993 Tissue distribution of somatostatin receptor subtype messenger ribonucleic acid in the rat. Endocrinology 133:2561–2567

Bruno JF, Xu Y, Song J, Berelowitz M 1994 Pituitary and hypothalamic somatostatin receptor subtype mRNA expression in the food-deprived and diabetic rat. Endocrinology 133:2561–2567

Collins S, Caron MG, Lefkowitz RI 1992 From ligand binding to gene expression: New insights into the regulation of G-protein-coupled receptors. Trends Biochem Sci 17:37–39

Katakami H, Berelowitz M, Marbach M, Frohman L 1985 Modulation of somatostatin binding to rat pituitary membranes by exogenously administered growth hormone. Endocrinology 117:557–560

O'Carroll A-M, Lolait SJ, König M, Mahan LC 1992 Molecular cloning and expression of a pituitary somatostatin receptor with preferential affinity for somatostatin-28. Mol Pharmacol 42:939–946

Olchovsky D, Bruno JF, Wood TL et al 1990 Altered pituitary growth hormone regulation in streptozotocin-diabetic rats: a combined defect of hypothalamic somatostatin and GH-releasing factor. Endocrinology 126:53–61

Presky DH, Schonbrunn A 1988 Somatostatin pretreatment increases the number of somatostatin receptors in GH_4 C_1 pituitary cells and does not reduce cellular responsiveness to somatostatin. J Biol Chem 263:714–721

Raynor K, Murphy WA, Coy DH et al 1993 Cloned somatostatin receptors: identification of subtype-selective peptides and demonstration of high affinity binding of linear peptides. Mol Pharmacol 43:838–844

Rens-Domiano S, Law SF, Yamada Y, Seino S, Bell GI, Reisine T 1992 Pharmacological properties of two cloned somatostatin receptors. Mol Pharmacol 42:28–34

Srikant CB, Patel YC 1981 Receptor binding of somatostatin-28 is tissue specific. Nature 294:259–260

Tannenbaum GS, Epelbaum J, Colle E, Brazeau P, Martin JB 1978 Antiserum to somatostatin reverses starvation-induced inhibition of growth hormone but not insulin secretion. Endocrinology 102:1909–1914

Tran VT, Beal MF, Martin JB 1985 Two types of somatostatin receptors differentiated by cyclic somatostatin analogs. Science 228:492–495

Walsh JB, Szabo M 1988 Impaired suppression of growth hormone release by somatostatin in cultured adenohypophyseal cells of spontaneously diabetic BB/W rats. Endocrinology 123:2230–2234

White JD, Stewart KD, McKelvy JF 1986 Measurement of neuroendocrine peptide mRNA in discrete brain regions. Methods Enzymol 124:548–560

Xu Y, Song JF, Berelowitz M, Bruno JF 1993 Ligand binding and functional properties of the rat somatostatin receptor SSTR4 stably expressed in Chinese hamster ovary cells. Mol Cell Neurosci 4:245–249

Yasuda K, Rens-Domiano S, Breder CD et al 1992 Cloning of a novel somatostatin receptor, SSTR3, coupled to adenylyl cyclase. J Biol Chem 267:20422–20428

DISCUSSION

Epelbaum: Did you include a preincubation step in your food deprivation studies to dissociate prelabelled endogenous somatostatin from pituitary receptors? We did these experiments in the pituitary of food-deprived rats using a different ligand and didn't observe a change in the number of receptors or the affinity of the somatostatin receptors for the ligand after preincubation (Hugues et al 1988).

Berelowitz: We washed the membranes in an acid medium beforehand to displace somatostatin from its binding sites.

Lightman: The levels of *sstr1*, *sstr2* and *sstr3* mRNA in food-deprived rats decreased dramatically in the pituitary, but were unchanged in the hypothalamus (Figs 1 and 2). We have found that the level of TRH (thyrotropin-releasing hormone) mRNA in the hypothalamus decreases to about 50% of its normal level within 48 h of food deprivation (Blake et al 1991). Have you looked at changes in hypothalamic releasing factors which also act on the pituitary?

Berelowitz: We've studied GHRH (growth hormone-releasing hormone) expression more extensively than somatostatin receptor expression (Bruno et al 1990). We can restore GHRH expression almost completely by refeeding the rats with normal food, dose-related amounts of protein or additions of a single amino acid (histidine). The addition of GH (growth hormone) or GHRH does not restore GHRH expression. We have not done similar experiments to look at *sstr* mRNA expression.

Lightman: We have similar results for TRH mRNA expression. It is the absence of the protein in the diet which reduces hypothalamic TRH mRNA levels (Shi et al 1993).

Reisine: Have you looked at somatostatin levels during GHRH-stimulated GH release?

Berelowitz: We have shown that dispersed anterior pituitary cells of diabetic rats are resistant to the inhibitory effect of somatostatin on GHRH-induced GH secretion (Olchovsky et al 1990).

Reisine: Were these experiments performed *in vivo*?

Berelowitz: The receptor number is taken from binding results obtained using the pituitaries of animals at the time of death and in those cases there was a 50–60% decrease in binding.

Reisine: Is sstr2 desensitized, so that you don't get a corresponding loss of inhibition of GH release?

Berelowitz: I think that the change in the responsiveness of these cells to somatostatin is due to a decrease in *sstr2* levels. We see a decrease in the levels of *sstr1*, *sstr2* and *sstr3* mRNA, but the decrease in *sstr2* is probably responsible for the decline in somatostatin binding.

Reisine: Do you support the idea that sstr2 is the receptor involved in regulation of GH release?

Berelowitz: Yes, this is what the results suggest.

Reisine: They suggest that sstr2 is knocked out in some way, causing desensitization of GH release. Other somatostatin receptors with normal sensitivity may be expressed in the pituitary, so that other functions of somatostatin are still operating. What are these functions? They could be regulation of TSH (thyroid-stimulating hormone) secretion, or some other hormone, or effects on proliferation or maintenance of somatotroph size.

Reichlin: During down-regulation, somatostatin receptors may be still present, but are just not working. How do you explain the mechanism of down-regulation during starvation?

Berelowitz: It could be due to a combined effect—when there are decreased levels of GHRH and increased levels of somatostatin, the levels of GH are decreased because there's little stimulation and much inhibition. Alternatively, the amount of receptor down-regulation that you see may depend on the level of spare receptors and whether receptor occupancy still results in a somatostatin-like effect. I have not looked at the spare somatostatin receptor population, but a similar situation exists for insulin. In some insulin-resistant states you can have substantial down-regulation of insulin receptors and still have an insulin effect, although you shift your dose–response curve to the right. This is also observed in diabetic rats.

Patel: But you are looking at two artificial situations—72 h of food deprivation or streptozocin-induced diabetic rats. These animals are characterized by high circulating levels of somatostatin that are capable of desensitizing some of the pituitary somatostatin receptors.

There are other more physiological situations where changing pituitary sensitivity to somatostatin could be quite important. For instance, the characteristic pulsatile pattern of GH secretion in the rat appears to be mediated by an interplay between somatostatin and GHRH released in surges from the hypothalamus. In addition, there may be changes in pituitary responsiveness to the two peptides through modulation of their receptors. A pulsatile burst of somatostatin would inhibit GH release, giving rise to a trough period, and desensitize pituitary somatostatin receptors, making the pituitary more sensitive to a subsequent stimulatory burst of GHRH release. This is a much more physiological setting for receptor regulation than the prolonged desensitization that occurs in starved or streptozocin-induced diabetic rats.

Berelowitz: Has anyone looked at the biological effect of decreasing the number of native plasma membrane receptors or receptors expressed in transfected cells?

Schonbrunn: Yes, we've studied the relationship between changes in somatostatin receptor number and response (Presky & Schonbrunn 1988). If we pretreat GH_4C_1 pituitary cells with somatostatin for 24 h, we observe a twofold increase in somatostatin receptor levels, measured by binding in whole cells. We carried out dose–response curves to detect a shift in the potency of somatostatin to inhibit hormone secretion from control and pretreated cells. We saw no desensitization of the response. We expected a twofold increase in the potency of somatostatin, because of the increase in receptor levels, but we could not detect any change in response. The sensitivity of the experiment may not have been sufficient to detect a twofold change in potency. Nonetheless, we showed clearly that there was desensitization produced by the somatostatin pretreatment.

Berelowitz: Is sstr5 the most abundant receptor in pancreatic islets?

Reisine: From pharmacological studies, sstr5 may be important for the release of insulin, but it is not known whether it is the most abundant receptor in pancreatic islets.

Berelowitz: Is there any evidence that insulin regulates the expression of *sstr5* mRNA in turn? Because *sstr5* is reduced in the pituitary and hypothalamus of diabetic rats and is restored by insulin therapy (Figs 3 and 4).

Reisine: Insulin receptors can desensitize just like somatostatin receptors, but there are no reports of the effect of insulin on somatostatin receptors.

Reichlin: GH regulation in the rat is very different from the human system (Reichlin 1974). In humans, hypersecretion of GH takes place in response to starvation. What happens to the somatostatin system in cases of human malnutrition?

Berelowitz: The final outcome of GH secretion may depend on the ambient levels of GHRH, somatostatin and IGF1 (insulin-like growth factor 1). In the rat, low levels of GHRH and high levels of somatostatin, plus whatever changes are taking place in the levels of IGF1, lead to a decrease in GH secretion. If the same changes took place in humans after food deprivation, but the balance of factors was different—less reduction in GHRH, less increase in somatostatin (which may be less physiologically dominant in humans) and a greater decrease in IGF1—the balance could be such that GH secretion takes place. My working hypothesis has always been to use the rat as a model to understand mechanisms, then apply these back to humans. The regulation depends on the balance of the systems; for example, the sheep has very little somatostatin function, so the balance is towards GHRH. In contrast, the rat has a very important somatostatin system, so the balance is towards GH inhibition.

Reichlin: Is there an animal species that is a good model for the human?

Berelowitz: The dog.

Reichlin: Does the starved dog have elevated GH levels?

Berelowitz: Yes, but dogs are too expensive to use.

Robbins: Michael Berelowitz looked at the entire hypothalamus and pituitary in his experiments. These both contain transcripts from a wide variety of cell types. It's possible that there could be changes in one group of cells that are offset by a change in a different group, resulting in no difference overall. A number of transformed cell lines, e.g. GH_3, express all five receptor transcripts; but does any one particular normal non-transformed cell make more than one receptor, or is there only one receptor per type of cell? For instance, does the somatotroph only express *sstr2*?

Patel: These experiments need to be done at the single cell level by *in situ* hybridization. We looked at the expression of somatostatin receptors in rat somatotrophs, purified by the method of Jack Kraicer (R. Panetta, M. Greenwood, C. B. Srikant, J. Kraicer & Y. C. Patel, unpublished work) and found a similar pattern of expression to that observed in GH_4C_1 cells. Using reverse transcriptase PCR (RT-PCR) methodology, we detected *sstr1*, *sstr2*, *sstr3* and a small amount of *sstr5*, but not *sstr4* mRNA. So a normal rat somatotroph cell expresses four somatostatin receptor mRNAs, *sstr2* being the most abundant.

We have also looked at the desensitization of *sstr2* in freshly isolated islets using the RT-PCR method (S. Grigorakis, R. Panetta & Y. C. Patel, unpublished work). We exposed rat islets to 10^{-7} M somatostatin-14 for 24 h and found a 10-fold decrease in the level of *sstr2* mRNA. We also saw a three- to fivefold stimulation of *sstr2* mRNA when the islets were exposed to $10\,\mu$M forskolin.

Reichlin: Is that true for all of the receptors or just *sstr2*?

Patel: sstr2 has only been investigated so far.

Reichlin: Have you found multiple receptors expressed in the same cell type?

Patel: Yes, I would say that there are multiple receptor types in normal cells.

Schonbrunn: Finally, I would like to add a note of caution. Decreases in receptor mRNA levels may not represent desensitization. When you measure changes in receptor mRNA levels at a particular point in time, it will not tell you what's happening to the receptor protein or to the biological response.

References

Blake NG, Eckland DJA, Foster OIF, Lightman SL 1991 Inhibition of hypothalamic TRH mRNA during food deprivation. Endocrinology 129:2714–2718

Bruno JF, Olchovsky D, White JD, Leidy JW, Song J, Berelowitz M 1990 Influence of food deprivation in the rat on hypothalamic expression of growth hormone-releasing factor and somatostatin. Endocrinology 127:2111–2116

Hugues JN, Epelbaum J, Voirol MJ, Modigliani E, Sebaoun J, Enjalbert A 1988 Influence of starvation on hormonal control of hypophyseal secretion in rats. Acta Endocrinol (Copenh) 119:195–202

Olchovsky D, Bruno JF, Wood TL et al 1990 Altered pituitary growth hormone regulation in streptozotocin-diabetic rats: a combined defect of hypothalamic somatostatin and GH-releasing factor. Endocrinology 126:53–61

Presky DH, Schonbrunn A 1988 Somatostatin pretreatment increases the number of somatostatin receptors in GH_4C_1 cells and does not reduce cellular responsiveness to somatostatin. J Biol Chem 263:714–721

Reichlin S 1974 Regulation of somatotrophic hormone secretion. In : Greep RO, Astwood EB (eds) American handbook of physiology. Williams and Wilkins Co Baltimore, vol IV:405–577

Shi Z-X, Levy A, Lightman SL 1993 The effect of dietary protein on thyrotropin-releasing hormone and thyrotropin gene expression. Brain Res 606:1–4

Transient expression of somatostatin receptors in the brain during development

Philippe Leroux, Corinne Bodenant, Evelina Bologna, Bruno Gonzalez and Hubert Vaudry

Laboratoire d'Endocrinologie Moléculaire, INSERM U 413, UA CNRS, Université de Rouen, 76821 Mont-Saint-Aignan, France

Abstract. The study of somatostatin receptors by means of autoradiography in tissue sections revealed high densities of binding sites in the immature central nervous system. In rat cerebral cortex, the receptors are present in the intermediate zone and in association with cells migrating through the cortical plate. Somatostatin receptors in the intermediate zone of fetuses and in the cortical plate of postnatal rats exhibit high and low affinities respectively for the somatostatin analogue MK 678. In the rat cerebellum, the external granule cell layer, a germinal matrix containing interneuron precursors, contains a high density of receptors. These receptors exhibit high affinity for MK 678 throughout the period of cell multiplication. In granule cell cultures from eight-day-old rats, MK 678, octreotide and somatostatin are able to inhibit cAMP formation induced by forskolin or pituitary adenylyl cyclase-activating polypeptide. Somatostatin reduces the intracellular Ca^{2+} concentration in cultured granule cells; this response desensitizes rapidly. These results suggest that the somatostatin receptors in the external granule cell layer are type 2 receptors (sstr2). A low density of receptors with low affinity for MK 678 was also detected in the external granule cell layer and in the granule cell layer of neonatal rats. In adult rats the cerebellum is devoid of somatostatin receptors. These observations indicate that somatostatin probably exerts morphogenetic activities through different receptor types in several structures of the central nervous system.

1995 Somatostatin and its receptors. Wiley, Chichester (Ciba Foundation Symposium 190) p 127–141

The development of the central nervous system (CNS) occurs as a succession of progressive and regressive phases. Active cell generation is followed by selection within the populations. Later, excessive fibre sprouting and formation of synaptic connections are regulated by competitive processes. In many areas of the CNS, somatostatin expression during ontogenesis appears to follow a similar developmental pattern, i.e. a succession of intense expression followed by down-regulation or switching off of the gene. Somatostatin is expressed very early during the development of the rat brain, as demonstrated by immunohistochemistry and *in situ* hybridization. Among the cells that synthesize somatostatin early, one can distinguish neuronal systems that express the somatostatin gene transiently, and those in which it is turned on permanently. The latter pattern is observed in the hypothalamus. Somatostatin mRNA becomes detectable at embryonic day 14 (E14) and shows a sixfold increase from E14 to adulthood. The gene is expressed before final neuronal differentiation (Almazan et al 1989) and long before somatostatin exerts any inhibitory effect on growth hormone secretion from anterior pituitary somatotrophs (Rieutort 1981). Somatostatin is detected in the hypothalamus as early as E12. This is earlier than when the mRNA is detected (E14), probably because of the relative insensitivity of mRNA detection (Daikoku et al 1983). *In situ* hybridization revealed the prevalence of mRNA in the periventricular nucleus, i.e. in the final location of somatostatin-expressing cells, from E14 onwards (Baram & Schultz 1991). The concentration of somatostatin rises by 30-fold progressively from E14 to adulthood (McGregor et al 1982) because somatostatin-synthesizing cells continue to be produced over a long time, rather than because of increased synthesis within the cells (Daikoku et al 1983). In keeping with the concept of progressive maturation of the somatostatin transmission system, it has been shown that in the hypothalamus, somatostatin receptors appear later than the neuropeptide (Heidet et al 1990).

Transient expression of somatostatin and somatostatin receptors in the rat brain

The onset of the physiological effects of somatostatin as a neuromodulator in the nervous system is difficult to determine. In many areas, the cell types immunoreactive for somatostatin during development are different from those found in the adult. For instance, it is present early in derivatives of the neural crest in quail and rat embryos. Neurons of the primordia of sympathetic and sensory ganglia express the somatostatin gene for short periods at the time of cell proliferation and synaptogenesis (Maxwell et al 1984, Katz et al 1992).

During the development of the fetal rat brain, transient innervation with somatostatin-positive neurons has been reported in fibre networks in the cerebral cortex (Naus et al 1988a), the hippocampus (Naus et al 1988b) and the cerebellum (Inagaki et al 1982). Transiently somatostatin-positive perikarya have been observed, corresponding to immature cells that will cease to express the peptide (Villar et al 1989). In the cortical plate of cats, somatostatin is stored in early differentiated cells committed to undergo apoptosis in the course of development (Chun et al 1987, Naus et al 1988a). These observations strongly suggest that somatostatin is involved in various developmental processes in different parts of the CNS. Thus, it is interesting to determine whether the neuronal systems that express somatostatin early are potentially functional, i.e. whether the immature brain contains somatostatin receptors coupled to effector systems at the right location and right time. We have studied the distribution and the binding characteristics of these receptors in some structures of the rat brain to examine the putative functions of somatostatin during ontogeny.

Physiological significance of somatostatin in the cerebral cortex: anatomical approaches

In the rat cerebral cortex, somatostatin is expressed mainly in a subclass of GABAergic (γ-aminobutyric acid) neurons (Somogyi et al 1984) located throughout layers II to VI, though generally less abundant in layer IV (Morrison et al 1983). In addition, it is present in some pyramidal and inverted pyramidal cells in the neocortex as well as in Cajal–Retzius-like neurons in layer I and layer VI. The latter cell types are particularly abundant during development and it is believed that they are involved in cortical maturation (Shatz et al 1988). Somatostatin mRNA has been detected from E16 onwards (Naus et al 1988a, Naus 1990) and somatostatin-immunoreactive neurons can be observed at E17 in the visual cortex (Eadie et al 1987). These neurons, located in the intermediate zone and subplate zone, innervate all the cortical plate; they belong to the neurons generated early in the cat visual cortex (Chun et al 1987, Chun & Shatz 1989).

Somatostatin receptors are expressed early in the immature somatosensory cortex and their localizations are distinct from those of the adult cortex. Autohistoradiographic studies using ($[^{125}I]$Tyr^0,DTrp^8]somatostatin-14 ((T0D8)somatostatin-14) as a tracer showed that the receptors are concentrated mainly in the intermediate zone and cortical plate of fetuses (Fig. 1). A clear spatial and temporal correlation is observed between the presence of the receptors and the occurrence of neuroblasts migrating through the cortical plate up to postnatal day 13 (P13). The receptors disappear from layers II and III when migration is completed. In contrast, somatostatin receptors do not disappear in the future layers V and VI. (Gonzalez et al 1991).

Somatostatin receptors in the intermediate zone are localized in the area containing the perikarya of transient neurons that arise early. Migrating neuroblasts that are likely the cells which express somatostatin receptors in cortical superficial layers in postnatal rats are innervated by somatostatin-immunoreactive fibres from intermediate zone neurons. The disappearance of receptors from these layers is concomitant with the regression of somatostatin innervation. From P8 to adulthood, receptors are also present in the future layers VI and V of the cerebral cortex. At the same time, interneurons located in these layers synthesize prosomatostatin, but do not process the precursor properly into the 14 amino acid peptide before P30 (Naus et al 1988a). Thus, there is no temporal overlap between the synthesis of somatostatin during morphogenesis and in the adult.

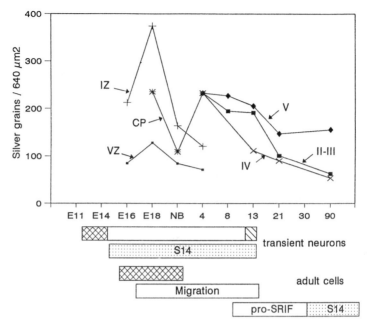

FIG. 1. Somatostatin receptor densities in the rat somatosensory cortex during development. Receptor densities were measured by grain counting in autohistoradiographic preparations of sections incubated with ($[^{125}I]$ Tyr^0,DTrp^8)somatostatin-14 as radioligand. CP, cortical plate; IZ, intermediate zone; NB, newborn; VZ, ventricular zone; II-III, layers II and III; IV, layer IV; V, layer V. Boxes indicate periods of cortical maturation: ☒ cell multiplication, ◩ apoptosis, ⊡ expression and processing of somatostatin (S14). pro-SRIF, prosomatostatin.

Using (Leu^8,DTrp^{22}, $[^{125}I]$ Tyr^{25})somatostatin-28 (abbreviated as (LDTT)somatostatin-28), a radioligand that labels more somatostatin receptor types than (T0D8)somatostatin-14 (Leroux et al 1985, Bucharles et al 1994), we have studied the affinity of cortical receptors for MK 678, a putative ligand of sstr2, in the somatosensory cortex during development. In fetal cortex, 95% of somatostatin receptors in the intermediate zone exhibit a high affinity for MK 678 (EC$_{50}$ = 0.25 nM). This ligand is totally unable, at 0.1 μM to compete with (LDTT)somatostatin-28 binding sites in layers II and III of P4 rats, an area within the cortical plate in which cell migration continues (Fig. 2). In P4 rats, MK 678 high affinity sites represent 60% of (LDTT)somatostatin-28 binding sites in the intermediate zone (Fig. 2). Later during development, from P8 to adulthood, the receptors with high affinity for MK 678 are restricted to layers V and VI of the cortex and represent fewer than 15% of somatostatin receptors (not shown).

Thus, at least three different populations of somatostatin receptors could be identified in the cerebral cortex on the basis of pharmacological and

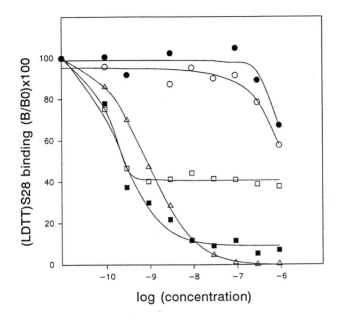

FIG. 2. Competition curves for binding of (LDTT)somatostatin-28 (S28) to somatostatin receptors by somatostatin-28 and MK 678 in different layers of developing rat somatosensory cortex: somatostatin-28 in cortical plate at P4 (△); MK 678 in intermediate zone at E21 (■); MK 678 in intermediate zone at P4 (□); MK 678 in cortical plate (including layers II and III) at P4 (○), MK 678 in layer V at P14 (●). Data are means of 12 determinations in three animals. B/B0, binding of ligand/initial binding.

spatiotemporal criteria. These observations support the view that somatostatin is involved in cortical morphogenesis.

Somatostatin receptors in the immature cerebellum

Localization of somatostatin

The cerebellum of adult rats contains very little somatostatin (Brownstein et al 1975). Somatostatin immunoreactivity has been described in ascending fibres (Inagaki et al 1982) and in a few intrinsic cerebellar neurons (Vincent et al 1985). In the perinatal period, a significant proportion of cerebellar neurons (Purkinje cells, Golgi cells) express the somatostatin gene (Inagaki et al 1989, Villar et al 1989, Naus 1990). At the same time, somatostatin-immunoreactive fibres arising from the pons and the medulla innervate the cerebellar cortex (Inagaki et al 1982). The density of somatostatin-containing neuronal processes and the intensity of expression of the peptide peak around P10 (Inagaki et al 1982, McGregor et al 1982, Hayashi 1987, Villar et al 1989, Naus 1990). This period corresponds to the maximum intensity of cell multiplication in the external granule cell layer (EGC), a germinative matrix in which cerebellar interneurons are generated during the first postnatal weeks. Somatostatin-immunoreactive fibres do not penetrate this layer, which is virtually devoid of synapses (Inagaki et al 1982, Villar et al 1989) and somatostatin is not synthesized within the EGC (Villar et al 1989, Naus 1990).

Localization of somatostatin receptors

Although the gene coding for sstr3 is expressed strongly in granule and Purkinje cells (Perez et al 1994), the concentration of somatostatin receptors in the cerebellar cortex of adult rat is very low. Conversely, high densities of somatostatin receptors were detected in cerebellar nuclei (Epelbaum et al 1985, Leroux et al 1985, Uhl et al 1985). These have been characterized as sstr1 according to pharmacological criteria using (LDTT)somatostatin-28 as a radioligand (Bucharles et al 1994). These sites were not labelled with (T0D8)somatostatin-14.

In immature rat cerebellum, binding sites for (T0D8)somatostatin-14 were detected in the EGC for as long as the layer was present (Gonzalez et al 1988), a short period corresponding to the maximum level of somatostatin in the cerebellum. At least two distinct sites could be resolved in membrane preparations from the cerebellum of P13 rats, using octreotide as a competitor (Gonzalez et al 1990). Autoradiographic studies were carried out using (LDTT)somatostatin-28 in cerebellum from rats aged 4–12 days. A large proportion of (LDTT)somatostatin-28 binding sites in the EGC showed high affinity for MK 678 ($IC_{50} = 0.2$ nM). MK 678 (30 nM) displaced 100% of (LDTT)somatostatin-28 binding in the EGC of rats at P4 (not shown), whereas at P8 (not shown) and P12 (Fig. 3A), biphasic competition curves were obtained. The sites with low affinity for MK 678 represent 26% and 12% of total (LDTT)somatostatin-28 binding sites in the EGC at P8 (not shown) and P12 (Fig. 3A), respectively. These sites exhibited low affinity for N-Ahep(7–10)SS14–Thr–bzl ($IC > 1$ μM), a ligand which exhibits higher affinity ($IC_{50} = 0.5$ μM) for sstr1 in transfected CHO cells (Raynor et al 1993) and in adult rat cerebellar nuclei (Bucharles et al 1994). The somatostatin analogue BIM 23056, a selective ligand for sstr3 in transfected CHO (Chinese hamster ovary) cells (Raynor et al 1993), bound weakly to (LDTT)somato-statin-28 binding sites in the EGC ($IC_{50} > 1$ μM). The granule cells generated in the EGC migrate across the molecular layer to settle in the granule cell layer (GCL). A low density of binding sites could be detected in the GCL from P8 to P12. MK 678 evoked a biphasic competition curve in GCL ($IC_{50} = 0.5$ nM and 0.1 μM, for high affinity and low affinity sites). In the GCL, MK 678 high affinity sites represent 60% and 54% of (LDTT)somatostatin-28 binding sites at P8 (not shown) and P12 (Fig. 3B), respectively.

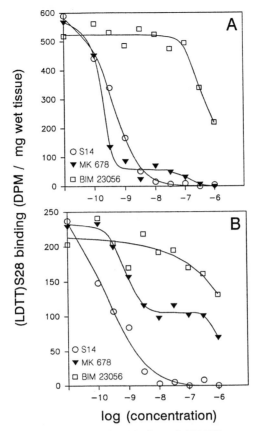

FIG. 3. Typical competition curves for binding of (LDTT)somatostatin-28 (S28) to somatostatin receptors by somatostatin-14 (S14), MK 678 and BIM 23056 in external granule cell layer (A) and granule cell layer (B) of P12 rat cerebellar cortex. Competitive effect of each molecule was measured in three definite areas of cerebellar lobules 6 and 7 in three animals. DPM, disintegrations per min.

These observations are difficult to reconcile with the mRNA localizations. Using a quantitative polymerase chain reaction (PCR), Wulfsen et al (1993) detected almost invariant levels of sstr2 mRNA in the cerebellum from birth to adulthood, whereas MK 678 high affinity sites disappear together with EGC around P21. On the other hand, sstr3 mRNA is abundant in the immature cerebellum, in both granule and Purkinje cells from P7 to adulthood (Meyerhof et al 1992, Wulfsen et al 1993, Perez et al 1994). Binding studies will show whether these mRNAs are translated in immature or adult cerebellar neurons. It is also possible that the analogue BIM 23056 does not have such a high affinity for sstr3 synthesized *in vivo* as for receptors synthesized in transfected cells. The binding sites of low affinity for MK 678 observed in the EGC and in the GCL may be relevant to the sstr1 and/or sstr4 mRNAs that have been detected by PCR in the immature cerebellum (Wulfsen et al 1993).

Somatostatin receptors in cerebellar granule cells in culture

Binding studies on cultures of granule cells from eight-day-old rats confirmed that cerebellar interneuron precursors express somatostatin receptors (Gonzalez et al 1992). In these cultures, somatostatin, octreotide (SMS 201–995, Sandostatin™) and MK 678 were able to inhibit cAMP formation induced by pituitary adenylyl cyclase activating-polypeptide by 85% ($ED_{50} = 3.0$, 4.0 and 0.6 nM, respectively). The effect was prevented by treatment of the cultures with 30 μM pertussis toxin for five hours. In addition, somatostatin decreased the intracellular Ca^{2+} concentration ($[Ca^{2+}]_i$) in granule cells in culture. Repeated pulses of somatostatin resulted in sequential inhibition of $[Ca^{2+}]_i$ with gradual attenuation of the response, which implies there is a desensitization mechanism (Gonzalez et al 1992). All together, these characteristics indicate the occurrence of sstr2 in cerebellar granule cell precursors.

The density of somatostatin receptors on granule cells in culture depends on the medium. In serum-free medium composed of Dulbecco's modified Eagle's medium (DMEM) and Ham's F12 in a ratio of 3:1, containing 5 mM K^+, a 50% decrease in receptor density was observed after four days *in vitro*. In cultures grown in the same medium supplemented with 10% serum and containing 25 mM K^+, cell survival was enhanced and morphological modifications indicated that neuronal differentiation occurred. In these conditions, the receptor density was reduced by 50% after two days *in vitro*. Thus, the expression of somatostatin receptors in granule cells in culture parallels that observed *in vivo*, i.e. their expression is inversely related to cell differentiation.

Effects of somatostatin in cerebellar explants

Cerebellar explants of rats at P8 were grown in DMEM:Ham's F12 (3:1) culture medium containing 10% fetal calf serum. In the presence of 1 μM insulin, granule cell precursors of the EGC incorporated pulse-labelled [^{3}H]thymidine for 36 h *in vitro*. The cells generated in the EGC completed their migration across the molecular layer and reached the GCL in eight hours. Somatostatin and octreotide did not affect the incorporation of [^{3}H]thymidine in these explants. It appears from this study that in the immature rat cerebellum, somatostatin is unlikely to modulate cell generation in the EGC, although its receptors are intensely expressed in the germinative epithelium.

Conclusion

Complete determination of the pattern of somatostatin receptor distribution in the different areas of immature CNS requires selective ligands for all five receptor types. It will also be necessary to identify the receptor proteins by biochemical or immunological approaches, in an attempt to reconcile the findings of *in situ* hybridization and binding studies. The present results suggest that somatostatin is involved in the development of the CNS. However, morphogenetic activities of the neuropeptide remain to be demonstrated.

Acknowledgements

This work has been supported by CNRS URA 650, INSERM grant 92.0808 and the Conseil Régional de Haute-Normandie.

References

Almazan G, Lefebvre DL, Zingg HH 1989 Ontogeny of hypothalamic vasopressin, oxytocin and somatostatin gene expression. Dev Brain Res 45:69–75

Baram TZ, Schultz L 1991 Ontogeny of somatostatin gene expression in rat diencephalon. Dev Neurosci 13:176–180

Brownstein M, Arimura A, Szabo H, Schally AV, Kizer JS 1975 The regional distribution of somatostatin in the rat brain. Endocrinology 96:1456–1461

Bucharles C, Vaudry H, Leroux P 1994 Pharmacological characterization of somatostatin receptors in rat cerebellar nuclei. Eur J Pharmacol 271:79–86

Chun JJM, Shatz CJ 1989 The earliest generated neurons of the cat-cerebral cortex: characterization by MAP2 and neurotransmitter immunohistochemistry during fetal life. J Neuroscience 9:1648–1667

Chun JJM, Nakamura MJ, Shatz CJ 1987 Transient cells of the developing mammalian telencephalon are peptide immunoreactive neurons. Nature 325:617–620

Daikoku S, Hisano S, Kawano K, Okamura Y, Tsuruo Y 1983 Ontogenetic studies of the topographical heterogeneity of somatostatin-containing neurons in rat hypothalamus. Cell Tissue Res 233:347–354

Eadie LA, Parnavelas JG, Franke E 1987 Development of the ultrastructural features of somatostatin-immunoreactive neurons in the rat visual cortex. J Neurocytol 16:445–459

Epelbaum J, Dussaillant M, Enjalbert A, Kordon C, Rostene W 1985 Autoradiographic localization of a non-reducible somatostatin analog (^{125}I–CGP) binding sites in the rat brain: comparison with membrane binding. Peptides 6:713–719

Gonzalez BJ, Leroux P, Laquerrière A, Coy DH, Bodenant C, Vaudry H 1988 Transient expression of somatostatin receptors in the rat cerebellum during development. Dev Brain Res 40:154–157

Gonzalez BJ, Leroux P, Bodenant C, Braquet P, Vaudry H 1990 Pharmacological characterization of somatostatin receptors in the rat cerebellum during development. J Neurochem 55:729–737

Gonzalez BJ, Leroux P, Bodenant C, Vaudry H 1991 Ontogeny of somatostatin receptors in the rat somatosensory cortex. J Comp Neurol 305:177–188

Gonzalez BJ, Leroux P, Lamacz M, Bodenant C, Balazs R, Vaudry H 1992 Somatostatin receptors are expressed by immature cerebellar granule cells: evidence for a direct inhibitory effect of somatostatin on neuroblast activity. Proc Natl Acad Sci USA 89:9627–9631

Hayashi M 1987 Ontogeny of glutamic acid decarboxylase, tyrosine hydroxylase, choline acetyl transferase, somatostatin and substance P in monkey cerebellum. Dev Brain Res 32:181–186

Heidet V, Faivre-Bauman A, Kordon C, Loudes C, Rasolonjanahary S, Epelbaum J 1990 Functional maturation of somatostatin neurons and somatostatin receptors during development of mouse hypothalamus in vivo and in vitro. Dev Brain Res 57:85–92

Inagaki S, Shiosaka S, Takatsuki K et al 1982 Ontogeny of somatostatin–containing neuron system of the rat cerebellum including its fiber connections: an experimental and immunohistochemical analysis. Dev Brain Res 3:509–527

Inagaki S, Shiozaka S, Sekitani M, Noguchi K, Shimada S, Takagi H 1989 In situ hybridization of the somatostatin containing neuron system in developing cerebellum of rats. Mol Brain Res 6:289–295

Katz DM, He H, White M 1992 Transient expression of somatostatin peptide is a widespread feature of developing sensory and sympathetic neurons in the embryonic rat. J Neurobiol 23:855–870

Leroux P, Quirion R, Pelletier G 1985 Localization and characterization of brain somatostatin receptors as studied with somatostatin-14 and somatostatin-28 receptor radioautography. Brain Res 347:74–84

McGregor GP, Woodhams PL, Chatey MA, Polak JM, Bloom SR 1982 Developmental changes in bombesin, substance P, somatostatin and vasoactive intestinal peptide in the rat brain. Neurosci Lett 28:21–27

Maxwell GD, Sietz PD, Chenard PH 1984 Development of somatostatin-like immunoreactivity in embryonic sympathetic ganglia. J Neurosci 4:576–584

Meyerhof W, Wulfsen I, Schonrock C, Fehr S, Richter D 1992 Molecular cloning of a somatostatin-28 receptor and comparison of its expression pattern with that of a somatostatin-14 receptor in rat brain. Proc Natl Acad Sci USA 89:10267–10271

Morrison JH, Benoit R, Magistretti PJ, Bloom FE 1983 Immunohistochemical distribution of prosomatostatin-related peptides in cerebral cortex. Brain Res 262:344–351

Naus CCG 1990 Developmental appearence of somatostatin in the rat cerebellum: in situ hybridization and immunohistochemistry. Brain Res Bull 24:583–592

Naus CCG, Miller FD, Morrison JH, Bloom FE 1988a Immunohistochemical and in situ hybridization analysis of the development of the rat somatostatin-containing neocortical neuronal system. J Comp Neurol 269:448–463

Naus CCG, Morrison JH, Bloom FE 1988b Development of somatostatin-containing neurons and fibers in the rat hippocampus. Dev Brain Res 40:113–121

Pérez J, Rigo M, Kaupmann K et al 1994 Localization of somatostatin (SRIF) SSTR1, SSTR2 and SSTR3 receptor mRNA in rat brain by in situ hybridization. Naunyn-Schmiedebergs Arch Pharmakol 349:145–160

Raynor K, Murphy WA, Coy DH et al 1993 Cloned somatostatin receptors: identification of subtype-selective peptides and demonstration of high affinity binding of linear peptides. Mol Pharmacol 43:838–844

Rieutort M 1981 Ontogenic development of the inhibition of growth hormone release by somatostatin in the rat: in-vivo and in-vitro (perifusion) study. J Endocrinol 9:355–363

Shatz CJ, Chun JJM, Luskin MB 1988 The role of the subplate in the development of the mammalian telencephalon. In: Peters A, Jones EG (eds) Cerebral cortex. Plenum Press, New York, p35–58

Somogyi P, Hodgsin A, Smith D, Nunzi MG, Gorio A, Wu AY 1984 Different populations of GABAergic neurons in the visual cortex and hippocampus of cat contain somatostatin or cholecystokinin-immunoreactive material. J Neurosci 4:2590–2603

Uhl GR, Tran V, Snyder SH, Martin JB 1985 Somatostatin receptors: distribution in rat central nervous system and human frontal cortex. J Comp Neurol 240:288–304

Villar MJ, Hökfelt T, Brown JC 1989 Somatostatin expression in the cerebellar cortex during post-natal development; an immunohistochemical study in the rat. Anat Embryol 179:257–267

Vincent SR, McIntosh CHS, Buchan AMJ, Brown JC 1985 Central somatostatin systems revealed with monoclonal antibodies. J Comp Neurol 238:169–186

Wulfsen I, Meyerhof W, Fehr S, Richter D 1993 Expression patterns of rat somatostatin receptor genes in pre- and postnatal brain and pituitary. J Neurochem 61:1549:1552

DISCUSSION

Robbins: Areas in the deep layers of the cerebral cortex that contain somatostatin receptors are critical for the proliferation and survival of neurons in that region. We have found that as early as E12 (embryonic day 12) or E13, the deepest layers of the neocortex (near the cortical plate) develop high levels of somatostatin receptors (S. Welsh & R. Robbins, unpublished observations). A variety of growth factors are important in the proliferation and subsequent differentiation of cells coming out of that plate; for instance, we have shown that IGF1 (insulin-like growth factor 1) is important in differentiation and proliferation, stimulating both protein kinase C and Ca^{2+} currents (Torres-Aleman et al 1990). The cells of the neocortex require a certain level of Ca^{2+}; if this level is too high the cell will undergo programmed cell death, if it is too low the cells will also die. Growth factors allow Ca^{2+} to enter the cell, but the regulatory mechanism is not understood. It is possible that somatostatin, by altering Ca^{2+} currents in these developing cells, could be a critical factor in deciding whether they continue to proliferate or begin to undergo differentiation during migration through the cortex, i.e. somatostatin may act as the switch between proliferation and differentiation in that region of the neocortex. The identification of somatostatin receptors in this region of the cortex at this stage of development supports this.

Leroux: Receptor levels are low in the proliferating ventricular zone, but are high in the non-proliferating intermediate layer (Gonzalez et al 1991).

Robbins: We haven't differentiated between immature cells and the cortical plate. Your results suggest that somatostatin is involved mostly in differentiation.

Meyerhof: We have used *in situ* hybridization to detect *sstr1* mRNA in the developing cortex and have found that it is present in postmigratory neurons, but not in premigratory or premitotic neurons (D. Hartmann, S. Fehr, W. Meyerhof & D. Richter, unpublished observations). Within the cortical plate, expression of the *sstr1* gene was first detected in parallel with the formation of the deep laminae V/VI at E16. Thereafter, *sstr1* expression spreads throughout the entire cortical plate, coincidental with the deposition of the supragranular neurons.

Robbins: The cells that extend long processes need cAMP. It would be interesting if sstr1 was linked to cAMP and sstr2 to Ca^{2+} currents.

Meyerhof: We have not observed coupling of rat sstr1 to the inhibition of adenylyl cyclase when sstr1 is stably expressed in NIH/3T3 cells (R. Nehring, C. Schönrock, W. Meyerhof & D. Richter, unpublished observations). Consequently, this receptor may be engaged in differentiation by affecting, either directly or indirectly, the intracellular Ca^{2+} concentration.

Robbins: Immature K^+ currents may also exist at this stage, but I don't think they are functional, even if sstr1 and sstr2 are present. It's possible that as the receptors mature they link up with other G proteins. It might be interesting to look at the relationship between the age of the cell and which function somatostatin can perform with a specific receptor.

Chesselet: One interesting area to look at the role of somatostatin in proliferation or differentiation would be the subventricular zone in the adult rat, because this region is one of the few in the brain that contains dividing cells (Smart 1961). Most of these cells do not differentiate into either mature glial cells or neurons (F. G. Szele & M.-F. Chesselet, unpublished results).

Reichlin: Has anyone made transgenic knockouts of somatostatin receptors to address the role of somatostatin in differentiation/proliferation?

Patel: There are only rumours that a somatostatin knockout exists.

Eppler: Did Philippe Leroux do a competition curve with David Coy's antagonist N–Ahep(7–10)S14–bzl (Fries et al 1982)? What was the IC_{50}?

Leroux: Yes, in adult rat cerebellar nuclei, the IC_{50} for N–Ahep(7–10)S14–bzl is 0.5 μM.

Coy: Which receptor is involved in the binding of this antagonist?

Leroux: The literature indicates that it has a low affinity for sstr4 (Raynor et al 1993). We use this argument to exclude the possibility that sstr4 functions in the cerebellar nuclei.

We have also used BIM 23052 in other areas, such as the anterior pituitary and the cerebral cortex, and found similar efficiency of this analogue everywhere we tested.

Reisine: We have done electrophysiological studies of cerebral cortical cultures and showed that the response to the same K^+ current, for example, was different for somatostatin-14 and somatostatin-28 (Wang et al 1989, 1990). If you look at total mRNA levels, you may not detect the low levels of mRNA that may actually make the relevant protein.

We've also done developmental studies in the brain and spinal cord using an antibody against sstr2. In the brain, the level of protein starts to rise after birth and then goes down. This is just the brain region as a whole, we made no attempt to separate the different regions. The interesting area is the spinal cord because there's a transient appearance of the sstr2 receptor before birth.

We have also done binding studies using MK 678 and found that it binds during the same period.

Leroux: Did you get the same result in the brain stem?

Reisine: We did not distinguish between the brain stem and the other regions of the brain.

Montminy: Anterior horn cells express somatostatin mRNA during development and this expression disappears at birth (M. Montminy, unpublished results). Has Philippe Leroux looked at this area to see if there are similar changes in receptor expression?

Leroux: No, not in the rat.

Montminy: Do you think that the same areas that express somatostatin also express the receptors?

Epelbaum: There is a temporal correlation between the appearance of somatostatin binding sites and somatostatin innervation in the ventral horn (the lower part of the spinal cord). We showed this using immunohistochemistry and binding studies during the period when neurons migrate from the ventral horn to the dorsal horn (Maubert et al 1994). The same pattern is also observed in the dorsal ganglia where sympathetic neurons are transiently somatostatinergic at the same time as somatostatin binding sites appear—between E13.5 and E15.5. A similar situation occurs in the adrenal lineage, where a transient appearance of the binding sites is observed.

Reisine: We've looked at sstr2 with an antibody and iodinated MK 678, and at somatostatin receptors in general (M. A. Theveniau & T. Reisine, unpublished results). Patterns of expression may be different for each receptor. sstr2 is present early in development, but then the levels decrease; however, a corresponding decrease is not observed in somatostatin binding overall. This suggests that sstr2 is not the only receptor present.

Montminy: Do you think that the same cells in the spinal cord express the receptor and somatostatin?

Epelbaum: No, not necessarily, although the temporal pattern of expression is the same in the dorsal root ganglion. It's very clear that sympathetic neurons which innervate the dorsal root ganglion express somatostatin and that somatostatin binding sites are present in the dorsal root.

Reichlin: There are innumerable differentiating brain growth factors that are operative during these stages; for example TGF1 (transforming growth factor 1), IGF2 and NGF (nerve growth factor). Have you looked at the effect of these factors in your isolated culture systems for the development of somatostatin receptors?

Leroux: No.

Reichlin: Does somatostatin influence the secretion of IGF2?

Robbins: We found that BDNF (brain-derived neurotrophic factor) and IGF1 increase the survival of somatostatin-expressing cells in the developing brain, but I don't know what the effects are on somatostatin target cells, the main output neurons in the cortex.

You suggest that by the time of birth or postnatal day 10 (P10) the transient somatostatin-expressing cell population has died off completely and that by P30, the prosomatostatin neurons begin to appear.

Leroux: Yes, this has been shown by Naus et al (1988) using different antibodies directed either against somatostatin-14, which recognize precursor forms of somatostatin, or against somatostatin-28$_{(1-12)}$ which attest that somatostatin-14 has been cleaved. The signal with the latter antibody disappears in the second postnatal week and does not reappear for 30 days.

Robbins: We have published that between P10 and P20, somatostatin-14 is present throughout the brain (Robbins & Reichlin 1983). We found, by chromatography, that somatostatin-14 is present from birth to adult; there doesn't seem to be a gap where the brain has no somatostatin-14 for 10 days.

Eppler: It would be interesting to repeat these studies in transgenic animals, for instance GH (growth hormone) or GHRH (growth hormone-releasing factor) transgenics. You might find different distributions of somatostatin receptors in these animals compared with normals. This might tell you something about feedback regulation in the hypothalamus.

Robbins: It would also be interesting if there were knockout mice for Pit-1, because at least one of the receptors has two Pit-1 response elements in the promotor.

Reichlin: There are mutant mouse strains that have neurological defects, for instance defects in the cerebellum. Have you looked at somatostatin in any of these mutants?

Leroux: We've looked at receptor expression in the external granule layer of *weaver* mice.

Reichlin: Do these mice develop normally?

Leroux: They develop ataxia because their granule cells do not migrate, but remain in the external granule cell layer and degenerate. Somatostatin receptors are present in these cells, so the migratory defect is not due to a lack of receptors. It is difficult to work with this mutant strain because the mutation is recessive and it is almost impossible to get homozygous adults. When you work with heterozygotes, a small proportion express the defect and ataxia is hard to detect at the time suitable to perform cell cultures.

References

Fries JL, Murphy WA, Sueiras-Diaz J, Coy DH 1982 Somatostatin antagonist analog increases GH, insulin and glucagon release in the rat. Peptides 3:811–814

Gonzalez BJ, Leroux P, Bodenant C, Vaudry H 1991 Ontogeny of somatostatin receptors in the rat somatosensory cortex. J Comp Neurol 305:177–188

Maubert E, Slama A, Ciofi P et al 1994 Developmental patterns of somatostatin binding sites and somatostatin immunoreactivity during early neurogenesis in the rat. Neuroscience 62:317–325

Naus CCG, Millar FD, Morrison JH, Bloom FE 1988 Immunohistochemical and in situ hybridization analysis of the development of the rat somatostatin-containing neocortical neuronal system. J Comp Neurol 269:448–463

Raynor K, O'Carroll A-M, Kong H et al 1993 Characterization of cloned somatostatin receptors sstr4 and sstr5. Molec Pharmacol 44:385–392

Robbins R, Reichlin S 1983 Somatostatin biosynthesis by cerebral cortex cells in monolayer culture. Endocrinology 113:574–581

Smart I 1961 The subventricular zone of the mouse brain and its cell production as shown by autoradiography after thymidine-3H injection. J Comp Neurol 116:325–338

Torres-Aleman I, Naftolin F, Robbins RJ 1990 Trophic effects of insulin-like growth factor I on fetal rat hypothalamic cells in culture. Neuroscience 35:601–608

Wang HL, Bogen C, Reisine T, Dichter M 1989 Somatostatin-14 and somatostatin-28 induce opposite effects on potassium currents in rat neocortical neurons. Proc Natl Acad Sci USA 86:9616–9620

Wang HL, Dichter M, Reisine T 1990 Lack of cross-desensitization of somatostatin-14 and somatostatin-28 receptors coupled to potassium channels in rat neocortical neurons. Mol Pharmacol 38:357–361

Expression of *sstr1* and *sstr2* in rat hypothalamus: correlation with receptor binding and distribution of growth hormone regulatory peptides

Alain Beaudet and Gloria S. Tannenbaum*

*Departments of Neurology & Neurosurgery and Pediatrics, McGill University and The Montreal Neurological Institute and *Montreal Children's Hospital Research Institute, Montreal, Quebec, Canada*

Abstract. With the aim of elucidating the role of individual somatostatin receptors in the central control of growth hormone secretion, we have examined the distribution of *sstr1* and *sstr2* mRNAs in the hypothalamus of the adult rat by *in situ* hybridization using ^{35}S-labelled antisense riboprobes. Both receptors were expressed strongly in the preoptic area, suprachiasmatic nucleus and arcuate nucleus. High *sstr1*, but low *sstr2*, expression was evident in the paraventricular and periventricular nuclei as well as in the ventral premammillary nucleus. Conversely, moderate to high *sstr2*, but low *sstr1*, mRNA levels were detected in the anterior hypothalamic nucleus, ventromedial and dorsomedial nuclei and medial tuberal nucleus. Within the arcuate nucleus, the distribution of cells expressing *sstr1* and *sstr2* was comparable to that of neurons which bind somatostatin-14 selectively, one third of which have been documented to contain growth hormone-releasing hormone. Within the periventricular nucleus, the distribution of cells expressing *sstr1* and, to a lesser extent, *sstr2* was reminiscent of that of both [^{125}I]somatostatin-labelled and somatostatin-immunoreactive cells. Taken together, these results imply a role for both sstr1 and sstr2 receptors in the central regulation of growth hormone-releasing hormone and somatostatin secretion, and hence of growth hormone release, by somatostatin.

1995 Somatostatin and its receptors. Wiley, Chichester (Ciba Foundation Symposium 190) p 142–159

Somatostatin was isolated originally from ovine hypothalamus on the basis of its ability to inhibit growth hormone (GH) secretion from cultured rat pituitary cells (Brazeau et al 1973). It was subsequently localized throughout the central nervous system and shown to exert a wide variety of neural effects in addition to being recognized as the major physiological inhibitor of GH secretion (see Reichlin 1983 for review). In mammals, two major forms of somatostatin are synthesized: somatostatin-14 and an N-terminally extended form of the tetradecapeptide, somatostatin-28 (Schally et al 1980). Like somatostatin-14, somatostatin-28 was found to exert potent, long-lasting inhibition of GH secretion (Tannenbaum et al 1982).

Both somatostatin-14 and somatostatin-28 can directly inhibit GH secretion from anterior pituitary somatotrophs through their release from tuberoinfundibular neurons into the portal circulation of the median eminence. Physiological (Katakami et al 1988, Lumpkin et al 1981, Tannenbaum et al 1990) and morphological (Bertherat et al 1992, Daikoku et al 1988, Liposits et al 1988, McCarthy et al 1992, Tannenbaum et al 1990) evidence suggests that somatostatin may also regulate pituitary GH release through its action within the central nervous system. Although the mechanisms underlying these central effects have not yet been fully elucidated, available data suggest that they may involve a direct action of somatostatin at the level of both somatostatin-containing neurons and growth hormone-releasing hormone (GHRH)-containing neurons in the mediobasal hypothalamus (Bertherat et al 1992, Daikoku et al 1988, Liposits et al 1988, Lumpkin et al 1981, McCarthy et al 1992, Richardson & Twente 1986). This is supported by the autoradiographic demonstration of high affinity somatostatin binding sites at the level of the periventricular and arcuate hypothalamic nuclei (see Table 1 for references) which have been shown to contain somatostatin- and GHRH-containing tuberoinfundibular neurons, respectively (Bloch et al 1983, Ishikawa et al 1987). Furthermore, within the arcuate nucleus, a proportion of these binding sites was shown to be directly associated with cells that both contain and express GHRH (McCarthy et al 1992, Bertherat et al 1992). However, the molecular identity of these binding sites is unknown.

Molecular biological studies have recently demonstrated the existence of at least five somatostatin receptors in the mammalian brain (Bruno et al 1992, O'Carroll et al 1992, Yamada et al 1992, Yasuda et al 1992). These receptors may be grouped into two families based on their structure and pharmacological profile: the sstr2/sstr3/sstr5 family which corresponds to the previously pharmacologically defined somatostatin-1 receptor and the sstr1/sstr4 family which is the equivalent of the previous somatostatin-2 receptor.

With the aim of elucidating the potential physiological role of these two receptor families in the transduction of somatostatin's neuroendocrine actions in the brain, we have examined the distribution of the mRNA coding for one receptor from each family (sstr1 and sstr2) in the hypothalamus of the adult rat. The results are reported in the context of our current knowledge of the distribution of somatostatin binding sites and of GH regulatory peptides in the mammalian hypothalamus.

Methods

In situ hybridization was used to visualize *sstr1* and *sstr2* mRNAs in serial coronal sections taken through the whole hypothalamus of male and female

TABLE 1 *sstr1* and *sstr2* localization and somatostatin binding in the adult rat hypothalamus

	In situ hybridization[a]		Receptor autoradiography[b]		
Region	*sstr1 mRNA*	*sstr2 mRNA*	*($[^{125}I]Tyr^0,D-Trp^8$)SS14*[c]	*($[^{125}I]Leu^8,D-Trp^{22},Tyr^{25}$)SS28*[d]	*$[^{125}I]MK\ 678$*[e]
Magnocellular pre-optic nucleus	+++	−	−	++	−
Medial pre-optic area	+++	+++	+++	++	+
Anterior hypothalamic nucleus	+	++	++	+	−
Paraventricular nucleus	+++	+	+	+++	−
Periventricular nucleus	++	+	−/+	+	−
Suprachiasmatic nucleus	+++	++	−	−	−
Arcuate nucleus	+++	+++	+	+	+
Ventromedial nucleus	+	+++	+	−	++
Dorsomedial nucleus	+	++	−	−	++
Mammillary complex	+	+	−	++	−

[a]Relative autoradiographic density values established separately for each probe. Data based on joint film and light microscopic observations and expressed as none (−) or in increasing order from (+) to (+++).
[b]Autoradiographic labelling densities based on densitometric values and/or qualitative descriptions in the text of references indicated.
[c]SS14, somatostatin-14. Epelbaum et al 1989, Krantic et al 1989, Leroux et al 1988.
[d]SS28, somatostatin-28. Leroux et al 1988, Uhl et al 1985.
[e]Martin et al 1991.

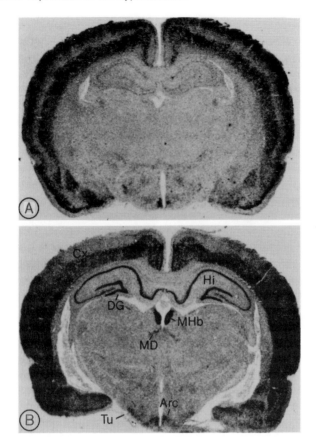

FIG. 1.   Regional distribution of *sstr1* (A) and *sstr2* (B) mRNA as detected by *in situ* hybridization in 20 μm thick coronal sections of the rat brain. Note the difference in the labelling patterns produced by the two antisense probes and the strong hybridization signal for both *sstr1* and *sstr2* in the arcuate nucleus of the hypothalamus (Arc). Abbreviations: Cx: cerebral cortex; DG: dentate gyrus; Hi: hippocampus; MD: medio-dorsal thalamic nucleus; MHb: medial habenular nucleus; Tu: medial tuberal nucleus.

rats. The brains were fixed by perfusion with 4% paraformaldehyde and cut frozen on a cryostat. Sections (20 μm thick) were hybridized with ³⁵S-labelled mouse *sstr1* and *sstr2* probes as described by Breder et al (1992). Original plasmids for both probes were generously provided by Dr Graeme Bell, Howard Hughes Medical Institute, University of Chicago. The *sstr1* probe is a 413 bp *Ban*II–*Acc*I fragment of the gene which encodes amino acids 214–352. The *sstr2* probe is a 458 bp *Bst*EII–*Xba*I fragment of the cDNA which encodes amino acids 254–369, the stop codon and 107 nucleotides of the 3′-untranslated region. To control for non-specific hybridization, additional sections were incubated

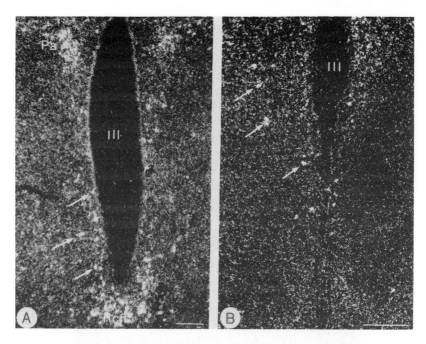

FIG. 2. (A) Cellular distribution of *sstr1* mRNA in the paraventricular (Pa) and periventricular (arrows) nuclei of the hypothalamus as detected by *in situ* hybridization using liquid emulsion autoradiographic processing. Darkfield. Note the small cluster of cells expressing *sstr1* in the retrochiasmatic area (Rch). Scale bar = 75 μm. (B) Distribution of ([^{125}I] Tyr^0,D–Trp^8)somatostatin-14 autoradiographically labelled cells in the periventricular nucleus (arrows) as seen in an enlarged frame of the same region. Note the similarity in the distribution of the two sub-ependymal labelled cell populations. Darkfield. III = third ventricle. Scale bar = 100 μm.

in parallel with equivalent concentrations of ^{35}S-labelled sense probe. To control for cross-hybridization, we added a 100-fold excess of cold *sstr2* or *sstr1* antisense probes to the ^{35}S-labelled *sstr2* and *sstr1* probes, respectively. The hybridization signal was revealed by exposing sections to Beta Max tritium-sensitive film (Amersham; Fig. 1) and/or by coating with NTB-2 nuclear emulsion (Kodak; Figs 2 and 3).

Distribution of *sstr1* mRNA

Both film (Fig. 1A) and light microscopic autoradiograms revealed a widespread distribution of *sstr1* mRNA throughout the hypothalamus of both male and female rats. No specific hybridization signal was apparent in sections incubated with the corresponding sense probe. Furthermore, there was no change in the hybridization pattern in sections exposed to an excess of dissimilar cold probe.

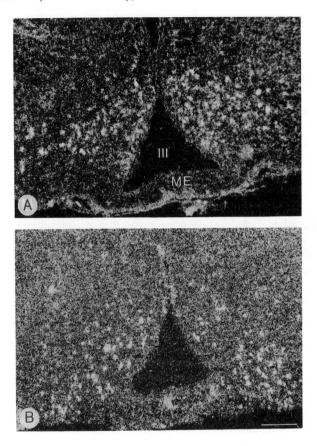

FIG. 3. Comparative distribution of *sstr1* mRNA (A) and ($[^{125}I]Tyr^0,D-Trp^8$) somatostatin-14 binding (B) in the arcuate nucleus of the hypothalamus. Note the similarity in the number and distribution of labelled cells and the absence of signal in the median eminence (ME) in both cases. Darkfield. III = third ventricle. Scale bar = 150 μm.

The distribution of *sstr1* hybridization signal is summarized in Table 1. Rostrally, moderate to high concentrations of *sstr1* mRNA were evident throughout the medial pre-optic area as well as within the magnocellular pre-optic nucleus. By contrast, the lateral pre-optic area and anterior hypothalamic nucleus were only labelled sparsely.

At mid-hypothalamic levels, a moderate to intense hybridization signal was apparent in the paraventricular nucleus dorsally and in the suprachiasmatic nucleus ventrally. In the former, *sstr1* hybridizing cells were concentrated mainly in the magnocellular division of the nucleus (Fig. 2A) and in the latter, within the ventrolateral segment. Scattered *sstr1* hybridizing cells were also apparent throughout the periventricular nucleus bordering the wall of the third ventricle from the paraventricular nucleus dorsally to the retrochiasmatic area ventrally (Fig. 2A).

The arcuate nucleus exhibited an intense *sstr1* hybridization signal throughout the entire rostrocaudal axis. Cells containing *sstr1* mRNA were most numerous in the caudal pole where they formed tight clusters on each side of the third ventricle (Fig. 3A). In contrast, only a weak hybridization signal was detected in the ventromedial and dorsomedial nuclei. Moderate concentrations of *sstr1* mRNA were detected in the more caudal ventral premammillary nucleus. A few cells containing *sstr1* were also observed in the dorsal premammillary, lateral mammillary and supramammillary nuclei.

The above distribution corresponds closely to that previously reported for *sstr1* mRNA in the mouse hypothalamus (Breder et al 1992), but is considerably more extensive than that described by Pérez et al (1994) in the rat, presumably because the latter study was based on the use of oligoprobes which are notoriously less sensitive than riboprobes for *in situ* hybridization studies.

The regional distribution of *sstr1* mRNA showed many similarities with that of somatostatin binding sites revealed by autoradiography using a variety of radioactive ligands (Table 1). Interestingly, the *sstr1* hybridization pattern resembled most closely that of the binding of somatostatin-28 which was shown in transfected cells to bind with greater affinity to sstr1 than to sstr2 (Raynor et al 1993). This was particularly apparent in the magnocellular pre-optic, paraventricular and periventricular nuclei which exhibited high concentrations of *sstr1* mRNA but low to undetectable levels of *sstr2* mRNA. These areas were reported previously to contain moderate to high concentrations of ($[^{125}I]$ Leu^8,D–Trp^{22},Tyr^{25})somatostatin-28 binding sites but not of other radioligands (Table 1). The congruence between *in situ* hybridization and binding data suggests that an important proportion of sstr1 receptors in the hypothalamus are directly associated with the perikarya and/or the dendrites of neurons that express them. Support for this interpretation also stems from the results of our high resolution receptor autoradiographic studies which have demonstrated a clear-cut association of high affinity somatostatin binding sites with neuronal perikarya in the periventricular and arcuate nuclei (Epelbaum et al 1989). The distribution of these binding sites bears a striking resemblance with that of the cells containing *sstr1* mRNA in these two regions (compare A and B in Figs 2 and 3).

Distribution of *sstr2* mRNA

Like *sstr1*, *sstr2* mRNA was distributed widely throughout the hypothalamus (Table 1). However, its nuclear patterning showed marked differences with that of *sstr1* mRNA (Fig. 1). The specificity of the hybridization signal was confirmed by the lack of hybridization of the corresponding sense riboprobe and by the fact that the hybridization signal was not affected by the addition of an excess of dissimilar cold probe.

Within the medial pre-optic area, the concentration of *sstr2* mRNA was moderate to high, although not quite as intense as that of *sstr1* mRNA. The *sstr2* hybridization signal was relatively dense in the anterior hypothalamic nucleus, whereas only scattered *sstr1* mRNA positive neurons were detected in the paraventricular nucleus (Table 1). The suprachiasmatic nucleus exhibited a moderate *sstr2* hybridization signal which was confined, like the *sstr1* signal, to the ventrolateral segment of the nucleus.

The *sstr2* hybridization pattern within the arcuate nucleus was comparable to that of *sstr1* in that the caudal pole exhibited a more intense concentration of *sstr2* mRNA than the rostral pole. However, cells containing *sstr1* mRNA tended to be concentrated medially, next to the borders of the third ventricle, whereas those containing *sstr2* were distributed mediolaterally more widely and diffusely (Fig. 1). *sstr2* mRNA was more concentrated than *sstr1* mRNA in both the ventromedial and the dorsomedial hypothalamic nuclei (Table 1). An intense *sstr2* hybridization signal was also detected in the medial tuberal nucleus along the basolateral edges of the hypothalamus (Fig. 1). In contrast to *sstr1* mRNA, which was observed in moderate to high concentrations in the ventral premammillary nucleus and in low concentrations in the dorsal premammillary nucleus, *sstr2* mRNA was observed rarely in either of these two nuclei. However, low concentrations of *sstr2* mRNA were present in both the lateral mammillary and supramammillary nuclei.

This pattern of *sstr2* expression is considerably more extensive than patterns described previously in the rat (Pérez et al 1994) and mouse (Breder et al 1992) hypothalamus. As mentioned above, the use of oligoprobes in the rat experiments may explain this discrepancy. Indeed, biochemical experiments (Bruno et al 1993, Kong et al 1994) show a higher expression of *sstr2* in the rat hypothalamus than can be accounted for by the *in situ* hybridization results of Pérez et al (1994). The discrepancy between our observations and those reported in the mouse, however, probably reflect true species differences since we have used the same riboprobes as Breder et al (1992) and our extrahypothalamic distributional patterns were comparable to theirs.

The topographical distribution of *sstr2* mRNA, like *sstr1* mRNA, is comparable to that of radiolabelled high affinity binding sites visualized by autoradiography (Table 1). A relatively high concentration of *sstr2* mRNA is present in areas such as the ventromedial and dorsomedial nuclei which bind sstr2-selective ligands such as MK 678 preferentially (Raynor et al 1993). It is also absent in structures such as the magnocellular pre-optic area where there are no binding sites for this ligand. It therefore appears that, as in the case of *sstr1*, neurons expressing *sstr2* integrate at least part of the receptors locally into their somatodendritic membrane. Furthermore, the overlap between the neuronal populations containing *sstr1* and *sstr2* mRNAs suggests that both receptors may coexist on the same cells. This interpretation is consistent with the recent demonstration by Patel et al (1994) that the same cells can express multiple somatostatin receptor genes.

Functional considerations

A major finding of this study was the overlapping expression of both *sstr1* and *sstr2* by subpopulations of neurons in the arcuate nucleus. Furthermore, the distribution of cells containing *sstr1* and *sstr2* within the arcuate nucleus was remarkably similar to that of neurons which bind ($[^{125}I]$ Tyr^0,D–Trp^8)somatostatin-14 (Epelbaum et al 1989). One third of these neurons contain (McCarthy et al 1992) and express (Bertherat et al 1992) GHRH. The rostrocaudal gradient of cells expressing *sstr1* and *sstr2* matched closely the one described previously for GHRH-containing cells (McCarthy et al 1992). These results imply, therefore, that the receptors present on the surface of arcuate GHRH neurons may represent either sstr1, sstr2 or both. Although this interpretation will need confirmation from double labelling experiments, our results suggest strongly that both sstr1 and sstr2 are involved in the central regulation of GHRH secretion, and hence of GH release, by somatostatin.

Similar conclusions may be drawn from the similarity between the distribution pattern of *sstr1* mRNA and that reported for somatostatin-immunoreactive neurons in the paraventricular and periventricular nuclei (Alpert et al 1976, Johansson et al 1984). Again, double labelling studies will be needed to determine whether *sstr1*, *sstr2* or both are expressed by somatostatin-containing cells. Our observations, however, suggest that a proportion of sstr1, and perhaps also sstr2, corresponds to somatostatin autoreceptors. Such receptors could account for the direct inhibitory action of somatostatin on somatostatin release demonstrated in cultured hypothalamic cells (Peterfreund & Vale 1984, Richardson & Twente 1986) and could therefore indirectly affect GH release (Lumpkin et al 1981).

In summary, these results demonstrate that neurons expressing *sstr1* and *sstr2* overlap extensively within the hypothalamus of the adult rat and share, in some hypothalamic nuclei, the same distribution pattern as those neurons that contain and express somatostatin or GHRH. They also suggest that both sstr1 and sstr2 are involved in the neuroendocrine regulation of GH secretion in both sexes of the rat.

Acknowledgements

The authors thank Beverley Lindsay for secretarial assistance and Helmut Bernhard for photographic work. The studies from the authors' laboratories were supported by grants from the Fonds de la Recherche en Santé du Québec and the Medical Research Council of Canada. Gloria S. Tannenbaum is the recipient of a Chercheur de Carrière award from the Fonds de la Recherche en Santé du Québec.

References

Alpert LC, Brawer JR, Patel YC, Reichlin S 1976 Somatostatinergic neurons in anterior hypothalamus: immunohistochemical localization. Endocrinology 98:255–258
Bertherat J, Dournaud P, Bérod A et al 1992 Growth hormone-releasing hormone-synthesizing neurons are a sub-population of somatostatin receptor-labelled cells in

the rat arcuate nucleus: a combined *in situ* hybridization and receptor light-microscopic radioautographic study. Neuroendocrinology 56:25–31

Bloch B, Brazeau P, Ling N et al 1983 Immunohistochemical detections of growth hormone releasing factor in brain. Nature 301:607–608

Brazeau P, Vale W, Burgus R et al 1973 Hypothalamic polypeptide that inhibits the secretion of immunoreactive pituitary growth hormone. Science 179:77–79

Breder CD, Yamada Y, Yasuda K, Seino S, Saper CB, Bell GI 1992 Differential expression of somatostatin receptor subtypes in brain. J Neurosci 12:3920–3934

Bruno JF, Xu Y, Song JF, Berelowitz M 1992 Molecular cloning and functional expression of a brain-specific somatostatin receptor. Proc Natl Acad Sci USA 89:11151–11155

Bruno JF, Xu Y, Song JF, Berelowitz M 1993 Tissue distribution of somatostatin receptor subtype messenger ribonucleic acid in the rat. Endocrinology 133:2561–2567

Daikoku S, Hisano S, Kawano H et al 1988 Ultrastructural evidence for neuronal regulation of growth hormone secretion. Neuroendocrinology 47:405–415

Epelbaum J, Moyse E, Tannenbaum GS, Kordon C, Beaudet A 1989 Combined autoradiographic and immunohistochemical evidence for an association of somatostatin binding sites with growth hormone-releasing factor-containing nerve cell bodies in the rat arcuate nucleus. J Neuroendocrinol 1:109–115

Ishikawa K, Taniguchi Y, Kurosumi K, Suzuki M, Shinoda M 1987 Immunohistochemical identification of somatostatin-containing neurons projecting to the median eminence of the rat. Endocrinology 121:94–97

Johansson O, Hökfelt T, Elde RP 1984 Immunohistochemical distribution of somatostatin-like immunoreactivity in the central nervous system of the adult rat. Neuroscience 13:265–339

Katakami H, Downs TR, Frohman LA 1988 Inhibitory effect of hypothalamic medial preoptic area somatostatin on growth hormone-releasing factor in the rat. Endocrinology 123:1103–1109

Kong H, DePaoli AM, Breder CD, Yasuda K, Bell GI, Reisine T 1994 Differential expression of messenger RNAs for somatostatin receptor subtypes SSTRl, SSTR2 and SSTR3 in adult rat brain: analysis by RNA blotting and *in situ* hybridization histochemistry. Neuroscience 59:175–184

Krantic S, Martel JC, Weissmann D, Quirion R 1989 Radioautographic analysis of somatostatin receptor sub-type in rat hypothalamus. Brain Res 498:267–278

Leroux P, Gonzalez BJ, Laquerrière A, Bodenant C, Vaudry H 1988 Autoradiographic study of somatostatin receptors in the rat hypothalamus: validation of a GTP-induced desaturation procedure. Neuroendocrinology 47:533–544

Liposits ZS, Merchenthaler I, Paull WK, Flerko B 1988 Synaptic communication between somatostatinergic axons and growth hormone-releasing factor (GRF) synthesizing neurons in the arcuate nucleus of the rat. Histochemistry 89:247–252

Lumpkin MD, Negro-Vilar A, McCann SM 1981 Paradoxical elevation of growth hormone by intraventricular somatostatin: possible ultrashort-loop feedback. Science 211:1072–1074

Martin J-L, Chesselet M-F, Raynor K, Gonzales C, Reisine T 1991 Differential distribution of somatostatin receptor subtypes in rat brain revealed by newly developed somatostatin analogs. Neuroscience 41:581–593

McCarthy GF, Beaudet A, Tannenbaum GS 1992 Colocalization of somatostatin receptors and growth hormone-releasing factor immunoreactivity in neurons of the rat arcuate nucleus. Neuroendocrinology 56:18–24

O'Carroll A-M, Lolait SJ, König M, Mahan LC 1992 Molecular cloning and expression of a pituitary somatostatin receptor with preferential affinity for somatostatin-28. Mol Pharmacol 42:939–946

Patel YC, Panetta R, Escher E, Greenwood M, Srikant CB 1994 Expression of multiple somatostatin receptor genes in AtT-20 cells. Evidence for a novel somatostatin-28 selective receptor subtype. J Biol Chem 269:1506–1509

Pérez J, Rigo M, Kaupmann K et al 1994 Localization of somatostatin (SRIF) SSTR-1, SSTR-2 and SSTR-3 receptor mRNA in rat brain by *in situ* hybridization. Naunyn-Schmiedebergs Arch Pharmakol 349:145–160

Peterfreund RA, Vale WW 1984 Somatostatin analogs inhibit somatostatin secretion from cultured hypothalamus cells. Neuroendocrinology 39:397–402

Raynor K, Murphy WA, Coy DH et al 1993 Cloned somatostatin receptors: identification of subtype-selective peptides and demonstration of high affinity binding of linear peptides. Mol Pharmacol 43:838–844

Reichlin S 1983 Somatostatin. N Engl J Med 309:1495–1501, 1556–1563

Richardson SB, Twente S 1986 Inhibition of rat hypothalamus somatostatin release by somatostatin: evidence for ultrashort loop feedback. Endocrinology 118:2076–2082

Schally AV, Huang W-Y, Chang RCC et al 1980 Isolation and structure of prosomatostatin: a putative somatostatin precursor from pig hypothalamus. Proc Natl Acad Sci USA 77:4489–4493

Tannenbaum GS, Ling N, Brazeau P 1982 Somatostatin-28 is longer acting and more selective than somatostatin-14 on pituitary and pancreatic hormone release. Endocrinology 111:101–107

Tannenbaum GS, McCarthy GF, Zeitler P, Beaudet A 1990 Cysteamine-induced enhancement of growth hormone-releasing factor (GRF) immunoreactivity in arcuate neurons: morphological evidence for putative somatostatin/GRF interactions within hypothalamus. Endocrinology 127:2551–2560

Uhl GR, Tran V, Snyder SH, Martin JB 1985 Somatostatin receptors: distribution in rat central nervous system and human frontal cortex. J Comp Neurol 240:288–304

Yamada Y, Post SR, Wang K, Tager HS, Bell GI, Seino S 1992 Cloning and functional characterization of a family of human and mouse somatostatin receptors expressed in brain, gastrointestinal tract, and kidney. Proc Natl Acad Sci USA 89:251–255

Yasuda K, Rens-Domiano S, Breder CD et al 1992 Cloning of a novel somatostatin receptor, SSTR3, coupled to adenylyl cyclase. J Biol Chem 267:20422–20428

DISCUSSION

Meyerhof: In the hypothalamic nuclei, is *sstr1* expressed only in neurons or is it also expressed in astrocytes? Because Krisch et al (1991) have reported that somatostatin binding sites are present in astrocytes in the suprachiasmatic nuclei. We have also observed *sstr1* in these nuclei (Meyerhof et al 1992).

Beaudet: I can't answer that question because the resolution of the technique is not sufficient to ascribe unequivocally the labelling to astrocytes. We do see what apears to be astrocytic labelling in cresyl violet-stained material, but we haven't performed glial fibrillary acidic protein (GFAP) immunohistochemistry to answer the question directly.

Chesselet: Your idea that sstr1 may be an autoreceptor is very interesting. Have you looked in brain regions where somatostatinergic neurons are preserved after making lesions (but their target cells are destroyed) to see whether there is preservation of *sstr4*?

Beaudet: No, we concentrated on the hypothalamus.

Robbins: You showed us a subpopulation of somatostatin interneurons in the hippocampus which expresses only *sstr2* (Fig. 1). In the hippocampus, somatostatin binds only in the CA1 region (Krantic et al 1992), whereas CA2 and CA3 are heavily labelled with *sstr2* (Beaudet et al 1994b). Do you think this is an example of neurons expressing the message but not the protein?

Beaudet: I don't know what the explanation is for this.

Reisine: We did *in situ* hybridization in the rat and found *sstr3* mRNA in the CA3 region. Somatostatin does not bind there, but it may be that the mRNA is in CA3 cell bodies which send collaterals to CA1 and that sstr3 is a presynaptic receptor (Kong et al 1994). The problem is that no radioligand is available that is selective for sstr3. The hippocampus is an interesting place to establish presynaptic or postsynaptic receptors rather than hypothalamus.

Robbins: Because the circuitry is well defined?

Reisine: Yes, and if you had selective ligands, you could also do electrophysiology to see whether you could separate the functions.

Berelowitz: In the arcuate nucleus you have co-localization of somatostatin binding with GHRH (growth hormone-releasing hormone)-containing neurons. Have you done the same for NPY (neuropeptide Y)?

Beaudet: No, we have only looked at GHRH. GHRH coexists with a variety of other peptides (e.g. neurotensin and galanin) as well as with tyrosine hydroxylase in neurons of the arcuate nucleus (reviewed by Sawchenko & Swanson 1990) but not, to my knowledge, with NPY.

Berelowitz: It's interesting that GHRH and NPY expression co-localize in the arcuate nucleus. These two peptides are co-regulated in food deprivation and diabetes (Berelowitz et al 1992). It is possible that there is a common signalling pathway that operates through the somatostatin receptors.

Epelbaum: We have also studied the co-localization of somatostatin receptors with GHRH in the arcuate nucleus. We found that receptors here are sensitive to octreotide, suggesting that they are sstr2 rather than sstr1 (Bertherat et al 1991). We haven't looked at co-localization of somatostatin receptors with NPY; but we found that somatostatin and somatostatin receptors are not co-localized in this region. NPY-containing neurons may well be the ones that bear somatostatin receptors in the periventricular portion of the arcuate nucleus.

Reichlin: Neurons containing specific peptides, e.g. GHRH, TRH (thyrotropin-releasing hormone) and somatostatin itself, also have somatostatin receptors on them (Toni & Lechan 1993, Bertherat et al 1992). Is the resolution of your method sufficient to identify which of these are somatostatin-innervated cells?

Beaudet: No. You need double labelling studies at the electron microscopic level to determine which type of peptidergic cell receives a somatostatin input. This has been done by Daikoku et al (1988) and Liposits et al (1988) in the case of GHRH neurons.

Epelbaum: I have a question concerning the 50% inhibitory effect of hypophysectomy on *sstr1* and *sstr2* mRNA levels. Would you expect such a dramatic effect if this was due to a decrease in GH (growth hormone) levels? Because the localization of GH receptors in the hypothalamus is restricted.

Beaudet: It is premature to attribute these effects to a decrease in GH levels. GH-replacement experiments would be needed to confirm this.

Reichlin: Has somatostatin receptor internalization been seen previously in the neuronal system? Is this part of a specific function or just a breakdown pathway?

Beaudet: This is a new finding for somatostatin (A. Beaudet & G. S. Tannenbaum, unpublished results), although we've seen the same phenomenon with neurotensin in neurons of the CNS (central nervous system) (Beaudet et al 1994a). Neurotensin internalization proceeds from the entire dendritic membrane of the cell as well as from the axon terminals. The ligand is internalized within endosomes which then travel to the cell body. The endosomes coalesce to form what we believe are first multivesicular bodies and then lysosomes, although further studies will be needed to confirm this. What's interesting is that these presumptive lysosomes cluster around the nuclear membrane and that even up to 2.5 h after internalization, a large proportion of the peptide present within them is still intact (Castel et al 1991).

I wouldn't be surprised if a similar situation occurred for somatostatin, given the similarity of the confocal images obtained with the two peptides. Furthermore, according to current models of endocytosis, ligand degradation should follow acidification of the endosomes. It is unlikely that the organelles containing internalized somatostatin are acidic, however, because of the lack of quenching in our FITC (fluorescein isothiocyanate) labelling (A. Beaudet & G. S. Tannenbaum, unpublished results). It would be interesting to look at the pH of the organelles in which the peptide is internalized.

Schonbrunn: We looked for ligand internalization following receptor binding in both GH pituitary cells and rat RINm5F insulinoma cells and did not see it (Presky & Schonbrunn 1986, Sullivan & Schonbrunn 1986).

Reichlin: Do you wash the cells in these studies?

Schonbrunn: We bind the ligand to the receptor then treat the cells with an acid wash to dissociate the ligand from plasma membrane receptors. If the ligand dissociates quickly (i.e. less than five minutes) then it must be on the outside of the cell, accessible to the acid. Viguerie et al (1987) did similar studies in pancreatic acinar cells and found that about 25% of the ligand was internalized.

Susini: In normal pancreatic cells the ligand is internalized into the nucleus and this ligand remains intact.

Schonbrunn: In our pituitary and pancreatic islet cells, intact ligand stays on the cell surface following receptor binding and we don't see receptor down-regulation. So, it is likely that not all somatostatin receptors internalize ligands in the same way—it may depend on the receptor or cell type. Also, what happens to the receptor and what happens to the ligand may be separate processes. Although Alain Beaudet sees intracellular ligand, he cannot say that the ligand was internalized with the receptor.

Beaudet: You are right; however, I cannot help but compare our results again with our earlier findings with neurotensin (Beaudet et al 1994a). We studied the internalization of a fluorescent photoactivatable analogue of neurotensin (azido-nitro-Na-fluoresceinyl-NT$_{2-13}$) after it had been cross-linked to cell surface receptors. We saw that internalization proceeded in the same way as dissociable ligands, suggesting that ligand–receptor complexes were internalized. In the case of somatostatin, this type of experiment remains to be done.

Reisine: If you treat sstr1-expressing cells with somatostatin, there's no change in receptor binding, although you cannot distinguish between internalized receptor and cell surface receptor. In contrast, if you treat sstr2-expressing cells with somatostatin, there is no high affinity ligand binding in membranes, suggesting that the receptors are desensitized and/or down-regulated. So, the effects of high affinity ligand binding are different between the two cells, despite a similar pattern of internalization.

Beaudet: Yes, our results suggest that both types of receptor are internalized.

Schonbrunn: With different somatostatin receptors you may or may not get ligand internalization—that's been established by studies in different cell types. The questions are which receptor does what and is internalization cell type specific? In terms of the receptor being internalized, again I would say that some are and some are not.

Reichlin: The internalized membrane is a small unit of the surface membrane. Is it likely to have a G protein associated with it?

Reisine: It may, or may not, have a G protein associated with it.

Reichlin: If you could establish that it was associated with a G protein, would that be a way to determine whether the internalized complex had all the elements required for an activated system?

Reisine: That's not an easy study to do. You need antibodies against G proteins to look at their internalization—there are few that are good enough to do this immunocytochemistry. You also need a good antibody against the receptor for the same reason. Even if you have these, you still have the problem of not having intracellular functions that you can link up with the internalized complex.

Berelowitz: Boris Draznin studied receptors in secretory granules, where receptors are present in vesicles together with their target.

Robbins: Yes, I was an author on that paper (Steiner et al 1986). We first permeabilized the membranes of primary cultures of rat pituitary cells with digitonin. We then introduced [^{125}I]somatostatin and found that secretory vesicles contained bound somatostatin before they reached the surface membrane and were released. When we exposed the cells to GHRH, we showed that the somatostatin receptors bound [^{125}I]somatostatin and were functional before they were vesicularized. When increased stores of [^{125}I]somatostatin were produced with chloroquine, the endocytotic somatostatin did not block the GHRH-induced release of GH, suggesting that somatostatin receptors and secretory vesicles are able to bind somatostatin before they reach the plasma membrane. But that's the opposite of what you were talking about, which is degradation.

Schonbrunn: One striking difference between the receptors that you were studying and the somatostatin receptors in other studies is that they had a much lower affinity—less than 10^{-8} M instead of 10^{-10} M. It is not clear that they are the same receptors. In fact, we did similar studies looking at high affinity somatostatin receptors in the RINm5F insulinoma cell line and saw no increase in somatostatin receptors following stimulated secretion (Sullivan & Schonbrunn 1988).

Epelbaum: We know that endogenous somatostatin is very prone to degradation, so what is the point of using ligands that are more resistant to degradation and may not be related physiologically to the action of the native peptide?

Schonbrunn: The $([^{125}I] Tyr^1)$somatostatin and $([^{125}I] Tyr^{11})$somatostatin ligands that we used are biologically active and, as they do not contain any D-amino acids, would not be expected to be more stable than the native peptide.

Epelbaum: Yes, but some analogues that are used are more resistant to degradation, so maybe the native peptide is degraded before it has time to internalize.

Beaudet: But internalization is extremely rapid—it's a matter of minutes. It is possible that once internalized, somatostatin is protected from degradation in the endosomal compartment, if the latter is not acidic.

Patel: We studied the internalization of $([^{125}I] Tyr^1)$somatostatin-14 in monolayer cultures of rat islet α, β and δ cells by quantitative electronmicroscopic autoradiography (Amherdt et al 1989). Our results agree with Alain Beaudet's, showing a time- and temperature-dependent internalization of radioligand into endocytotic vesicles, then to multivesicular bodies and finally into lysosomes. Internalization of radioligand occurred within minutes at 37 °C in all three cell types. Interestingly, after initial internalization, the intracellular progression occurred freely in β and α cells (i.e. cells heterologous for the radioligand) but poorly in δ cells (i.e. cells homologous for the radioligand).

Lamberts: Dr Hofland in our laboratory demonstrated a high degree of $([^{125}I] Tyr^3)$octreotide internalization in mouse AtT-20 pituitary tumour cells and cultured human GH-secreting pituitary tumour cells. However, ^{125}I-labelled native somatostatin was hardly ever internalized, probably because of its rapid degradation. From a clinical standpoint, this might be a way of getting radioactivity into somatostatin receptor-positive tumour cells. Indeed, we have also demonstrated considerable uptake of radioactivity in cultured human carcinoid cells (Lamberts et al 1995, this volume). When we administer $([^{111}In] DTPA)$octreotide (DTPA, diethylenetriaminepentaacetic acid) *in vivo*, we see uptake of radioactivity in human somatostatin receptor-positive tumours up to 0.2% of the initial radioactive dose. This observation can be used to carry out radiotherapy in patients with inoperable somatostatin receptor-positive tumours. The preliminary results in the first patients are promising (Krenning et al 1994). You can also take advantage of this internalization in patients with multiple metastases of somatostatin receptor-positive tumours, for example around the neck or in the abdomen, by administering ($[^{111}$In$]$DTPA)octreotide to the patient before the operation. The surgeon can then use a hand-held gamma-detecting probe to demonstrate radioactivity in the tumour and its lymph node metastases. This concept of radioguided surgery results in the localization of previously undetected somatostatin receptor-positive tumour cells (Schirmer et al 1993, Waddington et al 1994).

Schonbrunn: I would like to offer a different interpretation of those results. We have shown that analogues which are similar to natural somatostatin, e.g. ($[^{125}$I$]$Tyr^1)somatostatin and ($[^{125}$I$]$Tyr^{11})somatostatin, are not degraded rapidly after receptor binding (Sullivan & Schonbrunn 1986, Presky & Schonbrunn 1986, 1988). We bound these analogues to the receptor for various times, treated the cells with acid to dissolve the ligand–receptor complex, then looked at the nature of the ligand by HPLC (high performance liquid chromatography). We found that the receptor-bound peptide was intact. Hence, an alternative explanation for your observations is that internalization of the somatostatin analogue is both cell type specific and receptor specific.

Bruns: We have never observed receptor internalization in any of our cell lines. *In vivo* imaging studies in patients using ($[^{111}$In$]$DTPA)octreotide cannot tell us if internalization occurs. It is misleading to interpret the accumulation of radioactivity in the tumour as somatostatin receptor internalization. Terry Reisine, have you done binding experiments which demonstrate conclusively that receptor internalization occurs?

Reisine: We have only done binding studies with membranes and not with intact cells. We observe a decrease in binding if we expose sstr2 to an agonist for a long time, but we can't say conclusively that this decrease is because of internalization. If you have a receptor coupled to a G protein, loss of affinity towards an agonist could be due to uncoupling of the receptor from the G protein. sstr2 is a very unusual receptor—it doesn't degrade easily. We have treated cells for 24 h with agonist, and in conditions where there's no binding or function, we have shown, by Western blotting, that sstr2 is still present, although it's functionally inactive.

Meyerhof: The ability of neuronal cells to sort and target macromolecules is an important step in the generation of functionally different synapses. A new concept in neuroscience is that certain mRNAs, such as microtubule-associated protein 2 (MAP2, Garner et al 1988), the alpha subunit of Ca^{2+}/calmodulin-dependent protein kinase II (Burgin et al 1990) and brain-derived neurotrophic factor (Dugich-Djordjevic et al 1992) are not confined exclusively to the cell body but are targeted specifically to dendrites, probably to allow decentralized protein synthesis. Furthermore, several lines of evidence have shown that vasopressin and oxytocin mRNAs are located in axons of magnocellular hypothalamic neurons (Mohr et al 1991, Trembleau et al 1994).

Is it possible that dendritically localized somatostatin binding sites in the hypothalamus (A. Beaudet & G. S. Tannenbaum, unpublished results) could be due to decentralized translation of targeted *sstr* mRNA?

Beaudet: It is not known for somatostatin, however, there was recently a report describing axon localization of the δ-opiate receptor. Using peptide antibodies for the δ-opiate receptor, Dado et al (1993) found that all immunoreactive receptor proteins were presynaptic, i.e. they were associated exclusively with axon terminals. This would imply that the receptors targeted to axon terminals can be immunologically (and therefore presumably structurally) different from those targeted toward dendrites.

Chesselet: There are few examples of axonal transport of mRNA in adult neurons, except for vasopressin innervating the median eminence and the posterior pituitary (Trembleau et al 1994).

References

Amherdt M, Patel YC, Orci L 1989 Binding and internalization of somatostatin, insulin, and glucagon by cultured rat islet cells. J Clin Invest 84:412–417

Beaudet A, Mazella J, Nouel D et al 1994a Internalization and intracellular mobilization of neurotensin in neuronal cells. Biochem Pharmacol 47:43–52

Beaudet A, Greenspun D, Raelson J, Tannenbaum GS 1994b Patterns of expression of SSTR1 and SSTR2 somatostatin receptor subtypes in the hypothalamus of the adult rat: relationship to neuroendocrine function. Neuroscience, in press

Berelowitz M, Bruno JF, White JD 1992 Regulation of hypothalamic expression by peripheral metabolism. Trends Endocrinol Metab 3:127–133

Bertherat J, Slama A, Kordon C, Videau C, Epelbaum J 1991 Characterization of pericellular ([^{125}I] Tyr^{0},DTrp^{8}) somatostatin binding sites in the rat arcuate nucleus by a newly developed method: quantitative high-resolution light microscopic autoradiography. Neuroscience 41:571–579

Bertherat J, Dournaud P, Bérod A et al 1992 Growth hormone-releasing hormone-synthesizing neurons are a subpopulation of somatostatin receptor labelled cells in the rat arcuate nucleus: a combined *in situ* hybridization and receptor light-microscopic autoradiographic study. Neuroendocrinology 56:1125–1131

Burgin KE, Waxham MN, Rickling S, Westgate SA, Mobley WC, Kelly PT 1990 *In situ* hybridization histochemistry of Ca^{2+} calmodulin dependent protein kinase in developing rat brain. J Neurosci 10:1788–1798

Castel MN, Faucher D, Cuiné F, Dubédat P, Boireau A, Laduron PM 1991 Identification of intact neurotensin in the substantia nigra after its retrograde axonal transport in dopaminergic neurons. J Neurochem 56:1816–1818

Daikoku S, Hisano S, Kawano H et al 1988 Ultrastructural evidence for neuronal regulation of growth hormone secretion. Neuroendocrinology 47:405–415

Dado RJ, Law PY, Loh HH, Elde R 1993 Immunofluorescent identification of a delta (δ)-opioid receptor on primary afferent nerve terminals. Neuroreport 5:341–344

Dugich-Djordjevic MM, Tocco G, Willoughby DA et al 1992 BDNF messenger RNA expression in the developing rat brain following kainic acid-induced seizure activity. Neuron 8:1127–1138

Garner CC, Tucker RP, Matus A 1988 Selective localization of messenger RNA for cytoskeletal protein MAP2 in dendrites. Nature 336:674–677

Kong H, DePaoli AM, Breder CD, Yasuda K, Bell GI, Reisine T 1994 Differential expression of messenger RNAs for somatostatin receptor subtypes sstr1, sstr2 and sstr3 in adult rat brain—analysis by RNA blotting and in situ hybridization histochemistry. Neuroscience 59:175–184

Krantic S, Quirion R, Uhl G 1992 Somatostatin receptors. In: Bjorklund A, Hokfelt T, Kuhar MJ (eds) Handbook of chemical neuroanatomy, vol 11: Neuropeptide receptors in the CNS. Elsevier Science, Amsterdam, p321–346

Krenning EP, Kooy PPM, Bakker WHB et al 1994 Radiotherapy with radiolabelled somatostatin. Ann N Y Acad Sci, in press

Krisch B, Buchholz C, Mentlein R 1991 Somatostatin binding sites on rat diencephalic astrocytes—light microscopic study *in vitro* and *in vivo*. Cell Tissue Res 263:253–263

Lamberts SWJ, de Herder WW, van Koetsfeld PM et al 1995 Somatostatin receptors: clinical implications for endocrinology and oncology. In: Somatostatin and its receptors. Wiley, Chichester (Ciba Found Symp 190) p 222–239

Liposits ZS, Merchenthaler I, Paull WK, Flerko B 1988 Synaptic communication between somatostatinergic axons and growth hormone-releasing factor (GRF) synthesizing neurons in the arcuate nucleus of the rat. Histochemistry 89:247–252

Meyerhof W, Wulfsen I, Schinrock C, Fehr S, Richter D 1992 Molecular cloning of a somatostatin-28 receptor and comparison of its expression pattern with that of a somatostatin-14 receptor in rat brain. Proc Natl Acad Sci USA 89:10267–10271

Mohr E, Fehr S, Richter D 1991 Axonal transport of neuropeptides encoding messenger RNAs within the hypothalamophyseal tract of rats. EMBO J 10:2419–2424

Presky DH, Schonbrunn A 1986 Receptor-bound somatostatin and epidermal growth factor are processed differently in GH_4C_1 rat pituitary cells. J Cell Biol 102:878–888

Presky DH, Schonbrunn A 1988 Iodination of [Tyr[11]]somatostatin yields a super high affinity ligand for somatostatin receptors in GH_4C_1 pituitary cells. Mol Pharmacol 34:651–658

Sawchenko PE, Swanson LW 1990 Growth hormone-releasing hormone. In: Björklund A, Hökfelt T, Kuhar MJ (eds) Handbook of chemical neuroanatomy, vol 9: Neuropeptides in the CNS. Elsevier Science Publishers p131–163

Schirmer WJ, O'Dorisio TM, Schirmer TP et al 1993 Intraoperative localization of neuroendocrine tumors with ^{125}I-Tyr(3)-octreotide and a hand-held gamma-detecting probe. Surgery 114:745–752

Steiner C, Dahl R, Sherman N, Trowbridge M, Vatter A, Robbins R, Draznin B 1986 Somatostatin receptors are active before they are inserted into the plasma membrane. Endocrinology 118:766–772

Sullivan SJ, Schonbrunn A 1986 The processing of receptor-bound [^{125}I-Tyr[11]]somatostatin by RINm5F insulinoma cells. J Biol Chem 261:3571–3577

Sullivan SJ, Schonbrunn A 1988 Distribution of somatostatin receptors in RINm5F insulinoma cells. Endocrinology 122:1137–1145

Toni R, Lechan RM 1993 Neuroendocrine regulation of thyrotropin-releasing hormone (TRH) in the tuberoinfundibular system. J Endocrinol Invest 16:715–753

Trembleau A, Morales M, Bloom FE 1994 Aggregation of vasopressin messenger RNA in a subset of axonal swellings of the median eminence and posterior pituitary—light and electron microscopic evidence. J Neurosci 14:39–53

Viguerie N, Esteve J-P, Susini C, Vaysse N Ribet A 1987 Processing of receptor-bound somatostatin: internalization and degradation by pancreatic acini. Am J Physiol 252:G535–G542

Waddington WA, Kettle AG, Heddle RM, Coakley AJ 1994 Intraoperative localization of recurrent medullary carcinoma of the thyroid using indium-111 pentetreotide and a surgical probe. Nucl Med 21:363–364

Interaction of somatostatin receptors with G proteins and cellular effector systems

Terry Reisine, Donna Woulfe, Karen Raynor, Haeyoung Kong, Jennifer Heerding, John Hines, Melanie Tallent and Susan Law

Department of Pharmacology, University of Pennsylvania School of Medicine, 36th Street and Hamilton Walk, Philadelphia, PA 19104, USA

Abstract. Somatostatin induces its multiple biological actions by interacting with a family of receptors, referred to as sstr1–sstr5. To determine the molecular mechanisms of action of somatostatin, we have investigated the interaction of the different cloned receptors with G proteins and cellular effector systems. sstr2, sstr3 and sstr5 associate with pertussis toxin-sensitive G proteins and are able to mediate the inhibition of adenylyl cyclase activity by somatostatin. Two forms of sstr2, sstr2A and sstr2B, are generated by alternative splicing and differ in their C-terminal amino acid sequence. sstr2B couples to adenylyl cyclase whereas sstr2A does not. To investigate the basis for the differential coupling to adenylyl cyclase, we truncated sstr2B to the point of amino acid sequence divergence from sstr2A. The truncated sstr2B mediated the inhibition of cAMP formation by somatostatin, indicating that the C-terminus is not needed for coupling sstr2 to adenylyl cyclase. It is likely that the C-terminus of sstr2A hinders coupling to adenylyl cyclase. sstr2A associates with $G_i\alpha3$ and $G_o\alpha$ but does not effectively interact with $G_i\alpha1$, a G protein that is necessary for coupling somatostatin receptors to adenylyl cyclase. The differential association of the splice variants with $G_i\alpha1$ may explain their contrasting effects on adenylyl cyclase activity. sstr3 also couples to adenylyl cyclase. $G_i\alpha1$ links sstr3 to adenylyl cyclase and mutagenesis studies have shown that the C-terminus of $G_i\alpha1$ is necessary for this coupling. The C-terminus of the $G_i\alpha$ proteins differ by only a few amino acid residues and only $G_i\alpha1$ couples sstr3 to adenylyl cyclase. Our findings indicate that only a few amino acid residues in $G_i\alpha1$ are necessary for targeting sstr3 to adenylyl cyclase.

1995 Somatostatin and its receptors. Wiley Chichester (Ciba Foundation Symposium 190) p 160–170

Somatostatin is the major physiological inhibitor of growth hormone (GH) release from the pituitary and of insulin and glucagon secretion from the pancreas (Epelbaum 1986). It is also a neurotransmitter of the brain, where it has been shown to be involved in the control of locomotor activity and cognitive functions (Raynor & Reisine 1992a). It induces its biological actions by interacting with cell surface receptors. Recently, a family of somatostatin receptors has been cloned and referred to as sstr1–sstr5 (Bell & Reisine 1993).

The cloned somatostatin receptors are approximately 50% identical in amino acid sequence. Regions that diverge the most are the N- and C- termini. These regions may have critical roles in ligand binding and G protein coupling. Of the five receptors, sstr2, sstr3 and sstr5 have been shown clearly to associate with pertussis toxin-sensitive G proteins and to mediate the inhibition of cAMP formation by somatostatin (Bell & Reisine 1993, O'Carroll et al 1992, Rens-Domiano et al 1992, Yasuda et al 1992). These receptors may couple to other cellular effector systems besides adenylyl cyclase. Receptors in brain neurons in culture with similar pharmacological characteristics to sstr2 have been shown to couple to voltage-dependent Ca^{2+} and K^+ channels (Raynor et al 1992). Furthermore, in mouse pituitary tumour AtT-20 cells, which express *sstr2* mRNA, selective sstr2 agonists can inhibit an L-type Ca^{2+} channel and potentiate an inward-rectifying K^+ current (Reisine et al 1994). These cells also express *sstr5* mRNA and agonists that selectively interact with this receptor inhibit an L-type Ca^{2+} current. sstr2 has also been shown to mediate inhibition of cell proliferation, possibly by stimulating a protein tyrosine phosphatase (Buscail et al 1994). Thus, preliminary studies suggest that the somatostatin receptor subtypes couple to multiple cellular effector systems.

The somatostatin receptor subtypes may mediate distinct physiological actions of the hormone. sstr2 selectively mediates the inhibition of GH release by somatostatin (Raynor et al 1993a). The rank order of potencies of a large number of somatostatin analogues to inhibit GH release and to bind to sstr2 were highly correlated. No correlation existed for the other somatostatin receptors (Raynor et al 1993a,b). Recent studies by Rossowski & Coy (1993) indicate that sstr5 may be selectively involved in the inhibition of insulin release by somatostatin. These authors showed that sstr5-selective agonists inhibited insulin release as potently as somatostatin whereas agonists selective for sstr2 and sstr3 had minimal effects on insulin secretion. These studies suggest that sstr2 and sstr5 may have distinct biological roles.

Most cellular responses to somatostatin are pertussis toxin sensitive, indicating that the pertussis toxin-sensitive G proteins $G_i\alpha$ and $G_o\alpha$ play critical roles in coupling somatostatin receptors to different cellular effector systems. To investigate the G proteins that physically associate with somatostatin receptors, we initially investigated the interaction of the receptors with G proteins from rat brain and the pituitary cell line AtT-20 (Law et al 1991, Law & Reisine 1992). The approach that was employed involved solubilizing somatostatin receptor/G protein complexes with the mild detergent CHAPS (3-[(3-cholamidopropyl)-dimethylammonio]-1-propane-sulphonate) and then immunoprecipitating the complexes with antisera directed against different α and β subunits of G proteins. The immunoprecipitates were tested for the appearance of high affinity somatostatin agonist-binding sites and the supernatants were examined for the loss of agonist-binding sites. If high affinity agonist-binding sites were present in the immunoprecipitate, then we assumed that the receptor associated with the α or β subunit selectively immunoprecipitated with the peptide-directed antisera. If high affinity agonist-binding in the supernatant was reduced after the immunoprecipitation but no receptor was detected in the immunoprecipitate, we assumed that the antisera uncoupled the receptor from the α or β subunit to which the antisera was directed. If the solubilized somatostatin receptor was not immunoprecipitated or uncoupled by an antisera selectively directed against a particular α or β subunit, then we assumed that the G protein subunit did not stably associate with the receptor. Using this approach, we found that $G_i\alpha 1$, $G_i\alpha 3$ and $G_o\alpha$ formed stable complexes with the somatostatin receptors. Furthermore, the $G\beta 36$ subunit associated with these receptors, as did the $\gamma 3$ subunit. Thus, our biochemical studies indicate that rat brain and AtT-20 cell somatostatin receptors associate with the $G_i\alpha 1$, $G_i\alpha 3$, $G_o\alpha$, $G\beta 36$ and $\gamma 3$ subunits.

The association of the somatostatin receptors with different G proteins may be the basis of the functional diversity of these receptors. Recent functional studies have indicated that different G proteins link the receptors to different cellular effector systems. Tallent & Reisine (1992) showed that antisera against $G_i\alpha$ blocked the coupling of somatostatin receptors in AtT-20 cells to adenylyl cyclase. This effect was selective because the blockade by the antisera was prevented by a peptide to which the antisera was generated but not by an irrelevant peptide. Furthermore, antisera against $G_o\alpha$ did not block somatostatin receptor coupling to adenylyl cyclase. Peptide-directed antisera against $G_i\alpha 1$ prevented this coupling, whereas antisera against $G_i\alpha 2$ and $G_i\alpha 3$ did not, indicating that $G_i\alpha 1$ selectively couples the somatostatin receptor to adenylyl cyclase.

Knockout experiments by Kleuss et al (1991) have shown that $G_o\alpha$ couples somatostatin receptors in rat pituitary tumour GH_3 cells to voltage-dependent Ca^{2+} channels. These investigators injected antisense oligonucleotides into the cells, blocked expression of $G_o\alpha$ and prevented somatostatin from inhibiting the voltage-dependent Ca^{2+} current. In contrast, blockade of $G_i\alpha$ expression had no effect on the coupling of the receptor to the Ca^{2+} channel. These investigators also showed that blockade of $G\beta 36$ (Kleuss et al 1992) and $\gamma 3$ (Kleuss et al 1993) expression uncoupled the receptors from the Ca^{2+} channel, whereas blockade of expression of $G\beta 35$ and other γ subunits did not. These functional studies indicate that $G_o\alpha$, $G\beta 36$ and the $\gamma 3$ subunit are essential in coupling somatostatin receptors to Ca^{2+} channels.

In contrast, Yatani et al (1987) reported that $G_i\alpha 3$ couples somatostatin receptors in GH_3 cells to voltage-dependent K^+ channels. They inactivated endogenous G proteins with pertussis toxin and reconstituted somatostatin receptor K^+ channel coupling with purified $G_i\alpha 3$. Therefore, somatostatin receptors are coupled to adenylyl cyclase, Ca^{2+} channels and K^+ channels via different G proteins. These functional results are consistent with the biochemical studies that identified which G proteins physically associate with rat brain and AtT-20 cell somatostatin receptors.

To investigate the coupling of the cloned somatostatin receptors with G proteins and cellular effector systems, we have either stably expressed the receptors in CHO (Chinese hamster ovary) cells (DG-44) or transiently expressed them in African green monkey COS-7 kidney cells or human kidney fibroblast HEK293 cells. When the mouse sstr1 or the human sstr4 was expressed in COS or CHO cells, high affinity agonist-binding to these receptors was not affected by GTPγS or pertussis toxin treatment (Raynor et al 1993b, Rens-Domiano et al 1992, Yasuda et al 1992). Furthermore, no efficient coupling of these receptors to adenylyl cyclase was detected. The lack of coupling to G proteins and adenylyl cyclase differs from the reports of other groups (Kaupmann et al 1993, Patel et al 1994). These differences may be related to variations in the G protein composition or cellular environment of the cells, or to the levels of expression of the cloned receptors in the cells.

We have used the inability of sstr1 to couple to adenylyl cyclase to identify regions of other receptors involved in coupling to this enzyme. The δ opioid receptor has 40% overall amino acid sequence identity to sstr1 (Reisine & Bell et al 1993). It couples to adenylyl cyclase and mediates agonist inhibition of cAMP accumulation. The third intracellular loops of the δ opioid receptors and sstr1 are identical in size but differ in amino acid sequence. Exchanging the third intracellular loops of sstr1 and the δ receptor resulted in a chimeric sstr1/δ receptor that effectively coupled to adenylyl cyclase and mediated somatostatin inhibition of cAMP accumulation (Heerding et al 1994). In contrast, the δ receptor sstr1 chimera was not able to couple to adenylyl cyclase. These findings indicate that the third intracellular loop of the δ receptor contains the adenylyl cyclase coupling domain. In contrast, the sequence of the intracellular loop of sstr1 prevented the δ receptor from effectively interacting with adenylyl cyclase.

In contrast to sstr1 and sstr4, sstr3 and sstr5 have been clearly established to couple to adenylyl cyclase. With regards to sstr2, different results have been reported by various groups. We were unable to obtain functional coupling of the originally cloned mouse sstr2 (referred to as sstr2A) to adenylyl cyclase when expressed in COS-7 cells, CHO-DG44 cells or HEK293 cells (Law et al 1993, Rens-Domiano et al 1992 Yasuda et al 1992). Similar findings have been reported by Buscail et al (1994). However, other investigators have found coupling of this receptor to adenylyl cyclase (Kaupmann et al 1993, Patel et al 1994). Our inability to obtain coupling of mouse sstr2A to adenylyl cyclase was unexpected, because we have reported that sstr2-selective agonists could inhibit adenylyl cyclase activity in brain membranes (Raynor & Reisine 1992b). sstr2 exists in two forms, sstr2A and sstr2B, which are generated by alternative splicing of a common primary transcript (Vanetti et al 1992). The two forms differ only in their C-terminal amino acid sequences. We have reported that sstr2B couples to adenylyl cyclase and mediates the inhibition of cAMP accumulation by somatostatin (Reisine et al 1993). sstr2A does not efficiently couple to adenylyl cyclase. The differential coupling to adenylyl cyclase may be due to variations in the C-terminal sequences of the receptor. To test this hypothesis, we truncated sstr2B at the point of sequence divergence of the two receptors, so that the remaining receptor was identical with that region of sstr2A (Woulfe & Reisine 1994). This truncated receptor effectively coupled to adenylyl cyclase, indicating that the C-terminus of sstr2 is not necessary for this coupling and that other intracellular domains are involved. One likely region is the third intracellular loop. This region has been reported for other receptors to be critical for the coupling to adenylyl cyclase (Dohlman et al 1987). In recent studies, we have found that the third intracellular loop of the δ opioid receptor is necessary for coupling the receptor to adenylyl cyclase (Heerding et al 1994). The size of the third intracellular loop of sstr2B is identical to that of the δ receptor and sstr1, although the amino acid sequences differ. Chimeras between sstr1 and sstr2 should enable us to address the question of whether the third intracellular loop of sstr2 contains the adenylyl cyclase-coupling domain.

The finding that the C-terminus of sstr2 is not essential for coupling the receptor to adenylyl cyclase suggests that the inability of sstr2A to mediate inhibition of cAMP formation may be due to the C-terminus of this receptor disrupting association of the receptor with adenylyl cyclase and the G proteins involved in coupling the receptor to adenylyl cyclase. Biochemical studies have shown that sstr2A forms stable complexes with $G_i\alpha3$ and $G_o\alpha$ (Law et al 1993). However, previous studies have suggested that $G_i\alpha1$ couples somatostatin receptors, with similar characteristics of sstr2, to adenylyl cyclase (Tallent & Reisine 1992). The inability of sstr2A to form stable complexes with $G_i\alpha1$ may be the molecular basis for its lack of coupling to adenylyl cyclase.

Our findings indicate that the splice variants of sstr2 differentially couple to adenylyl cyclase. It is possible that the molecular basis for this variation in response is due to an inhibitory effect of the C-terminus of sstr2A on the ability of the receptor to associate with $G_i\alpha1$ to couple to adenylyl cyclase. To our knowledge this is the first example of a region of a receptor hindering coupling to adenylyl cyclase. This may direct the receptor to selective effector systems.

sstr3 also couples to adenylyl cyclase and mediates agonist inhibition of cAMP formation (Yasuda et al 1992). To investigate the molecular basis of the coupling of sstr3 to adenylyl cyclase, we initially expressed sstr3 in COS-7, HEK293 and CHO-DG44 cells. COS-7 cells express all forms of $G_i\alpha$ and sstr3 in those cells coupled to adenylyl cyclase (Yasuda et al 1992). HEK293 cells express $G_i\alpha1$ and $G_i\alpha3$ but no detectable $G_i\alpha2$ (Law et al 1993). sstr3 couples to adenylyl cyclase when expressed in these cells (Law et al 1993); however, it did not effectively mediate the inhibition of cAMP formation by somatostatin when expressed in CHO cells, which primarily express $G_i\alpha3$ and $G_o\alpha$ (Law et al 1994). These results suggested that $G_i\alpha1$ was critical for sstr3 to couple to adenylyl cyclase. To test this hypothesis directly, we coexpressed sstr3 with either $G_i\alpha1$ or $G_i\alpha2$

in CHO-DG44 cells. sstr3-mediated agonist inhibition of cAMP formation in cells expressing $G_i\alpha1$ but not in cells expressing $G_i\alpha2$ (Law et al 1994). Thus, $G_i\alpha1$ is necessary for sstr3 to couple to adenylyl cyclase.

To investigate the regions of $G_i\alpha1$ needed for coupling sstr3 to adenylyl cyclase, we constructed chimeric proteins consisting of the N-terminal two-thirds of $G_i\alpha2$ and the C-terminal third of $G_i\alpha1$ (Law et al 1994). In cells coexpressing this chimera and sstr3, somatostatin was able to inhibit forskolin-stimulated cAMP formation. In contrast, in cells expressing a $G_i\alpha2/G_i\alpha3$ chimera, sstr3 did not couple to adenylyl cyclase. The findings of these studies indicate that the C-terminal third of $G_i\alpha1$ is involved in the coupling of sstr3 to adenylyl cyclase. Comparison of the predicted amino acid sequences of the C-terminal third of $G_i\alpha1$ with $G_i\alpha3$, which does not couple sstr3 to adenylyl cyclase, reveal only six amino acids that are different between the two proteins. Two in particular are in strategic positions. The residue of the C-terminus and the amino acid five residues from the C-terminus differ between $G_i\alpha1$ and $G_i\alpha3$. Exchange mutagenesis of these residues might reveal the limited structural differences of these two G proteins necessary for coupling sstr3 to adenylyl cyclase.

Conclusion

Somatostatin induces multiple cellular responses by interacting with a family of receptors. Individual somatostatin receptor subtypes are capable of coupling to several effector systems. Thus, activation of a particular receptor could result in the simultaneous lowering of cAMP and Ca^{2+} in a cell as well as hyper-polarization of the cell membrane owing to K^+ efflux. G proteins may provide the molecular basis for the diversity of action of somatostatin. One of the somatostatin receptors, sstr2, is capable of coupling to $G_i\alpha1$, $G_i\alpha3$ and $G_o\alpha$. $G_i\alpha1$ links the receptor to adenylyl cyclase. $G_i\alpha3$ couples the receptor to a K^+ channel and $G_o\alpha$ associates the receptor with Ca^{2+} channels. Depending on which G proteins are expressed in a given cell, sstr2 will be capable of regulating different arrays of effector systems. Since $G_i\alpha1$ and $G_i\alpha3$ are 94% identical in amino acid sequence, subtle structural differences must be essential for creating the specificity in coupling sstr2 to adenylyl cyclase and K^+ channels. For sstr3 coupling to adenylyl cyclase, only a few amino acids in the C-terminus of $G_i\alpha1$ are essential for specifying the coupling of the receptor to adenylyl cyclase.

The cloning of the somatostatin receptors will now allow for structure–function analysis to identify the regions of the receptors involved in associating with particular G proteins and cellular effector systems. By combining such analysis with mutagenesis of the G proteins, we will be able to identify the surface topology of the receptors and G proteins needed for association and coupling to adenylyl cyclase and ion channels. Such studies should reveal the molecular basis of action of somatostatin.

Acknowledgements

We thank Yuan-Jiang Yu and Frederick Livingston for their expert technical assistance in these studies. The work was supported by NIH grants MH45533, MH48518 and GM34781, a predoctoral NIMH fellowship to H. K. and a grant from the Scottish Rite Foundation to K. R.

References

Bell GI, Reisine T 1993 Molecular biology of somatostatin receptors. Trends Neurosci 16:34–38

Buscail L, Delesque N, Estève J-P et al 1994 Stimulation of tyrosine phosphatase and inhibition of cell proliferation by somatostatin analogues: mediation by human somatostatin receptor subtypes sstr1 and sstr2. Proc Natl Acad Sci USA 91:2315–2319

Dohlman H, Caron M, Lefkowitz R 1987 A family of receptors coupled to guanine nucleotide regulatory proteins. Biochemistry 26:2657–2664

Epelbaum J 1986 Somatostatin in the central nervous system: physiology and pathological modifications. Prog Neurobiol 27:63–100

Heerding J, Raynor K, Reisine T 1994 The third intracellular loop of the delta opioid receptor is involved in coupling to adenylyl cyclase. Soc Neurosci Abstr 20:907

Kaupmann K, Bruns C, Hoyer D, Seuwen K, Lubbert H 1993 Distribution and second messenger coupling of four somatostatin receptor subtypes expressed in brain. FEBS Lett 331:53–59

Kleuss C, Hescheler J, Ewel C, Rosenthal W, Schultz G, Wittig R 1991 Assignment of G-protein subtypes to specific receptors inducing inhibition of calcium currents. Nature 353:43–48

Kleuss C, Scherübl H, Hescheler J, Schultz G, Wittig B 1992 Different β-subunits determine G-protein interaction with transmembrane receptors. Nature 358:424–426

Kleuss C, Scherübl H, Hescheler J, Schultz G, Wittig B 1993 Selectivity in signal transduction determined by α subunits of heterotrimeric G proteins. Science 259:832–834

Law S, Reisine T 1992 Agonist binding to rat brain somatostatin receptors alters the interaction of the receptor with G proteins. Mol Pharmacol 42:398–402

Law SF, Manning D, Reisine T 1991 Identification of the subunits of GTP-binding proteins coupled to somatostatin receptors. J Biol Chem 266:17885–17897

Law S, Yasuda K, Bell G, Reisine T 1993 $G_{i\alpha3}$ and $G_{o\alpha}$ selectively associate with the cloned somatostatin receptor subtype sstr2. J Biol Chem 268:10721–10727

Law S, Zaina S, Sweet R et al 1994 $G_{i\alpha1}$ selectively couples the somatostatin receptor subtype sstr3 to adenylyl cyclase: identification of the functional domains of this alpha subunit necessary for mediating somatostatin's inhibition of cAMP formation. Mol Pharmacol 45:587–590

O'Carroll A-M, Lolait SJ, König M, Mahan LC 1992 Molecular cloning and expression of a pituitary somatostatin receptor with preferential affinity for somatostatin–28. Mol Pharmacol 42:939:946

Patel YC, Greenwood MT, Warszynska A, Panetta R, Srikant CB 1994 All five cloned human somatostatin receptors (hsstr1–5) are functionally coupled to adenylyl cyclase. Biochem Biophys Res Commun 198:605–612

Raynor K, Reisine T 1992a Somatostatin receptors. Crit Rev Neurobiol 14:273–289

Raynor K, Reisine T 1992b Differential coupling of somatostatin₁ receptors to adenylyl cyclase in the rat striatum vs. the pituitary and other regions of the rat brain. J Pharmacol Exp Ther 260:841–848

Raynor K, Wang H, Dichter M, Reisine T 1992 Subtypes of somatostatin receptors couple to multiple effector systems. Mol Pharmacol 40:248–253

Raynor K, Murphy WA, Coy DH et al 1993a Cloned somatostatin receptors: identification of subtype-selective peptides and demonstration of high affinity binding of linear peptides. Mol Pharmacol 43:838–844

Raynor K, O'Carroll A-M, Kong HY et al 1993b Characterization of cloned somatostatin receptors sstr4 and sstr5. Mol Pharmacol 44:385–392

Reisine T, Bell G 1993 Molecular biology of opioid receptors. Trends Neurosci 16:506–510

Reisine T, Kong H, Raynor K et al 1993 Splice variant of the somatostatin receptor 2 subtype, sstr2B couples to adenylyl cyclase. Mol Pharmacol 44:1008–1015

Reisine T, Tallent M, Dichter M 1994 Somatostatin receptor subtypes endogenously expressed in AtT-20 cells couple to three different ionic currents. Soc Neurosci Abstr 20:908

Rens-Domiano S, Law SF, Yamada Y, Seino S, Bell GI, Reisine T 1992 Pharmacological properties of two cloned somatostatin receptors. Mol Pharmacol 42:28–34

Rossowski WJ, Coy DH 1993 Potent inhibitory effects of a type four receptor-selective somatostatin analog on rat insulin release. Biochem Biophys Res Commun 197:366–371

Tallent M, Reisine T 1992 $G_{i\alpha1}$ selectively couples somatostatin receptors to adenylyl cyclase in the pituitary-derived cell line AtT-20. Mol Pharmacol 41:452–455

Vanetti M, Kouba M, Wang XM, Vogt G, Höllt V 1992 Cloning and expression of a novel mouse somatostatin receptor sstr2B. FEBS Lett 311:290–294

Woulfe D, Reisine T 1994 Splice variants of sstr2 differentially couple to adenylyl cyclase. Soc Neurosci Abstr 20:907

Yasuda K, Rens-Domiano S, Breder CD et al 1992 Cloning of a novel somatostatin receptor, sstr3, coupled to adenylyl cyclase. J Biol Chem 267:20422–20428

Yatani A, Codina J, Sekura R, Birnbaumer L, Brown A 1987 Reconstitution of somatostatin and muscarinic receptor-mediated stimulation of K^+ channels by isolated G_k protein in clonal rat anterior pituitary cell membranes. Mol Endocrinol 1:283–289

DISCUSSION

Stork: Hershberger et al (1994) have also shown coupling of sstr1 and sstr2 to the inhibition of adenylyl cyclase using a different system. Discrepancies between groups may be due to differences in G protein components. CHO (Chinese hamster ovary)-K1 cells, for example, express $G_i\alpha2$ and $G_i\alpha3$, whereas $G_i\alpha2$ is not expressed in the CHO-DG44 cells that you used. The $G_i\alpha1$ is not expressed in either cell line. Can these cell lines that have different G protein α subunits be used productively to determine which G protein subunits couple to the inhibition of adenylyl cyclase?

Reisine: Our work on sstr3 addresses this question. sstr3 couples with adenylyl cyclase in African green monkey COS-7 kidney cells that have all forms of $G_i\alpha$ and in human kidney HEK293 fibroblasts that have $G_i\alpha1$ and $G_i\alpha3$, but not in CHO-DG44 cells which lack $G_i\alpha1$ and $G_i\alpha2$. Only in cells that have $G_i\alpha1$ do we get coupling of adenylyl cyclase to sstr3. We then localized the part of $G_i\alpha1$ involved in coupling to sstr3 to the C-terminal third of $G_i\alpha1$, by constructing chimeric proteins with the N-terminus of $G_i\alpha 2$. In this region there's only about four amino acids that are different between $G_i\alpha 1$ and $G_i\alpha 3$, so we're now making point mutations of these residues. One of these residues is the fifth residue from the C-terminus and is next to the cysteine that is ADP-ribosylated by pertussis toxin.

Eppler: It is possible that it is system specific. I certainly wouldn't rule out that $G_i\alpha 1$ could, in some circumstances, couple somatostatin receptors to the inhibition of adenylyl cyclase, even though there's evidence that $G_i\alpha 2$ is involved. For example, when $G_i\alpha 2$ antisense mRNA is transfected into cultured cells, all somatostatin-mediated inhibition of adenylyl cyclase is lost (Liu et al 1994, Moxham et al 1993). We have also purified sstr2A from rat GH_4C_1 pituitary tumour cells (Eppler et al 1992) and confirmed its identity by sequencing (Hulmes et al 1992). Luthin et al (1993) then showed, by Western blotting, that $G_i\alpha 2$ was the main G protein co-purifying with the receptor.

Reisine: Yes, but this is a different approach. You used an affinity column to purify the receptors—so you looked at agonist-bound receptors, whereas we looked at agonist-free receptors. We have also studied agonist-bound receptors and we do get an association between sstr2 and $G_i\alpha 2$ (we do not observe this association when the agonist is not present), however, the predominant G proteins that we find associated are still $G_i\alpha 1$, and $G_i\alpha 3$ and $G_o\alpha$.

With regards to the antisense mRNA experiment that you mentioned, when you knock out particular G proteins, cells don't grow very well. It is also very difficult to make stable cell lines that express mutant G proteins, so a number of other aspects may explain the discrepancies. I'm not saying that there's no possibility that $G_i\alpha 2$ is involved in the coupling, but the approaches we used indicate that it is not the predominant one coupled to sstr2 or sstr3.

Berelowitz: Patel et al (1994) found that adenylyl cyclase is inhibited by sstr1, whereas you don't. Why is this? Have you exchanged materials to see if this is the reason?

Reisine: In 1992, when we first got sstr1 from Graeme Bell, we tried to get it to couple to adenylyl cyclase, but we couldn't. We have used this result to our advantage to look at regions of other receptors to see if they are involved in coupling. We have not exchanged material with a number of other people because there are more interesting things to do in the field than to try and understand why we can't get coupling and other people can.

Eppler: It is important in the GH_4C_1 antisense experiments that Liu et al (1994) performed that adenylyl cyclase inhibition by somatostatin was retained when G_o, rather than $G_i\alpha 2$, was knocked out.

Reisine: But it's also known that different G proteins have different roles in both the cell cycle and cell growth, so it's possible that G_o and $G_i\alpha 2$ have different roles in this.

Eppler: They must've had good cell growth, otherwise they wouldn't have been able to select drug-resistant transfectants.

Schonbrunn: If you knock out $G_i\alpha2$ you prevent somatostatin inhibition of adenylyl cyclase, but you do not affect somatostatin reduction of intracellular Ca^{2+}.

Stork: The studies presented here (Reisine et al 1995, this volume) define G proteins in the absence of somatostatin. It is possible that G proteins coupled to somatostatin receptors in the presence of somatostatin are distinct from those that are coupled in the absence of somatostatin. When we looked at sstr1 in CHO cells we had difficulty in finding GTP effects on binding (Hershberger et al 1994). Our CHO cell line had $G_i\alpha2$ present and we saw inhibition of adenylyl cyclase.

Reisine: How do you explain a lack of GTP effect on high affinity agonist binding with apparent coupling to adenylyl cyclase?

Stork: We weren't able to see the GTP shift of binding. It is possible that it's a technical problem on our part because other groups have been able to do so. We were able to show a GTP shift of effector coupling, which confirmed that sstr1 was coupled to a G protein. Since the ability of sstr1 to couple to adenylyl cyclase correlates with the expression of a $G_i\alpha2$, we suggested that $G_i\alpha2$ might play a role in coupling; however, we did not look at the β and γ subunits in these cell lines.

Reichlin: Are all other coupling data in agreement? Is this the only exception?

Reisine: Few people have looked at the coupling of the cloned receptors to channels.

Reichlin: Both Yogesh Patel and Christian Bruns (Patel et al 1994, Bruns et al 1995, this volume) claim that all five receptors are coupled to adenylyl cyclase.

Meyerhof: We have also shown coupling of sstr3, but not sstr1, in NIH/3T3 mouse fibroblasts (R. Nehring, C. Schönrock, W. Meyerhof & D. Richter, unpublished results).

Reisine: Christiane Susini has not observed coupling of sstr1 or sstr2A to adenylyl cyclase (Buscail et al 1994).

Eppler: We didn't observe co-purification of G_o with sstr2 (Luthin et al 1993). It is possible that it is coupled in the intact cell, but falls off during purification. The G_i proteins are the only cases I know where G proteins co-purify with receptors such as dopamine, μ-opioid, sstr2 and adenosine A1 (Eppler et al 1993). Indeed, we agree only that coupling occurs in the case of $G_i\alpha2$ when preformed complexes of receptor, G protein and ligand are analysed.

I don't know of any good examples where G_s proteins co-purify with receptors, so it is possible that these interactions are more fragile.

Reisine: In contrast, our immunoprecipitation studies show that the $\alpha2$ adrenergic receptor, the somatostatin receptor, the δ receptor and $\varkappa$ opioid receptor all associate strongly with $G_o\alpha$. Functional evidence linking the receptors to Ca^{2+} channels also supports this coupling. It's possible that the variations are due to differences in technique.

Patel: I think we should exchange cells to see if the differences are due to different techniques or cell lines. For example, in our hands, all five human somatostatin receptors are functionally coupled to adenylyl cyclase in CHO-K1 and COS-7 cells (Patel et al 1994). Perhaps we would observe similar coupling if we used Terry Reisine's cell lines. Alternatively, maybe there's something missing in his cell line that is required for coupling.

Reisine: We are all using very artificial systems. The number of receptors may pose a problem if the presence of a large number ultimately results in the coupling to any signal transduction pathway. Also, different cell lines have different G proteins. It is important for us to resolve whether all five receptors can, if required, couple to adenylyl cyclase or not. In all of our systems, we have put in receptors that do, or do not, allow coupling. It's not something about the particular cells we're using that doesn't allow coupling, it's the expression of particular receptors.

References

Bruns C, Weckbecker G, Raulf H, Lübbert H, Hoyer D 1995 Characterization of somatostatin receptor subtypes. In: Somatostatin and its receptors. Wiley, Chichester (Ciba Found Symp 190) p 89–110

Buscail L, Delesque N, Estève JP et al 1994 Stimulation of tyrosine phosphatase and inhibition of cell proliferation by somatostatin analogues: mediation by human somatostatin receptor subtypes SSTR1 and SSTR2. Proc Natl Acad Sci USA 91:2315–2319

Eppler CM, Zysk JR, Corbett M, Shieh HM 1992 Purification of a pituitary receptor for somatostatin. The utility of biotinylated somatostatin analogs. J Biol Chem 267:15603–15612

Eppler CM, Hulmes JD, Wang JB et al 1993 Purification and partial amino acid sequence of a μ-opioid receptor from rat brain. J Biol Chem 268:26447–26451

Hershberger RE, Newman BL, Florio T et al 1994 The somatostatin receptors SSR1 and SSR2 are coupled to inhibition of adenylyl cyclase in CHO cells via pertussis toxin sensitive pathways. Endocrinology 134:1277–1285

Hulmes JD, Corbett M, Zysk JR, Bohlen P, Eppler CM 1992 Partial amino acid sequence of a somatostatin receptor isolated from GH_4C_1 pituitary cells. Biochem Biophys Res Comm 184:131–136

Liu YF, Jakobs KH, Rasenick MM, Alberts PR 1994 G protein specificity in receptor-effector coupling. Analysis of the roles of G_o and G_{i2} in GH_4C_1 pituitary cells. J Biol Chem 269:13880–13886

Luthin DR, Eppler CM, Linden J 1993 Identification and quantification of G_i type GTP-binding proteins that copurify with a pituitary somatostatin receptor. J Biol Chem 268:5990–5996

Moxham CM, Hod Y, Malbon CC 1993 Induction of $G_{\alpha i2}$-specific antisense RNA *in vivo* inhibits neonatal growth. Science 260:991–995

Patel YC, Greenwood MT, Warszynska A, Panetta R, Srikant CB 1994 All five cloned human somatostatin receptors (hsstr1–5) are functionally coupled to adenylyl cyclase. Biochem Biophys Res Commun 198:605–612

Somatostatin modulates voltage-dependent Ca^{2+} channels in GH$_3$ cells via a specific G$_o$ splice variant

C. Kleuss

Institut für Molekularbiologie und Biochemie, Freie Universität Berlin, Arnimallee 22, D-12197 Berlin, Germany

Abstract. In rat pituitary GH$_3$ cells Ca^{2+} current through L-type channels is reduced by somatostatin. This modulation of channel activity by somatostatin receptors is mediated by a guanine nucleotide-binding regulatory protein (G protein). It is sensitive to pertussis toxin, indicating the involvement of a G$_o$- or G$_i$-type G protein in this pathway. The identity of this G protein was determined by suppressing the expression of endogenous G proteins individually via intranuclear injection of antisense oligonucleotides. This method was applied to GH$_3$ cells to screen several G protein α, β and γ subunits for their roles in the defined signal transduction pathway. The loss of somatostatin's modulating activity on the voltage-dependent Ca^{2+} channel after oligonucleotide injection revealed the involvement of G$_o\alpha2\beta1\gamma3$ to the exclusion of other closely related subtypes.

1995 Somatostatin and its receptors. Wiley, Chichester (Ciba Foundation Symposium 190) p 171–186

Heterotrimeric G proteins represent a family of homologous regulatory proteins, each composed of one α, β and γ subunit (Hepler & Gilman 1992). G proteins were first identified and purified by monitoring activity and names were assigned with subscripts to evoke functional roles, for example G$_s$ which stimulates adenylyl cyclase activity and G$_i$ which inhibits its activity. Further discoveries, based mostly on similarities during purification of proteins or cloning of cDNAs, revealed an increasing number of G proteins but their functions were not immediately known. These 'newer' G proteins were named at the whim of the researcher, for example G$_o$ (other G protein), which was originally purified from brain. G proteins are grouped into subfamilies according to sequence homology of their α subunits. About 20 different cDNA sequences (including splice variants) encoding α subunits are known; at least five different cDNAs coding for G protein β subunits and six cDNAs for γ subunits have been identified (Simon et al 1991). Their functions are highly diverse and for some subfamilies (G_s, G_i, G_o, G_q and G_t), receptors stimulating the G protein and the corresponding effectors have been identified. The functions of others, e.g. G_{12}, G_{16}, G_z and additional pertussis toxin-insensitive G proteins remain largely unknown.

Assignment of a function to specific G protein subunits and $\alpha\beta\gamma$ subunit heterotrimers is a current challenge in studies of G protein-mediated signal transduction. In general, the specificity of interactions between receptors, G proteins and effectors appears greater in intact cells than in reconstitution experiments. The heterotrimeric composition of the G protein that couples the somatostatin receptor to the voltage-operated Ca^{2+} channel in intact rat pituitary GH_3 cells was determined using antisense oligonucleotides which transiently and selectively knock out expression of G protein subunits. An observed loss of function, i.e. the inability of the voltage-gated Ca^{2+} channel to be modulated by somatostatin, indicates the involvement of the targeted G protein subunit in this signal transduction pathway.

The selective repression of individual genes by nucleic acids in antisense orientation, i.e. complementary to the respective mRNA, has already been demonstrated in numerous applications (Izant & Weintraub 1984). When an antisense DNA oligonucleotide is delivered to recipient cells, it is supposed to hybridize to a complementary region in the mRNA and render it susceptible to cellular RNase H activity (Minshull & Hunt 1986, Shuttleworth & Colman 1988). The high degree of sequence similarity among subtypes of G protein subunits disallows the use of conventional recombinant eukaryotic expression systems to produce antisense mRNA for selective suppression of G protein subunit subtypes. Such an approach could address the involvement of a G protein subfamily in a certain signal pathway, e.g. α_i or α_o (Goetzl et al 1994, Liu et al 1994), but not of a specific member, e.g. α_{i1}, α_{i2}, α_{i3} (Gollasch et al 1993) or α_{o1}, α_{o2}. This problem can be partially overcome by targeting the more diverse, but species-specific 5'-untranslated sequences adjacent to the translational start site of the mRNA (Watkins et al 1992). Alternatively, short DNA oligonucleotides can be designed that are highly selective for each part of a known gene sequence. They may be delivered to the cytoplasm of mammalian cells simply by adding the oligonucleotide to the culture medium at a very high concentration. However, the efficiency of uptake and the amount of oligonucleotide finally surviving passage through degradative organelles is influenced by ingredients of the culture medium, the cell type, stage of cell cycle, length of the oligonucleotide and probably even the oligonucleotide sequence (Loke et al 1989). To avoid these variabilities, I microinjected DNA oligonucleotides into nuclei. The advantages of this technique are that the delivery of the antisense nucleic acid is virtually independent of oligonucleotide length and sequence, the oligonucleotides are applied to the assumed site of action (thus the amount of oligonucleotide actually reaching the nucleus is well controlled) and microinjection does not change the optimum culture environment.

Ca^{2+} currents in GH$_3$ cells are probably due to activation of L-type channels (Rosenthal et al 1988) and are inhibited to 70–75% of control currents by somatostatin (Fig. 1, top panel). Inhibition is abolished by treating the cells with pertussis toxin. This component of the Ca^{2+} current and its inhibition are not affected by cAMP applied intracellularly (Offermanns et al 1991). Therefore, a G$_i$-mediated inhibition of adenylyl cyclase is not involved.

Antisense oligonucleotides that hybridized selectively with the mRNA of a particular G protein subunit were microinjected into the nuclei of single cells.

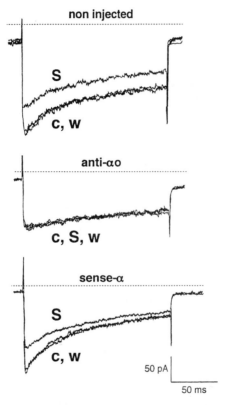

FIG. 1. Time-current recordings of the voltage-sensitive Ca^{2+} channel in GH$_3$ cells. Whole-cell Ca^{2+} currents were recorded under voltage-clamp conditions by depolarizing pulses from −80 to 0 mV. c, control currents obtained before application of somatostatin; S, currents recorded during superfusion of cells with 1 μM somatostatin; w, currents recorded after removal of somatostatin. Upper panel, uninjected cell; middle panel, cell injected with oligonucleotide anti-αo; lower panel, cell injected with oligonucleotide sense-α. An oligonucleotide that can hybridize with the mRNA is denoted as 'anti', whereas 'sense' refers to an oligonucleotide whose sequence is identical to the gene for the specified G protein subunit. A detailed description of the methods used and sequences of the oligonucleotides is given elsewhere (Kleuss et al 1991, 1992).

These cells were incubated for a certain time to wait for reduction in the concentration of the target subunit. The overall concentration is the amount of already synthesized protein at the time of injection plus the amount newly made by protein synthesis (which is affected by antisense oligonucleotides) minus the amount degraded during the incubation. The resulting biological effects were measured electrophysiologically by the whole-cell modification of the patch-clamp technique (Hamill et al 1981).

In cells that had been microinjected with an antisense oligonucleotide directed against the 5'-translated part of the mRNA coding for the α_o subunit of G proteins (anti-αo), somatostatin was no longer effective (Fig. 1, middle panel). The injection of a sense oligonucleotide affected neither the somatostatin modulation nor the kinetics of the Ca^{2+} current (Fig. 1, bottom panel).

A time course indicated that substantial inhibition of the somatostatin effect occurred between 24 h and 50 h after injection of the oligonucleotide (Fig. 2). The time required for the Ca^{2+} channel to regain sensitivity to somatostatin varied between 10 h and 30 h, depending on which G protein subunit was targeted

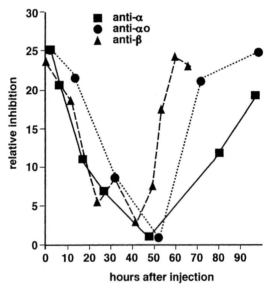

FIG. 2.   Time course of Ca^{2+} current inhibition by somatostatin (1 μM) in GH_3 cells injected with antisense oligonucleotides. Oligonucleotides anti-α (squares, solid line), anti-αo (circles, dotted line) and anti-β (triangles, dashed line) were injected at time zero. Mean values are shown (n $\geqslant$ 10; SEM $\leqslant$ 5% of means). Oligonucleotide anti-α is able to hybridize with mRNAs of all known variants of α_i, α_o, α_s and α_t. Oligonucleotide anti-αo selectively hybridizes with mRNAs of all known variants of α_o. Oligonucleotide anti-β can hybridize with the mRNAs of β1, β2, β3 and β4. The detailed sequences and presumed hybridization sites of the used oligonucleotides are described elsewhere (Kleuss et al 1991, 1992).

by the antisense oligonucleotide. Differences in the time course are probably due to different half-lives of the targeted subunits. The successful repression of G protein function implies that the inhibition of synthesis lasts longer than the life-time of already expressed subunits. Biochemical studies revealed half-lives of α_o subunits from 21 h to 35 h under steady-state conditions (Silbert et al 1990). In immunohistochemical experiments, G protein α subunits were reduced to barely detectable levels 24 h after injection of oligonucleotides (Kleuss et al 1991). An impaired function was observed 19 h after injection of oligonucleotides.

To determine whether the loss of Ca^{2+} current inhibition is observed only when α_o is suppressed, I tested the ability of other antisense oligonucleotides to interfere with the receptor-induced inhibition of Ca^{2+} currents (Fig. 3). As shown above (Fig. 1), oligonucleotide anti-αo largely reduced the somatostatin-induced inhibition of Ca^{2+} currents. In contrast, antisense oligonucleotides complementary to translated 5'-terminal regions of α_s and α_i mRNAs (anti-αs and anti-αi, respectively) had no effect. The sense oligonucleotide (sense-α) corresponding to oligonucleotide anti-α, or an oligonucleotide with an unrelated sequence (nonsense), was likewise without effect.

The mammalian genome contains a single gene coding for the α subunit of G_o (Tsukamoto et al 1991). Alternative splicing of the transcript yields two mRNAs, α_{o1} and α_{o2} (two more splice variants differ only in the 3'-non-coding regions; they encode the same protein and therefore are not addressed in this investigation) (Bertrand et al 1990, Strathmann et al 1990, Murtagh et al 1994). The two α_o subunits differ only in their C-terminal regions, which are assumed to be a recognition site for activated receptors (Masters et al 1986), because pertussis toxin-mediated ADP-ribosylation occurs at the C-termini of α_o and α_i subunits and uncouples these proteins from receptors. In GH_3 cells injected with oligonucleotide anti-$\alpha o2$ (that can hybridize to a 3'-translated sequence of α_{o2} mRNA), Ca^{2+} current inhibition by somatostatin was largely reduced (Fig. 3), as was observed after injection of oligonucleotide anti-$\alpha o2$. In contrast, the inhibition was not affected in cells that had been injected with oligonucleotide anti-$\alpha o1$, which has antisense orientation to the 3'-coding region of α_{o1} mRNA and cannot hybridize with α_{o2} mRNA. This demonstrates that the α_{o2} subunit of the trimeric G protein complex selectively mediates the action of the somatostatin receptor, whereas the α_{o1} subunit is not involved. These results allow, for the first time, assignment of different functions to alternatively spliced gene products of G proteins. A similar approach was used successfully to demonstrate that $G_o\alpha1$ couples to the muscarinic M_4 receptor and transmits inhibitory modulation to the same Ca^{2+} channel (Kleuss et al 1991). As α_{o1} and α_{o2} proteins differ only in their last two exon sequences E7 and E8 (a third exon, E9, is probably not translated into protein) (Tsukamoto et al 1991), their distinct functions are obviously defined by their C-terminal regions.

To obtain information about the mechanism of antisense oligonucleotide-mediated suppression of G protein subunits, I used additional α_o selective oligonucleotides complementary to sequences located immediately upstream from the translation start. As is shown in Fig. 3, the effect of an antisense oligonucleotide that can hybridize with the 5'-untranslated region of α_o mRNA (5' anti-αo, common to α_{o1} and α_{o2}) was similar to the effect of the previously studied oligonucleotide (anti-αo) that can hybridize with the translated region of the α_o mRNA. Oligonucleotide anti-αo1/2, which is complementary to the most 3'-coding region common to both α_{o1} and α_{o2} mRNAs, also reduced the response of cells to somatostatin. Antisense oligonucleotides 3' anti-αo1 and 3' anti-αo2 (which target the 3'-non-coding sequences of α_{o1} and α_{o2} mRNA,

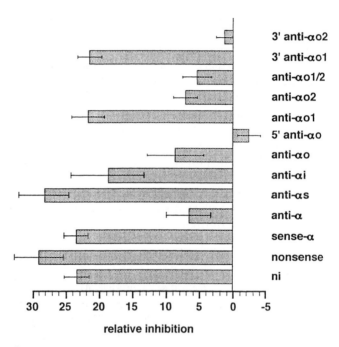

FIG. 3. Ca^{2+} current inhibition by somatostatin in GH$_3$ cells injected with antisense oligonucleotides directed against mRNAs encoding different G protein α subunits. Whole-cell Ca^{2+} currents were measured between 24 and 40 h after injection of the oligonucleotide. Relative inhibition denotes the Ca^{2+} currents induced by somatostatin (1 μM) as a percentage of the currents observed in cells not exposed to the agonist (mean values with SEM). ni, non injected cells. Targets of injected oligonucleotides: nonsense, sequence unrelated to G protein subunits; sense α and anti-α cover all known variants of α_i, α_o, α_s and α_t; anti-αs, anti-αi, anti-αo, 5' anti-αo and anti-αo1/2 are selective for all known variants of α_s, α_i and α_o, respectively; anti-αo1, anti-αo2, 3' anti-αo1 and 3' anti-αo2 are selective for the two splice variants of α_o. 5' and 3' denote hybridization of the oligonucleotide to the 5'- and 3'-untranslated regions of the α_o mRNA, respectively. The detailed sequences and presumed hybridization sites of the oligonucleotides are described elsewhere (Kleuss et al 1991).

respectively) had similar effects as their counterparts for the translated region: $3'$ anti-αo2 abolished the somatostatin-induced inhibition of Ca^{2+} currents, whereas oligonucleotide $3'$ anti-αo1 was ineffective.

These results help specify the level at which the antisense oligonucleotide acts (for review see Hélène & Toulmé 1990). A translational block caused by a hybridization complex between the antisense DNA oligonucleotide and the mRNA is highly unlikely to be the mechanism because, depending on the mRNA region, this complex would either interfere with the translation initiation or lead to premature termination owing to stalling of ribosomes. The presented results show that repression is independent of assumed hybridization positions: oligonucleotides against untranslated sequences upstream or downstream from coding sequences were equally effective at blocking the action of α subunits. Antisense oligonucleotides hybridizing to $3'$-untranslated regions of α_o mRNA should not cause premature termination. It is also obvious from the selective suppression of α_o splice variants that antisense oligonucleotides do not prevent transcription by hybridizing to the genomic DNA. All the results presented here are consistent with a mechanism by which RNase H destroys the α_o mRNA in a hybrid formed by the mRNA and the antisense oligonucleotide.

The involvement of β-subunits in the G protein-mediated Ca^{2+} current inhibition was demonstrated by microinjecting the oligonucleotide anti-β, which is complementary to the $5'$-coding regions of β1, β2, β3 and β4 mRNAs. After injection of anti-β, the inhibition of Ca^{2+} currents by somatostatin was abolished (Figs 2 and 4). The results, together with those of Blumer & Thorner (1990), emphasize the requirement for β subunits in the signal transduction pathway at the level of receptor recognition.

To study whether there is a preferential interaction between the activated receptor and G_o heterotrimers composed of a given β subunit subtype, I injected GH_3 cells with antisense oligonucleotides that can hybridize specifically with the mRNA of one particular β subunit subtype (Fig. 4). Ca^{2+} currents of cells that had been injected with antisense oligonucleotide anti-β1.1 were no longer inhibited by somatostatin. In contrast, neither β2-, β3- nor β4-specific antisense oligonucleotides prevented the inhibitory effects of somatostatin on the Ca^{2+} channel, although the respective mRNAs were expressed in GH_3 cells (Kleuss et al 1992). This suggests that a G protein containing the β1 subtype (apparent molecular mass of 36 kDa on a denaturing polyacrylamide gel) couples to the somatostatin receptor. This is consistent with co-precipitation of the β36 subunit and the somatostatin receptor solubilized from rat brain (Law et al 1991). The exclusive receptor coupling of $G_o\alpha$2 containing the β1 subunit was confirmed by injection of two other specific antisense oligonucleotides, anti-β1.2 and anti-β3.2 (β3 was chosen because this subunit has been shown to mediate the M_4 receptor-mediated Ca^{2+} current inhibition in GH_3 cells; Kleuss et al 1992).

The role of individual γ subunit subtypes in selective somatostatin receptor–Ca^{2+} channel-coupling was also studied using the antisense oligonucleotide

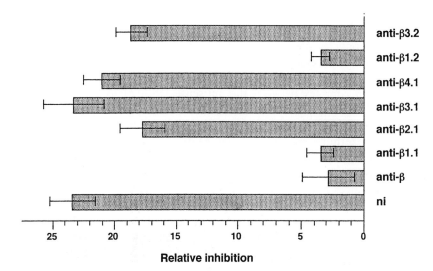

FIG. 4. Ca^{2+} current inhibition by somatostatin in GH_3 cells injected with antisense oligonucleotides directed against mRNAs encoding different G protein β subunits. Whole-cell Ca^{2+} currents were measured about 40 h after injection of the oligonucleotide indicated. Relative inhibition as for Fig. 3 (mean values with SEM). Targets of injected oligonucleotides: anti-β covers all β subunits (β1, β2, β3 and β4); anti-β2 (and anti-β4) are selective for subunits β2 and β4, respectively; anti-β1.1, anti-β1.2 and anti-β3.1, anti-β3.2 are selective for subunits β1 and β3, respectively, hybridizing at different sites on the respective mRNA. The detailed sequences and presumed hybridization sites of the oligonucleotides are described elsewhere (Kleuss et al 1992).

method. Somatostatin was ineffective in cells that had been microinjected with an antisense oligonucleotide (anti-γ) directed against four known G protein γ subunit mRNA sequences (Fig. 5). Cells that were injected with a γ1-selective oligo-nucleotide (anti-γ1) responded to somatostatin in the same manner as uninjected cells. The ineffectiveness of the injected γ1-selective antisense oligonucleotide agrees with the absence of any detectable γ1 mRNA in uninjected GH_3 cells (Kleuss et al 1993): the G protein subunit γ1 is expressed exclusively in retinal cells and is part of the rod-specific G protein heterotrimer.

The Ca^{2+} influx was no longer inhibited by somatostatin in those cells that had been injected with γ3-selective antisense oligonucleotides (anti-γ3.1, anti-γ3.2). In cells that had been injected with γ2 or γ4-selective antisense oligonucleotides (anti-γ2, anti-γ4.1, γ4.2), somatostatin was still effective despite the presence of γ2 and γ4 mRNA in untreated GH_3 cells. These results establish a third line of evidence that heterotrimeric G proteins composed of particular α, β, and γ subunits transduce signals from somatostatin receptors to voltage-sensitive Ca^{2+} channels.

In summary, these results verify the previous hypothesis that G_o rather than G_i couples inhibitory receptors to voltage-sensitive Ca^{2+} channels (Hescheler

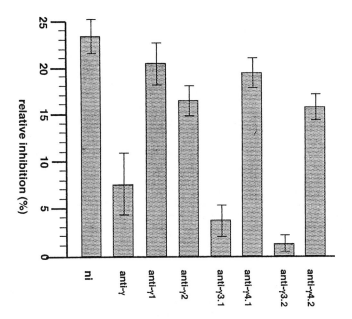

FIG. 5. Ca^{2+} current inhibition by somatostatin in GH_3 cells injected with antisense oligo-nucleotides directed against mRNAs encoding different types of G protein γ subunits. Whole-cell Ca^{2+} currents were measured about 40 h after injection of the oligo-nucleotide indicated. Because of its high degeneracy of base sequence, oligonucleotide anti-γ was microinjected at a concentration 10 times greater than that for γ subtype-selective oligonucleotides (50 μM instead of 5 μM). Relative inhibition as for Fig. 3 (mean values with SEM). Targets of injected oligonucleotides: anti-γ covers all γ subunits (γ_1, γ_2, γ_3 and γ_4); anti-γ1 and anti-γ2 are selective for subunits γ_1 and γ_2, respectively; anti-γ3.1, anti-γ3.2 and anti-γ4.1, anti-γ4.2 are selective for subunits γ_3 and γ_4, respectively, hybridizing at different sites on the respective mRNA. The detailed sequences and presumed hybridization sites of the oligonucleotides are described elsewhere (Kleuss et al 1993).

et al 1987) and also specify which isoform of G_o is responsible. Injection of antisense oligonucleotides into nuclei of rat pituitary GH_3 cells largely inhibited the expression of selected G protein subunits. Oligonucleotides complementary to the mRNA encoding α_{o2} (but not α_{o1}) abolished the inhibition of voltage-dependent Ca^{2+} currents by somatostatin. Intranuclear injection of oligonucleotides complementary to the mRNAs encoding G protein $\beta1$ and $\gamma3$ subunits (but not others) also abolished the inhibition of Ca^{2+} currents by somatostatin. As the cytosolic Ca^{2+} concentration is regulated by Ca^{2+} channel activity, the receptor-induced inhibition of Ca^{2+} channels by somatostatin may be crucial for the control of secretion. The results obtained with GH_3 cells are consistent with the observation made in rat pituitary tumour cells (Collu et al 1988), in which the expression of α_o correlated with the inhibition of prolactin secretion by dopamine.

The contribution of each G protein subunit α, β and γ to selective receptor–effector coupling is now substantiated. Although we cannot exclude a direct interaction between $\beta\gamma$ complexes and the Ca^{2+} channel, there are no indications for such effects in this system. The role of β and γ subunits in ablating the somatostatin-induced inhibition of Ca^{2+} channel activity is likely due to their role in G protein binding to activated receptors. Since the G protein binds to the activated receptor in its heterotrimeric form (Gilman 1987), the loss of either the α, β or γ subunit caused by an injected antisense oligonucleotide prevents formation of the proper heterotrimer and thus transduction of the signal across the membrane. As β or γ subunit subtypes other than those targeted by the injected antisense oligonucleotides apparently do not substitute for the suppressed ones, these results do not support the hypothesis of interchangeable β subunits or $\beta\gamma$ complexes in intact cells (Birnbaumer et al 1990), but rather suggest that the structural diversity of G proteins (Simon et al 1991) corresponds to a functional one. The presented results indicate that the receptor discriminates exquisitely between individual G protein subtypes containing subtly different α, β and γ subunits. They point out that the somatostatin receptor functionally couples via $G_o\alpha2$ containing the β_1 and γ_3 polypeptide to the voltage-dependent Ca^{2+} channel.

Acknowledgements

I thank Drs J. Hescheler and H. Scherübl for electrophysiological measurements, Drs B. Wittig and G. Schultz (all FU Berlin, Germany) for helpful advice and discussions, and Dr S. M. Mumby (UT Southwestern, Texas) for critical reading of the manuscript. The work was supported by grants from Deutsche Forschungsgemeinschaft, Fonds der Chemischen Industrie and from Trude-Goerke-Stiftung.

References

Bertrand P, Sanford J, Rudolph U, Codina J, Birnbaumer L 1990 At least three alternatively spliced mRNAs encoding two α subunits of the G_o GTP-binding protein can be expressed in a single tissue. J Biol Chem 265:18576–18580
Birnbaumer L, Abramowitz J, Brown AM 1990 Receptor–effector coupling by G proteins. Biochim Biophys Acta 1031:163–224
Blumer KJ, Thorner J 1990 β and γ subunits of a yeast guanine nucleotide-binding protein are not essential for membrane association of the α subunit but are required for receptor coupling. Proc Natl Acad Sci USA 87:4363–4367
Collu R, Bouvier C, Lagace G et al 1988 Selective deficiency of guanine nucleotide-binding protein G_o in two dopamine-resistant pituitary tumors. Endocrinology 122:1176–1178
Gilman AG 1987 G proteins: transducers of receptor-generated signals. Annu Rev Biochem 56:615–649

Goetzl EJ, Shames RS, Yang J et al 1994 Inhibition of Human HL-60 cell responses to chemotactic factors by antisense messenger RNA depletion of G proteins. J Biol Chem 269:809–812

Gollasch M, Kleuss C, Heschler J, Wittig B, Schultz G 1993 G_{i2} and protein kinase C are required for thyrotripin-releasing hormone-induced stimulation of voltage-dependent Ca^{2+} channels in rat pituitary GH_3 cells. Proc Natl Acad Sci USA 90:6265–6269

Hamill OP, Marty A, Neher E, Sakmann B, Sigworth FJ 1981 Improved patch-clamp techniques for high-resolution current recording from cells and cell-free membrane patches. Pflügers Arch Eur J Physiol 391:85–100

Hélène C, Toulmé J-J 1990 Specific regulation of gene expression by antisense, sense and antigene nucleic acids. Biochim Biophys Acta 1049:99–125

Hepler JR, Gilman AG 1992 G proteins. Trends Biochem Sci 17:383–387

Hescheler J, Rosenthal W, Trautwein W, Schultz G 1987 The GTP-binding protein, G_o, regulates neuronal calcium channels. Nature 325:445–447

Izant JG, Weintraub H 1984 Inhibition of thymidine kinase gene expression by antisense RNA: a molecular approach to genetic analysis. Cell 36:1007–1015

Kleuss C, Hescheler J, Ewel C, Rosenthal W, Schultz G, Wittig B 1991 Assignment of G-protein subtypes to specific receptors inducing inhibition of calcium currents. Nature 353:43–48

Kleuss C, Scherübl H, Hescheler J, Schultz G, Wittig B 1992 Different β-subunits determine G-protein interaction with transmembrane receptors. Nature 358:424–426

Kleuss C, Scherübl H, Hescheler J, Schultz G, Wittig B 1993 Selectivity in signal transduction determined by γ subunits of heterotrimeric G proteins. Science 259:832–834

Law SF, Manning D, Reisine T 1991 Identification of the subunits of GTP-binding proteins coupled to somatostatin receptors. J Biol Chem 266:17885–17897

Liu YF, Jakobs KH, Rasenick MM, Albert PR 1994 G protein specificity in receptor–effector coupling. J Biol Chem 269:13880–13886

Loke SL, Stein CA, Zhang XH et al 1989 Characterization of oligonucleotide transport into living cells. Proc Natl Acad Sci USA 86:3474–3478

Masters SE, Strout RN, Bourne HR 1986 Family of G protein α chains: amphipathic analysis and predicted structure of functional domains. Protein Eng 1:47–54

Minshull J, Hunt T 1986 The use of single-stranded DNA and RNase H to promote quantitative 'hybrid arrest of translation' of mRNA/DNA hybrids in reticulocyte lysate cell-free translations. Nucleic Acid Res 14:6433–6451

Murtagh JJ Jr, Moss J, Vaughan M 1994 Alternative splicing of the guanine nucleotide-binding regulatory protein $G_{o\alpha}$ generates four distinct mRNAs. Nucleic Acids Res 22:842–849

Offermanns S, Gollasch M, Hescheler J et al 1991 Inhibition of voltage-dependent Ca^{++} currents and activation of pertussis toxin-sensitive G-proteins via muscarinic receptors in GH_3 cells. Mol Endocrinol 5:995–1002

Rosenthal W, Hescheler J, Hinsch K-D, Spicher K, Trautwein W, Schultz G 1988 Cyclic AMP-independent, dual regulation of voltage-dependent Ca^{++} currents by LHRH and somatostatin in a pituitary cell line. EMBO J 7:1627–1633

Shuttleworth J, Colman A 1988 Antisense oligonucleotide-directed cleavage of mRNA in Xenopus oocytes and eggs. EMBO J 7:427–434

Silbert S, Michel T, Lee R, Neer EJ 1990 Differential degradation rates of the G protein α_o in cultured cardiac and pituitary cells. J Biol Chem 265:3102–3105

Simon MI, Strathmann MP, Gautam N 1991 Diversity of G proteins in signal transduction. Science 252:802–808

Strathmann M, Wilkie TM, Simon MI 1990 Alternative splicing produces transcripts encoding two forms of the α subunit of GTP-binding protein G_o. Proc Natl Acad Sci USA 87:6477–6481

Tsukamoto T, Toyama R, Itoh H, Kozasa T, Matsuoka M, Kaziro Y 1991 Structure of the human gene and two rat cDNAs encoding the α chain of GTP-binding regulatory protein G_o: two different mRNAs are generated by alternative splicing. Proc Natl Acad Sci USA 88:2974–2978

Watkins DC, Johnson GL, Malbon CC 1992 Regulation of the differentiation of teratocarcinoma cells into primitive endoderm by $G_{\alpha i2}$. Science 258:1373–1375

DISCUSSION

Reisine: We have also looked at the selectivity of particular G protein α subunits with particular β and γ subunits (Reisine et al 1995, this volume). We tried immunoprecipitation with antisera against different α subunits to determine what β or γ subunit was co-precipitated, and also immunoprecipitation with a selective β subunit antiserum. We found that any α subunit could couple with any β subunit and any γ subunit. It is possible that there's no pre-set association, but that the receptor interacts with a particular α subunit which is then selective for particular β and γ subunits; however, this is difficult to prove—you would have to have an antibody against the receptor and then look at subunit selectivity.

Kleuss: I have not done immunoprecipitation studies on the somatostatin receptor. We have looked at the G protein-coupled galanin receptor which interacts with two different $\beta\gamma$ complexes and one α_o subunit (F. Kalkbrenner, C. Kleuss & G. Schultz, unpublished results). It is possible that all G protein triplets immunoprecipitate with the receptor, but you cannot tell which specific triplet is involved in a pathway with a particular effector.

Reichlin: Terry Reisine, are you saying that the subunits are not specific?

Reisine: No, I'm saying that the specificity is not pre-set, but is created by the interaction of a receptor with an α subunit. A particular conformation is thus set up, the α subunit recognizes a particular β and γ subunit and the formation of this complex creates stability.

Reichlin: Are your antisera entirely specific for every G protein subtype?

Reisine: No antiserum is entirely specific.

Reichlin: Is the antisense approach more specific?

Reisine: Our antisera are selected on the basis of having a five to tenfold higher affinity for a particular subunit versus any other subunit and also by their ability to immunoprecipitate known ^{35}S-labelled α, β or γ subunits from particular cell lines.

Kleuss: Iniguez-Lluhi et al (1992) have shown that the effector cannot distinguish between the three different β subunits *in vitro* and that α subunit functions are influenced equally by $\beta 1$, $\beta 2$ and $\beta 3$ subunits. On the other hand, receptors and receptor kinases show preferences for specific $\beta\gamma$ complexes, effectors show specificity for particular α subunits and not all conceivable complexes are formed by the β and γ subunits. For every step in signal transduction, there is at least one example of a specific interaction between two subunits, but we do not know if these examples are exceptions and that selectivity is achieved instead by the formation of the multicomponent complex.

Reisine: Yes, most people think that the α subunit and the receptor interact, then the β associates with the receptor independently of the α subunit. In fact, there is no evidence for this.

Stork: Terry Reisine, do your chimeric receptors that couple adenylyl cyclase (Reisine et al 1995, this volume) associate with new α subunits? If they do, could you determine whether they still associate with the same β and γ subunits?

Reisine: We have not done this, but it can be done. We would have to make stable cell lines which express the chimeric receptors, then carry out co-immuno-precipitation experiments. Our previous studies of chimeric receptors were in transient African green monkey COS-7 kidney cells.

We have made chimeric γ receptors that knock out coupling to adenylyl cyclase, and we still get GTP-regulated agonist binding. It still associates with G proteins, but the effector pathway is different.

Montminy: Christiane Kleuss, how long did you wait after microinjection of the antisense oligonucleotides to measure changes in Ca^{2+} conductance? Also, did you measure the levels of the proteins that you were trying to eliminate?

Kleuss: We waited for at least one day, but not longer than two days, before we measured the Ca^{2+} current. This was because three days after injection, the effect of somatostatin was restored. We showed by immunofluorescence that the amount of α subunit was below the level of detection after injection of the respective antisense oligonucleotide.

Reichlin: Did you perform these experiments on single cells?

Kleuss: Cells were seeded on a glass plate and all cells in a circled area were injected. After one or two days, the plate was transferred to the patch–clamp device and Ca^{2+} currents were measured.

Reichlin: How many cells did you inject?

Kleuss: About 500–1000 cells were injected per plate; of these, about 10 were measured electrophysiologically. These measurements take a long time and the cells deteriorate in the bathing solution, so you have to use a fresh plate every hour or so.

Stork: It has been debated whether or not antisense oligonucleotides directed to start codons, stop codons or other intron junctions are better suited for inhibition (Falker et al 1994). What does your experience suggest?

Kleuss: In our system any region is suitable. I am aware of the reports where only one target site is effective for antisense oligonucleotides, however, these authors used a different set-up—they did *in vitro* translation experiments with different proteins, like globin and *myc* protein, which were inhibited by antisense mRNA.

Montminy: Did you try microinjecting your antisense oligonucleotides into the cytoplasm to see if you could identify the site at which these oligonucleotides are acting?

Kleuss: This is almost impossible in GH_3 rat pituitary tumour cells because the nucleus is so large. If you inject oligonucleotides into PC12 rat adrenal phaeochromocytoma cells, only about 50% is injected into the cytoplasm. In these cells, we can only knock out G protein subunits successfully if I use exonuclease-protected oligonucleotides to stabilize the oligonucleotides until they are transported into the nucleus. We don't have to use protected oligonucleotides for studies in GH_3 cells.

Coy: Does it make any difference whether a particular receptor is coupled to its effector system or not? Will this change the ligand binding profile?

Schonbrunn: A complex can exist between an effector system and the receptor, for instance in the case of phospholipase C (Mah et al 1992).

Reisine: If the receptor is not coupled to a G protein you would predict that it is in a low affinity state for the agonist. The problem with somatostatin is that the radioligands available are all agonists, so you could not label G protein-uncoupled receptors.

Epelbaum: Unless you had an antagonist.

Reisine: But there aren't any.

Reichlin: Christiane Kleuss, your marker is an electrophysiological signal that enables you to study a single cell after intranuclear injection. Have you used this technique for other types of second messengers?

Kleuss: Microinjection experiments are limited to techniques, such as channel current measurements or immunofluorescence, that detect the behaviour of one cell, or at least a very small population of cells. Usually, second messenger effects are studied in biochemical assays starting with several thousand cells or whole tissues—numbers that are too large to be microinjected. Alternatively, you can use the reverse transcriptase polymerase chain reaction method to study small numbers of cells. We used this to look for a reduction of G protein mRNA levels after microinjection, but this does not give any information on the function of the targeted protein.

Reichlin: Can you use your approach to define the specific coupling protein responsible for protein synthesis regulation or cAMP regulation?

Kleuss: There are techniques which can measure cAMP stimulation in a single cell.

Goodman: Yes, for example Riabowol et al (1988) injected CREs (cAMP response elements), linked to the β-galactosidase gene, into single cells and looked at the generation of cAMP signals. It is a similar system to the one Marc Montminy used (Montminy et al 1995, this volume) to look at signals mediated by CREB (CRE binding) protein and CBP (CREB protein binding protein).

Stork: I have also examined whether somatostatin inhibition of cAMP can be visualized using a β-galactosidase reporter plasmid hooked up to CREs (P. J. S. Stork, unpublished results). The cells turn blue when β-galactosidase expression is stimulated by the CRE. In the presence of forskolin, the field of cells can be stained blue. A sigmoidal dose–response curve of somatostatin's inhibition of adenylyl cyclase can be generated by examining the per cent of blue cells at various somatostatin concentrations. The EC_{50} estimated by this approach is similar to that seen using assays for adenylyl cyclase activity. Christiane Kleuss could use this approach with her microinjecting technique, although one limitation is that the cells are killed during the β-galactosidase assay, so you can't do anything with them afterwards.

Kleuss: Somatostatin regulation of the voltage-dependent Ca^{2+} channel is independent of cAMP. In the presence of forskolin, we get the same pattern of regulation. To study the inhibition of adenylyl cyclase, you may have to look at another receptor, probably sstr3, but this is not present in GH_3 cells.

Montminy: There's also an antibody which our lab has developed that distinguishes between phosphorylated and dephosphorylated CREB protein (Hagiwara et al 1993). You could therefore use immunofluorescence to look at the protein kinase A pathway in single cells.

Patel: Have you tried direct uptake of antisense oligonucleotides?

Kleuss: We tried and failed. Brown et al (1992) reported that it is possible to apply antisense oligonucleotides directly into the medium; but oligonucleotide uptake is inefficient. We need several oligonucleotides targeting the same, or related, G protein subunits to evaluate the effect observed with a single antisense oligonucleotide; but if these oligonucleotides do not reach their targets at equal concentrations over a similar time period, such experiments are meaningless.

Epelbaum: Interactions between M-type muscarinic receptors, somatostatin receptors and K^+ channels were recently reported (Jaçquin et al 1988). How is this feasible, since your experiments in GH_3 cells show that the α β and γ subunits linked to muscarinic channels are completely different from those linked to somatostatin channels?

Kleuss: It seems that the K^+ channel interacts with either the α_i subunit, the $\beta\gamma$ complex or both. This area is controversial—one lab sees an α_i effect, another a $\beta\gamma$ effect. This α_i subunit is distinct from α_o, which interacts with the Ca^{2+} channel. It is conceivable that a different G_o triplet couples the muscarinic or somatostatin receptor to the Ca^{2+} channel, but the same G_i triplet couples both receptors to the K^+ channel. The results you mentioned could also be explained by the interaction of α subunits activated by one receptor with the $\beta\gamma$ complex released by the other receptor.

References

Brown KE, Kindy MS, Sonenshein GE 1992 Expression of the *c-myb* protooncogene in bovine vascular smooth muscle cells. J Biol Chem 267:4625–4630

Falker B, Herlitze S, Amthor S, Zenner H-P, Ruppersberg J-P 1994 Short antisense oligonucleotide-mediated inhibition is strongly dependent on oligo length and concentration but almost independent of location of the target sequence. J Biol Chem 269:16187–16194

Hagiwara M, Brindle P, Marootunian A et al 1993 Coupling of hormonal stimulation and transcription via the cyclic AMP-responsive factor CREB is modified by nuclear entry of protein kinase A. Mol Cell Biol 13:4852–4859

Iñiguez-Lluhi JA, Simon MI, Robishaw JD, Gilman AG 1992 G protein beta–gamma subunits synthesized in SF9 cells: functional characterization and the significance of prenylation of gamma. J Biol Chem 267:23409–23417

Jaçquin T, Champagnat J, Madamba S, Denavitsaubie M, Siggins GR 1988 Somatostatin depresses excitability in neurons of the solitary tract complex through hyperpolarization and augmentation of IM, a non-inactivating voltage-dependent outward current blocked by muscarinic agonists. Proc Natl Acad Sci USA 85:948–952

Mah SJ, Ades AM, Mir R, Siemens JR, Williamson JR, Fluharty SJ 1992 Association of solubilized angiotensin II receptors with phospholipase C-alpha in murine neuroblastoma NIE-115 cells. Mol Pharmacol 42:217–226

Montminy M, Brindle P, Arias J, Ferreri K, Armstrong R 1995 Regulation of somatostatin gene transcription by cAMP. In: Somatostatin and its receptors. Wiley, Chichester (Ciba Found Symp 190) p 7–25

Reisine T, Woulfe D, Raynor K et al 1995 Interaction of somatostatin receptors with G proteins and cellular effector systems. In: Somatostatin and its receptors. Wiley, Chichester (Ciba Found Symp 190) p 160–170

Riabowol KT, Fink JS, Gilman MZ, Walsh DA, Goodman RH, Feramisco JZ (1988) The catalytic subunit of cAMP-dependent protein kinase is sufficient to induce expression of genes containing the cAMP-regulated enhancer. Nature 336:83–86

A tyrosine phosphatase is associated with the somatostatin receptor[1]

N. Delesque, L. Buscail, J. P. Estève, I. Rauly, M. Zeggari, N. Saint-Laurent, G. I. Bell*, A. V. Schally†, N. Vaysse and C. Susini

INSERM U151, Institut Louis Bugnard, CHU Rangueil, 1 Avenue Jean Poulhès, 31054 Toulouse, France, Howard Hughes Medical Institute, University of Chicago, Chicago, IL 60637, USA, †Veterans Affairs Medical Center and Tulane University School of Medicine, New Orleans, LA 70112, USA*

Abstract. Regulation of tyrosine phosphorylation is thought to be an essential step in signal transduction mechanisms that mediate cellular responses. In pancreatic tumour cells we demonstrated that somatostatin analogues inhibited cell proliferation and stimulated a membrane protein tyrosine phosphatase (PTP) activity at concentrations at which they bind to the somatostatin receptor. To elucidate the role of PTP in the signal transduction pathway activated by somatostatin receptors we first studied the interaction of PTP with the somatostatin receptor at the membrane. We purified somatostatin receptors by immunoaffinity from pancreatic membranes that strongly expressed the type 2 somatostatin receptor sstr2. We identified the receptor as an 87 kDa protein. We demonstrated that a PTP activity co-purified with somatostatin receptors. The PTP was identified as a 66 kDa protein immunoreactive to antibodies against SHPTP1. These antibodies immunoprecipitated somatostatin receptors either occupied or unoccupied by ligand indicating that SHPTP1 is associated with somatostatin receptors. We then expressed *sstr2A* in monkey kidney COS-7 cells and mouse NIH/3T3 fibroblasts and demonstrated that somatostatin analogues (RC 160, octreotide and BIM 23014) which exhibited high affinity for sstr2 stimulated a PTP activity and inhibited cell proliferation in proportion to their affinities for sstr2. Under the same conditions these analogues have no effect on the growth of cells expressing *sstr1*. All these results suggest that a PTP related to SHPTP1 is associated with somatostatin receptors and may be involved in the negative growth signal promoted by sstr2.

1995 Somatostatin and its receptors. Wiley, Chichester (Ciba Foundation Symposium 190) p 187–203

The protein tyrosine phosphatases (PTP)s constitute a family of enzymes that catalyse the dephosphorylation of proteins at tyrosine residues. About 40 PTPs have been identified so far; the structure, function and regulation of these enzymes have only recently been appreciated (Walton & Dixon 1993). Evidence

---

[1]This paper was presented by C. Susini

is accumulating to implicate PTPs in the control of signal transduction mechanisms triggered by hormones, growth factors and oncogenes that lead to cellular processes such as growth, differentiation and transformation. If one of the roles of PTPs is to counteract protein tyrosine kinase activities that are involved in growth factor-induced mitogenic responses, it might be expected that stimulation of PTP activity is implicated in the signal transduction pathways of negative regulators of cell growth, such as somatostatin.

It has been well documented that somatostatin negatively controls growth *in vivo* and *in vitro* in a variety of cell types, including normal and tumour epithelial cells (Lamberts et al 1991). Somatostatin may affect cell growth via indirect effects by inhibiting the release of trophic factors but also by direct interaction with specific receptors on target cells. Specific somatostatin receptors have been characterized on many cell types and shown to be coupled to a variety of signal transduction pathways, including adenylyl cyclase, ion conductance channels and protein dephosphorylation on serine/threonine and tyrosine residues (Reubi et al 1990, Lewin 1992, Patel et al 1990, Liebow et al 1989). Molecular cloning has recently identified five types of somatostatin receptor (Bell & Reisine 1993, O'Carroll et al 1992, Bruno et al 1992). The antiproliferative effect of somatostatin may be mediated by different somatostatin receptor types via multiple signal transduction systems depending on the target cell, its growth promoters and its cellular environment. Here, we have focused our attention on the role of PTPs in the signal transduction pathway initiated by somatostatin in regulating cell proliferation.

A PTP is involved in the negative growth signal promoted by somatostatin in pancreatic cells

In pancreatic tumour cells, we demonstrated that somatostatin analogues antagonized the mitogenic effect of growth factors acting on tyrosine kinase receptors, such as epidermal growth factor (EGF) and fibroblast growth factor (FGF) (Liebow et al 1989, Viguerie et al 1989, Bensaïd et al 1992). Furthermore, somatostatin or its analogues stimulated a membrane PTP activity in normal and tumour cells (Liebow et al 1989, Pan et al 1992, Tahiri-Jouti et al 1992, Colas et al 1992). The ability of somatostatin analogues to stimulate PTP activity correlates with their inhibitory effect on cell growth and their affinity for somatostatin receptors. We characterized the PTP activity in pancreatic cells and demonstrated that the amount of activity eluting with a molecular mass of 70 kDa in control membranes was increased after treatment of cells with a somatostatin analogue. The enzyme was then partially purified by ion-exchange chromatography and the two peaks generated by high performance liquid chromatography on DEAE (diethylaminoethyl)-cellulose contained an immunoreactive protein of apparent molecular mass 67 kDa revealed by antibodies directed against a conserved sequence of PTP (Colas et al 1992). All these findings support the involvement of PTP in the antiproliferative signal mediated by somatostatin receptors.

A PTP related to SHPTP1 is associated with the somatostatin receptor

To gain insight into the mechanism of action of somatostatin, we purified solubilized complexes of somatostatin bound to its receptor from rat pancreatic membranes pretreated with somatostatin-28 by immunoaffinity chromatography using immobilized polyclonal antibodies raised against the N-terminal half of somatostatin-28, somatostatin-28 (1–14), which is not involved in receptor binding site recognition. Analysis by SDS polyacrylamide gel electrophoresis and silver staining of immunopurified proteins revealed a band at 87 kDa specific to the somatostatin receptor. This band was not observed when somatostatin receptors were solubilized from pancreatic membranes pretreated in the absence of somatostatin-28 or with somatostatin-14 or when preimmune serum replaced the somatostatin-28 (1–14) antiserum. Somatostatin-14 inhibited the appearence of the 87 kDa protein in the same range of concentrations as those that inhibited radioligand binding on pancreatic membranes. Rat pancreatic acinar cells strongly express *sstr2* mRNA (B. Colas, personal communication 1994) and have somatostatin receptors with similar pharmacological characteristics to sstr2, suggesting that sstr2 is a candidate for the 87 kDa protein.

We then tested the PTP activity of immunopurified proteins containing somatostatin receptors using two phosphorylated substrates: ^{32}P-poly (Glu, Tyr) and ^{32}P-labelled EGF receptors. After treatment of membranes with somatostatin-28, purified somatostatin receptor preparations exhibited a PTP activity that dephosphorylated phosphorylated EGF receptors and poly (Glu, Tyr). This activity was related to the presence of somatostatin receptors in the purified material. It was increased by dithiothreitol, a known stimulator of PTPs, and inhibited by orthovanadate, a potent inhibitor of PTPs (Fig. 1).

Among the PTPs that have been cloned, the 68 kDa SHPTP1 that possesses two Src homology 2 (SH2) domains (Shen et al 1991) has the potential to interact with tyrosine kinase growth factor receptors (Uchida et al 1994). The demonstration that SHPTP1 inhibits growth factor-induced tyrosine phosphorylation (Yi et al 1993, Vogel et al 1993) suggests a role for this enzyme in the negative control of growth factor receptor-mediated signals. To test whether PTP activity detected in immunopurified material is immunologically related to SHPTP1, we generated polyclonal anti-SHPTP1 antibodies. In purified material containing somatostatin receptors, anti-SHPTP1 polyclonal antibodies identified a protein of 66 kDa that was not detected in the absence of somatostatin receptor in the eluted fractions, indicating that the PTP co-purified with somatostatin receptors (Fig. 2).

Furthermore, the anti-SHPTP1 antibodies immunoprecipitated specific ([^{125}I] Tyr^3)octreotide binding from pancreatic membranes prelabelled with ([^{125}I] Tyr^3)octreotide. (Fig. 3A). Anti-SHPTP1 antibodies immunoprecipitated

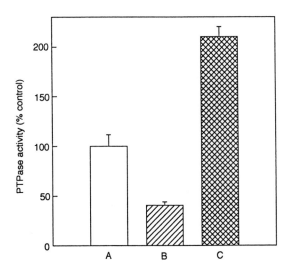

FIG. 1.   Effect of dithiothreitol and orthovanadate on tyrosine phosphatase activity of immunoaffinity purified somatostatin receptors. Solubilized somatostatin receptor complexes from pancreatic membranes treated with somatostatin-28 were incubated with an antibody against the N-terminal half of somatostatin-28. The bound material was eluted with the peptide and the tyrosine phosphatase (PTPase) activity of the eluted proteins was assayed. Purified somatostatin receptors were incubated with ^{32}P-poly (Glu, Tyr) for 10 min at 30 °C in the presence of 0.1 mM orthovanadate (B) or 5 mM dithiothreitol (C) and the amount of released ^{32}P$_i$ was measured. Values are corrected from that obtained in eluate derived from untreated membranes. Results are expressed as per cent of control (without agent, A) and are mean ± SEM of two experiments.

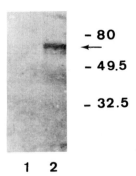

FIG. 2.   Co-purification of SHPTP1 with somatostatin receptors. Immunopurified somatostatin receptors from pancreatic membranes pretreated (lane 2) or not (lane 1) with 2 nM somatostatin-28 were subjected to electrophoresis on a 10% polyacrylamide gel. Proteins were transferred and blotted with anti-SHPTP1 antibodies. The 66 kDa immunoreactive protein is indicated by an arrow.

functional somatostatin receptors from solubilized pancreatic membranes as shown by specific binding of ($[^{125}I]$ Tyr³)octreotide to somatostatin receptors in the anti-SHPTP1 immunoprecipitates (Fig. 3B). Specific binding corresponds to the difference between total binding (measured in the presence of labelled ligand) and non-specific binding (measured in the presence of labelled ligand and an excess of cold ligand [1 μM octreotide]). These results indicate that SHPTP1 or an SHPTP1–related protein is able to associate with ligand-occupied and ligand-unoccupied somatostatin receptors and that the somatostatin receptor–SHPTP1 complexes exist in the membrane in the resting state.

A PTP is involved in signal transduction by sstr2

In an attempt to discover which somatostatin receptor stimulates a PTP, two human cloned somatostatin receptors, *sstr1* and *sstr2*, were expressed

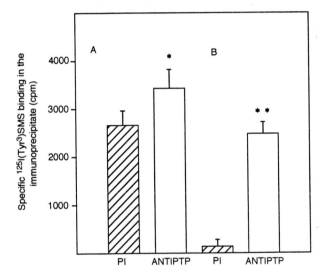

FIG. 3. Anti-SHPTP1 antibodies immunoprecipitate somatostatin receptors. (A) Pancreatic membranes were prelabelled with 1 nM ($[^{125}I]$ Tyr³)octreotide (^{125}I [Tyr³] SMS), in the presence or absence of 1 μM octreotide, solubilized and soluble ($[^{125}I]$ Tyr³octreotide-receptor complexes were immunoprecipitated with anti-SHPTP1 antibodies (ANTIPTP) or preimmune serum (PI). Specific bound ($[^{125}I]$ Tyr³)octreotide, corresponding to the difference between binding in the presence or absence of 1 μM octreotide, was detected in the immunoprecipitates. (B) Soluble somatostatin receptors from pancreatic membranes were immunoprecipitated with anti-SHPTP1 antibodies or preimmune serum. The presence of somatostatin receptors in the immunoprecipitates was detected using ($[^{125}I]$ Tyr³)octreotide binding assay. Values are the mean $\pm$ SEM of three different experiments. Statistical significance between preimmune serum and anti-SHPTP1 is indicated as *p < 0.05, **p < 0.01.

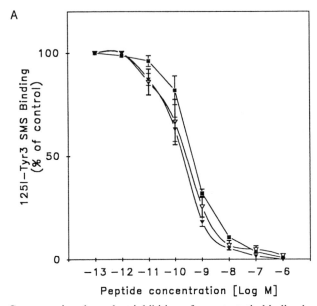

FIG. 4A. Concentration-dependent inhibition of somatostatin binding by somatostatin analogues in COS-7 cells expressing *sstr2*. Membranes from COS-7 cells transiently expressing *sstr2* were incubated with ($[^{125}I]Tyr^3$) octreotide (^{125}I-Tyr^3SMS) and the indicated concentrations of RC 160 (▼), octreotide (▽) and BIM 23014 (■). Results are expressed as per cent of specific binding observed without addition of competitor and are the mean ± SEM of three experiments.

transiently and stably in monkey kidney COS-7 and mouse NIH/3T3 fibroblasts, respectively. Binding experiments performed with membranes prepared from COS-7 cells expressing *sstr2* revealed that somatostatin analogues, RC 160, octreotide (generous gift from C. Bruns, Sandoz, Basle) and BIM 23014 (generous gift from F. Thomas, Ipsen Biotech, Paris), bound to sstr2 with high affinity ($IC_{50} = 0.27 \pm 0.12$ nM, 0.19 ± 0.06 nM and 0.5 ± 0.05 nM, respectively) (Fig. 4A). In contrast, in COS-7 cells expressing *sstr1*, analogues inhibited the binding of $[^{125}I]Tyr^{11}$ somatostatin with a low affinity ($IC_{50} = 0.43 \pm 0.07$ μM, 1.5 ± 0.2 μM, 2.9 ± 0.4 μM for RC 160, octreotide and BIM 23014, respectively). Treatment of COS-7 cells expressing *sstr2* with analogues for 15 min induced a rapid stimulation of PTP activity in a dose-dependent manner (Fig. 4B). Maximal stimulation (40–60% over basal value) occurred at 1 nM and half-maximal stimulation at 2 ± 1 pM RC 160, 6 ± 3 pM octreotide and 22 ± 2 pM BIM 23014. A similar dose-dependent stimulation of PTP activity was observed when NIH/3T3 cells expressing *sstr2* were treated with octreotide. In COS-7 cells expressing *sstr1*, weak stimulation of PTP activity was observed only at 1 μM RC 160.

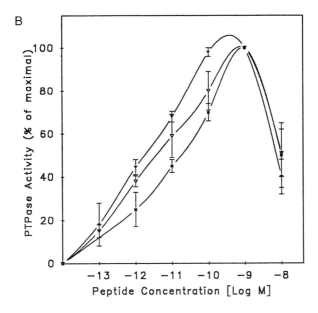

FIG. 4B. Stimulation of PTP activity by somatostatin analogues in COS-7 cells expressing *sstr2*. Cells were incubated for 15 min at 25°C in the presence of RC 160 (▼), octreotide (▽), or BIM 23014 (■). Cells were lysed in liquid N_2, centrifuged at 500 *g* and the homogenate was assayed for PTP activity as described (Colas et al 1992). Results are expressed as per cent of maximal stimulation over basal value observed at 1 nM analogue and are mean ± SEM of three experiments.

Treatment of NIH/3T3 cells expressing *sstr2* with analogues for 24 h resulted in inhibition of cell proliferation induced by serum or FGF. The inhibition was dose-dependent with a maximal effect at 1 nM RC 160 or octreotide and a half-maximal effect at 6.3 and 12 pM, respectively (Buscail et al 1994). In the same conditions, the analogues did not affect the proliferation of NIH/3T3 cells expressing *sstr1*. The good correlation between the ability of analogues to inhibit somatostatin binding, to stimulate PTP activity and to inhibit cell proliferation argues for the involvement of a PTP in the growth inhibition mediated by sstr2. Furthermore, both the stimulation of PTP activity and the inhibition of cell proliferation in NIH/3T3 cells induced by RC 160 were suppressed by 1 μM orthovanadate (Fig. 5). Whether the somatostatin-stimulated PTP cells expressing *sstr2* is related to SHPTP1 is currently being addressed.

In conclusion, all these findings suggest that SHPTP1 or an SHPTP1-related PTP is associated with somatostatin receptors at the membrane and could be part of the negative growth signal promoted by sstr2.

Delesque et al

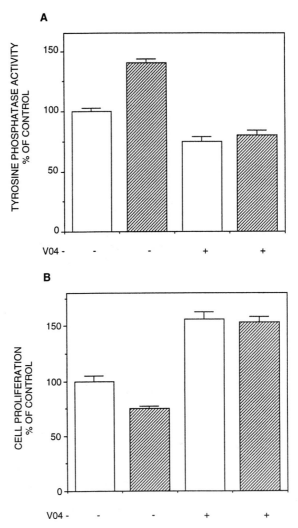

FIG. 5.   Effect of orthovanadate on stimulation of PTP activity and inhibition of cell
proliferation induced by RC 160 in NIH/3T3 cells expressing *sstr2*. Cells were treated
with (hatched bar) or without (open bar) 1 nM RC 160 in the presence or not of 1 μM
orthovanadate (VO^{4-}) for 15 min before PTP assay (A) or for 24 h (B) before
assessment of cell growth as described by Buscail et al (1994). Results are mean $\pm$ SEM
of two experiments.

Acknowledgements

These studies were supported in part by grants: Règion Midi-Pyrénées nb 9208647;
Association pour la Recherche sur le Cancer nb 6755; European Economic Community
nb SC1CT91–0632.

References

Bell GI, Reisine T 1993 Molecular biology of somatostatin receptors. Trends Neurosci 16:34–38

Bensaïd M, Tahiri-Jouti N, Cambillau C et al 1992 Basic fibroblast growth factor induces cell proliferation of a rat pancreatic cancer cell line. Inhibition by somatostatin. Int J Cancer 50:796–799

Bruno JF, Xu Y, Song JF, Berelowitz M 1992 Molecular cloning and functional expression of a brain-specific somatostatin receptor. Proc Natl Acad Sci USA 89:11151–11155

Buscail L, Delesque N, Estève JP et al 1994 Stimulation of tyrosine phosphatase and inhibition of cell proliferation by somatostatin analogues: mediation by human somatostatin receptor subtypes SSTR1 and SSTR2. Proc Natl Acad Sci USA 91:2315–2319

Colas B, Cambillau C, Buscail L et al 1992 Stimulation of a membrane tyrosine phosphatase activity by somatostatin analogues in pancreatic acinar cells. Eur J Biochem 207:1017–1024

Lamberts SWJ, Krenning EP, Reubi JC 1991 The role of somatostatin and its analogs in the diagnosis and treatment of tumors. Endocr Rev 12:450–482

Lewin MJM 1992 The somatostatin receptor in the GI tract. Annu Rev Physiol 54:455–468

Liebow C, Reilly V, Serrano M, Schally AV 1989 Somatostatin analogues inhibit growth of pancreatic cancer by stimulating a tyrosine phosphatase. Proc Natl Acad Sci USA 86:2003–2007

O'Carroll A-M, Lolait SJ, König M, Mahan LC 1992 Molecular cloning and expression of a pituitary somatostatin receptor with preferential affinity for somatostatin-28. Mol Pharmacol 42:939–946

Pan MG, Florio T, Stork PJS 1992 G protein activation of a hormone-stimulated phosphatase in human tumor cells. Science 256:1215–1217

Patel YC, Murthy KK, Escher EE, Banville D, Spiess J, Srikant CB 1990 Mechanism of action of somatostatin: an overview of receptor function and studies of the molecular characterization and purification of somatostatin receptor proteins. Metabolism (suppl 2) 39:63–69

Reubi J-C, Kvols LK, Waser et al B 1990 Detection of somatostatin receptors in surgical and percutaneous needle biopsy samples of carcinoids and islet cell carcinomas. Cancer Res 50:5969–5977

Shen SH, Bastien L, Posner BI, Chrétien P 1991 A protein-tyrosine phosphatase with sequence similarity to the SH2 domain of the protein-tyrosine kinase. Nature 352:736–739

Tahiri-Jouti N, Cambillau C, Viguerie N et al 1992 Characterization of a membrane tyrosine phosphatase in AR42J cells: regulation by somatostatin. Am J Physiol 262:G1007–G1014

Uchida T, Matozaki T, Noguchi T et al 1994 Insulin stimulates the phosphorylation of Tyr538 and the catalytic activity of PTP1C, a protein tyrosine phosphatase with Src homology-2 domains. J Biol Chem 269:12220–12228

Viguerie N, Tahiri-Jouti N, Ayral AM et al 1989 Direct inhibitory effects of a somatostatin analog, SMS 201–995, on AR42J cell proliferation via pertussis toxin-sensitive guanosine triphosphate-binding protein-independent mechanism. Endocrinology 124:1017–1025

Vogel W, Lammers R, Huang J, Ullrich A 1993 Activation of a phosphotyrosine phosphatase by tyrosine phosphorylation. Science 259:1611–1614

Walton KM, Dixon JE 1993 Protein tyrosine phosphatases. Annu Rev Biochem 62:101–120

Yi T, Mui ALF, Krystal G, Ihle JN 1993 Hematopoietic cell phosphatase associates with the interleukin-3 (IL-3) receptor β-chain and down-regulates IL-3-induced tyrosine phosphorylation and mitogenesis. Mol Cell Biol 13:7577–7586

DISCUSSION

Schonbrunn: Have you determined whether pertussis toxin blocks the effect of somatostatin either on phosphatase stimulation or on cell proliferation?

Susini: We tested whether pertussis toxin could block the effect of somatostatin on cell proliferation, but pertussis toxin itself decreases the proliferation of CHO (Chinese hamster ovary) cells by about 50%, so it is not possible to use this agent to find out if G proteins are involved. However, now that we have isolated the complex, we can look for G protein subtypes with different antisera.

In *sstr2*-expressing cells, pertussis toxin inhibits the effect of RC 160 on tyrosine phosphatase activity, indicating that a pertussis toxin-dependent G protein is involved.

Stork: We looked at tyrosine phosphatase stimulation by somatostatin in human MIA PaCa-2 pancreatic cells and found that it was pertussis toxin sensitive (Pan et al 1992). Using cloned sstr1 and sstr2 receptors in the CHO cells that I discussed earlier (Reisine et al 1995, this volume), where we showed inhibition of adenylyl cyclase with sstr1 and sstr2 (Hershberger et al 1994), we also examined whether somatostatin could stimulate phosphatase activity (Florio et al 1994). In agreement with your published work (Buscail et al 1994), we showed that somatostatin could stimulate the phosphatase activity of sstr1. In our hands, this stimulation was pertussis toxin sensitive. We were not able to show stimulation of phosphotyrosine activity associated with sstr2 in this system (Florio et al 1994).

Susini: But you used 1 μM somatostatin. We do not observe stimulation of tyrosine phosphatase activity associated with sstr2 at this concentration either.

Stork: However, this concentration is sufficient to stimulate tyrosine phosphatase activity associated with sstr1. This may suggest that each receptor is coupled to a different phosphatase. This may also explain why rat AR42J pancreatic tumour cells, which express *sstr2*, are insensitive to pertussis toxin (Viguerie et al 1989).

The SHPTP1 (Src homology protein tyrosine phosphatase 1) phosphatase that you identified was cloned and characterized originally as a lymphocyte-specific phosphatase (Shen et al 1991, Matthews et al 1992, Zhao et al 1993). Have you examined the expression of that phosphatase in your pancreatic cells as well as CHO cells and fibroblast lines in which you have demonstrated somatostatin's stimulation of phosphatase?

Susini: Yes, we have observed expression of SHPTP1 in human tumour pancreatic cell lines, CHO cell lines and mouse NIH/3T3 fibroblasts.

Stork: This is in contrast to the published literature on the expression of PTP1C (Shen et al 1991, Matthews et al 1992, Zhao et al 1993).

Susini: It is expressed in hematopoietic cells and also in epithelial cells.

Bruns: You stated that the inhibitory effect of somatostatin analogues is independent of cAMP in FGF (fibroblast growth factor)-stimulated or EGF (epidermal growth factor)-stimulated cell proliferation of AR42J cells. You proved this by pretreating the cells with pertussis toxin (Viguerie et al 1989). Furthermore, you stated that sometimes, the phosphatase activity is sensitive to pertussis toxin treatment (up to 50%). If the somatostatin receptor-dependent effect on cAMP and phosphatase activity are both affected by pertussis toxin, how can you conclude that the antiproliferative effect of octreotide is cAMP independent and that the somatostatin-activated tyrosine phosphatase is the mechanism important for antiproliferative effects?

Susini: We didn't measure the effect of pertussis toxin on tyrosine phosphatase activity in AR42J cells. We only looked at the inhibitory effect of somatostatin on adenylyl cyclase activity which was suppressed by pertussis toxin treatment.

Reisine: Does sstr2 couple to adenylyl cyclase in NIH/3T3 cells?

Susini: No (Buscail et al 1994).

Patel: Does this phosphatase have a transmembrane domain?

Susini: No, it's cytoplasmic.

Patel: This is the first time that a G protein-coupled receptor has been shown to associate with a phosphatase. Can you speculate how this may occur?

Susini: No. There is increasing evidence that stimulatory G protein-coupled receptors, such as bombesin, CCK (cholecystokinin) and gastrin receptors can stimulate tyrosine kinase activity. Alternatively, it is possible that an inhibitory receptor is coupled to a tyrosine phosphatase.

Patel: Does ligand binding induce a conformational change?

Susini: This phosphatase can bind to phosphorylated receptors via its SH2 domain and to insulin receptors via its C-terminal domain (Uchida et al 1994), so it's possible that there is an association between the C-terminal domain of the phosphatase and a G protein-dependent receptor. Coupling between the tyrosine phosphatase and somatostatin receptor is observed in the presence and absence of the ligand, so the complex is present on the membrane.

Patel: Are ligand binding, stimulation of phosphatase and inhibition of cell proliferation functionally coupled in the case of sstr2 and sstr5?

Susini: Only sstr2, not sstr5.

Montminy: How much phosphatase activity do you recover from the cell lysate when you purify it on your column?

Susini: The specific activity in the immunopurified material is about 9 nmol/mg per min. This corresponds to an increase of about 50-fold over that observed in solubilized material.

Montminy: What per cent of total phosphatase activity is in your column fraction?

Susini: I don't know. There is another SH2 tyrosine phosphatase called PTP1D (Kazlauskas et al 1993) that can also bind to growth factor receptors, like PDGF (platelet-derived growth factor) and IGF (insulin-like growth factor). Only 0.5–1% of total PTP1D activity is coupled to the receptors, so it is possible that only a small amount of phosphatase couples to the somatostatin receptor.

Stork: Our studies indicate that somatostatin can stimulate about a 50% increase in phosphatase activity (Pan et al 1992, Florio et al 1994), suggesting that it is recruiting more than a small percentage of SH2 phosphatase.

Your gel filtration studies showed that the major peak of somatostatin phosphatase activity elutes as a single species (Colas et al 1992). Does your SHPTP1 antibody detect anything in this fraction?

Susini: We haven't tried this antibody. We have only tested the immunoreactivity of the fractions using a polyclonal antibody raised against a conserved sequence of PTP.

Stork: Does somatostatin stimulate the activity of the phosphatase that is immunoprecipitated with SHPTP1 antiserum?

Susini: We haven't tried this.

Berelowitz: Kahn (1994) has established that insulin receptor targets have an SH2 binding domain consensus sequence. Does the somatostatin receptor sstr2 have that consensus sequence?

Susini: No.

Kleuss: Does your phosphatase have a pleckstrin homology domain?

Susini: As far as I know, this protein–protein interaction domain is not present in SHPTP1.

Schonbrunn: Have you looked at the effect of somatostatin on tyrosine phosphorylation in whole cells?

Susini: We observe stimulation on homogenates or crude membranes, so we stimulate cells for 15 min first and then prepare crude membranes to measure tyrosine phosphatase activity.

Schonbrunn: Can you stimulate your membranes without stimulating the cells and still see an increase in phosphatase activity?

Susini: We can stimulate the tyrosine phosphatase activity in membranes, but it's more difficult.

Schonbrunn: One problem is whether phosphorylation is something that you see because of your *in vitro* conditions or whether it actually occurs in a whole cell. That is why I asked whether you had looked at changes in phosphorylation of whole cells. Can you prelabel cells with $^{32}PO_4$, stimulate them with, for example, EGF, with and without somatostatin, and then look at the phosphorylation state of the EGF receptor?

Susini: Yes, we've done this. We observed a decrease in the phosphorylation state of the EGF receptor after pretreatment with somatostatin analogues.

Reichlin: One of the aims of the people working on somatostatin analogues is to find one analogue which is selectively antiproliferative, so that high doses can be used for cancer suppression. Can you dissociate the tyrosine phosphatase activity of any somatostatin analogue from its other actions?

Susini: Stimulation of tyrosine phosphatase activity is observed at a somatostatin analogue concentration in the picomolar range and can be dissociated from the cAMP inhibition that is observed at about 0.1–1 nM. However, all the analogues we tested have the same effect.

Taylor: One problem is that all the analogues that were made historically by David Coy and Biomeasure are all selective for sstr2. Only the two linear peptides, BIM 23052 and BIM 23054, exhibit other selectivities; for sstr5 and sstr3 respectively (Raynor et al 1993a,b). However, the development of different selective analogues might not help in the case of cell proliferation if sstr2 is the only receptor that's involved.

Patel: Your results show that sstr2 is the principle, if not the only, receptor that is coupled to stimulation of phosphatase and may also mediate the antiproliferative response of somatostatin. *sstr2* is also expressed in large amounts in many tumours. Can you explain the variability in the antiproliferative response of somatostatin and its analogues (somatostatin-14, octreotide, RC 160, BIM 23014)? Because all these analogues bind to sstr2 with comparable potency, at least in our hands.

Susini: It may be a difference in the level of phosphatases or in the level of sstr2, since not all tumours express *sstr2*. The efficiency of the coupling between sstr2 and phosphatase may also be important.

Patel: What is the situation in pituitary tumours or neuroendocrine tumours?

Lamberts: Kubota et al (1994) recently investigated the expression of the five somatostatin receptors in human neuroendocrine tumours and found an unpredictable variation of expression. One of these tumours was a carcinoid that did not respond to octreotide *in vivo*. About 95% of carcinoids respond to octreotide, so this was an unusual case. This was also the only tumour which did not express *sstr2*, although it did express *sstr1*, *sstr3* and *sstr4*. This suggests that the efficacy of octreotide may depend, at least in part, on the expression of *sstr2* in tumours.

We have also investigated the differential coupling of antimitotic and anti-hormone secretory actions of octreotide in the pituitary cell line 7315b (Hofland et al 1992). *In vitro*, if these cells have less than about 1000 fmol of binding sites/mg protein, there is an inhibitory effect of octreotide on hormone release, but no antimitotic effect. However, if we upregulate somatostatin receptor expression to about 3000–5000 fmol/mg protein, somatostatin exerts significant antimitotic effects as well. It is possible, therefore, that the different effects of the analogues are dependent on the number of receptors and not on the type of receptor.

Epelbaum: What was the concentration of octreotide?

Lamberts: It was in the nanomolar region for both the inhibitory effects on secretion and mitosis. The maximum inhibitory effect on DNA synthesis was observed with a higher octreotide concentration.

Taylor: We looked at established tumour cell lines and observed, by both binding assays and mRNA detection, that every cell line had sstr2 (Taylor et al 1994). There were a few small-cell lung carcinomas that also had sstr1. We believe, therefore, that the sstr2 is the primary tumour-associated receptor. The receptor concentration fell into three groups: the AR42J pancreatic tumour cell line that had about 2500 fmol/mg protein; the small-cell lung carcinoma that had about 1000 fmol/mg protein and everything else that had about 5–50 fmol/mg protein. Therefore, receptor concentration may be important and you may have to have a certain threshold level of receptor to see an inhibitory effect.

Reichlin: Are you arguing that the response of any cell type to a given stimulation depends on the number of receptors?

Taylor: It is possible. We haven't analysed these cell lines for tyrosine phosphatase activity, only antiproliferative effects.

Reichlin: In malignant melanoma, the MSH (melanocyte-stimulating hormone) receptor can be up-regulated by treating the cells with γ-interferon (Kameyama et al 1989). Is it possible to up-regulate the somatostatin receptors in a similar way?

Lamberts: We are trying to induce a transient up-regulation of somatostatin receptor expression in human tumours in order to carry out radioguided surgery or radiotherapy with isotopes coupled to somatostatin analogues. There are a couple of examples; for instance, somatostatin receptors can be induced in prolactinomas after exposure to oestrogens. Also, ACTH (adenocorticotropic hormone)-secreting pituitary tumours do not normally express somatostatin receptors if the patients have high levels of cortisol, as in the case of Cushing's Disease. However, if you pretreat the patients with the glucocorticoid receptor blocking agent RU486 (200 mg/day), somatostatin receptors are expressed (Lamberts et al 1994).

I also believe that *sstr2*, expressed on the human tumours, mediates both antihormonal and antiproliferative effects.

Bruns: We need a different approach other than just looking at the expression of somatostatin receptors. We need to find specific markers that indicate somatostatin receptor-mediated antiproliferative effects, for example the activation of a phosphatase.

Patel: The mitogenic signal is another variable. Presumably, the somatostatin receptor attenuates the mitogenic signal by stimulating tyrosine phosphatase activity.

Reubi et al (1989) have shown an inverse relationship between the number of somatostatin receptors and EGF receptors in tumours. They suggest that the more malignant tumours are characterized by an increase in the ratio of EGF receptors to somatostatin receptors. Conversely, the presence of more somatostatin receptors in a tumour carries a more favourable prognosis.

Montminy: There is no induction of the classical EGF pathway in cAMP-dependent cell lines, such as the rat thyroid cell line FRTL-5, suggesting that the mitogenic pathways are different in tumours that arise presumably from cAMP induction.

Lamberts: Has anyone studied tyrosine phosphatase activity in pituitary tumour cell lines?

Stork: RNase protection assays show that rat GH_4C_1 pituitary tumour cells express predominantly *sstr1* and little *sstr2*. Conversely, mouse AtT-20 pituitary tumour cells express *sstr1* and not *sstr2* (Florio et al 1994). We found that GH_4C_1 cells, but not AtT-20 cells, were able to sustain a stimulation of phosphatase activity with somatostatin. It would now be interesting to determine whether AtT-20 cells express the SH2-containing phosphatase, or whether some other components are missing. It is possible that the antiproliferative effects of sstr2 are not mediated via a phosphatase since these effects in AR42J pancreatic cells are independent of pertussis toxin (Viguerie et al 1989) and therefore may be acting via a different mechanism. Also, somatostatin was shown by Schally's group (Hierowski et al 1985, Lee et al 1991) to have no effect on phosphatase activity whereas RC 160 does.

Reisine: We can't compare GH_4C_1 cells and AtT-20 cells because AtT-20 somatostatin receptors desensitize quickly, whereas the GH_4C_1 are more resistant to regulation. These two cell lines also have a different morphology— AtT-20 cells are neuronal shaped, whereas GH_4C_1 cells are spherical.

Stork: cAMP may be stimulating a cascade that is independent of growth pathways. The targets of these phosphatases have remained elusive. For instance, when we compare immunoblots of cells with or without somatostatin treatment, using a phosphotyrosine antiserum, we do not detect the loss of specific bands following somatostatin treatment. Somatostatin receptors have also been linked recently to MAP (mitogen-activated protein) kinase activation (Bito et al 1994). This connection links up growth factor pathways and G protein-coupled pathways, providing multiple opportunities for cross-talk between the two pathways via a phosphotyrosine phosphatase.

Berelowitz: In somatotrophs, IGF1 inhibits the secretion of GH (growth hormone) through the IGF1 receptor, and somatostatin inhibits the secretion of GH through sstr2 (Berelowitz et al 1981). Have you looked at the effect of phosphorylation of the IGF1 receptor in pituitary cells?

Susini: No.

Schonbrunn: Do you have an explanation for the biphasic nature of the dose–response of somatostatin to inhibit cell growth? Why is there an effect only within a narrow window of hormone concentration?

Susini: sstr2 can also couple to $InsP_3$ (inositol trisphosphate) formation and high doses of somatostatin can stimulate $InsP_3$ formation. It's possible that high concentrations of Ca^{2+} inhibit tyrosine phosphatase activity.

Reisine: This is a classic response for many peptides, e.g. CCK stimulation of amylase secretion and stimulation of locomotor activity by somatostatin are probably related to desensitization (Raynor et al 1993c).

Schonbrunn: It is always more puzzling to see a biphasic effect when the endpoint being measured is linked directly to the receptor, than when it's complicated by simultaneous activation of multiple signalling pathways. There are not many situations where there is one receptor on a cell giving a biphasic effect on a single signal transduction pathway such as stimulation of adenylyl cyclase.

References

Berelowitz M, Szabo M, Frohman LA, Firestone S, Chu L 1981 Somatomedin-C mediates growth hormone negative feedback by effects on both the hypothalamus and pituitary. Science 212:1279–1281

Bito H, Sakanaka C, Honda Z-I, Shimizu T 1994 Functional coupling of SSTR4, a major hippocampal somatostatin receptor, to adenylate cyclase inhibition, arachadonate release and activation of the mitogen-activated protein kinase cascade. J Biol Chem 269:12722–12730

Buscail L, Delesque N, Estève JP et al 1994 Stimulation of tyrosine phosphatase and inhibition of cell proliferation by somatostatin analogues: mediation by human somatostatin receptor subtypes SSTR1 and SSTR2. Proc Natl Acad Sci USA 91:2315–2319

Colas B, Cambillau C, Buscail L et al 1992 Stimulation of a membrane tyrosine phosphatase activity by somatostatin analogues in pancreatic acinar cells. Eur J Biochem 207:1017–1024

Florio T, Rim C, Pan MG, Stork PJS 1994 The somatostatin receptor SSTR1 is coupled to phospho-tyrosine phosphatase activity in CHO-K1 cells. Mol Endocrinol 8:1289–1297

Hershberger RE, Newman BL, Florio T et al 1994 The somatostatin receptors SSR1 and SSR2 are coupled to inhibition of adenylyl cyclase in CHO cells via pertussis toxin sensitive pathways. Endocrinology 134:1277–1285

Hierowski MT, Liebow C, du Sapin K, Schally 1985 Stimulation by somatostatin of dephosphorylation of membrane proteins in pancreatic cancer MIA PaCa-2 cell lines. FEBS Lett 179:252–256

Hofland LJ, van Koetsfeld PM, Wouters N, Reubi J-C, Lamberts SWJ 1992 Dissociation of antiproliferative and antihormonal effects of the somatostatin analog octreotide on 7315b pituitary tumor cells. Endocrinology 131:571–577

Kahn CR 1994 Insulin action, diabetogenes, and the cause of type II diabetes. Diabetes 43:1066–1084

Kameyama K, Tanaka S, Ishida Y, Hearing VJ 1989 Interferons modulate the expression of hormone receptors on the surface of murine melanoma cells. J Clin Invest 83:213–221

Kazlauskas A, Feng GS, Pawson T, Valius M 1993 The 64 kDa protein that associates with the platelet-derived growth factor receptor beta subunit via Tyr-1009 is the Sh2-containing phosphotyrosine phosphatase Syp. Proc Natl Acad Sci USA 90:6939–6942

Kubota A, Yamada Y, Kagimoto S et al 1994 Identification of somatostatin receptor subtypes and an implication for the efficacy of somatostatin analogue SMS 201-995 in treatment of human endocrine tumours. J Clin Invest 93:1321–1325

Lamberts SWJ, Deherder WW, Krenning EP, Reubi J-C 1994 A role of (labeled) somatostatin analogs in the differential diagnosis and treatment of Cushing's syndrome. J Clin Endocrinol & Metab 78:17–19

Lee MT, Liebow C, Kamer AR, Schally AV 1991 Effects of epidermal growth factor and analogues of luteinizing hormone-releasing hormone and somatostatin on phosphorylation and dephosphorylation of tyrosine residues of specific protein substrates in various tumors. Proc Natl Acad Sci USA 88:1656–1600

Matthews RJ, Bowne DB, Flores E, Thomas ML 1992 Characterization of hematopoietic intracellular protein tyrosine phosphatases: description of a phosphatase containing an SH2 domain and another enriched in proline-, glutamic acid-, serine-, and threonine-rich sequences. Mol Cell Biol 12:2396–2405

Pan MG, Florio T, Stork PJS 1992 G protein activation of a hormone-stimulated protein-tyrosine phosphatase in human tumor cells. Science 256:37–39

Raynor K, Murphy W, Coy DH et al 1993a Cloned somatostatin receptors: identification of subtype-selective peptides and demonstration of high-affinity binding of linear peptides. Mol Pharmacol 43:838–844

Raynor K, O'Carroll A-M, Kong HY et al 1993b Characterization of cloned somatostatin receptors sstr4 and sstr5. Mol Pharmacol 44:385–392

Raynor K, Lucki I, Reisine T 1993 Somatostatin (1) receptors in the nucleus accumbens selectively mediate the stimulatory effect of somatostatin on locomotor activity in rats. J Pharmacol Exp Ther 265:67–73

Reisine T, Woulfe D, Raynor K et al 1995 Interaction of somatostatin receptors with G proteins and cellular effector systems. In: Somatostatin and its receptors. Wiley, Chichester (Ciba Found Symp 190) p 160–170

Reubi J-C, Horisberger U, Lang W 1989 Coincidence of EGF receptors and somatostatin receptors in meningiomas but inverse differentiation-dependent relationship in glial tumours. Am J Pathol 134:337–344

Shen S, Bastien L, Posner B, Chretien P 1991 A protein-tyrosine phosphatase with sequence similarity to the SH2 domain of the protein tyrosine kinases. Nature 352:736–739

Taylor JE, Theveniau M, Bashirzadeh R, Reisine T, Eden PA 1994 Detection of somatostatin receptor subtype 2 (sstr2) in established tumors and tumor cell lines: evidence for sstr2 heterogeneity. Peptides 15:1229–1236

Uchida T, Matozaki T, Noguchi T et al 1994 Insulin stimulates the phosphorylation of Tyr538 and the catalytic activity of PTP1C, a protein tyrosine phosphatase with Src homology-2 domains. J Biol Chem 269:12220–12228

Viguerie N, Tahiri-Jouti N, Cambillau AM et al 1989 Direct inhibitory effects of a somatostatin analog, SMS 201-995, on AR4-2J cell proliferation via pertussis toxin-sensitive guanosine triphosphate-binding protein-independent mechanism. Endocrinology 124:1017–1025

Zhao Z, Bouchard P, Diltz CD, Shen S, Fisher EH 1993 Purification and characterization of a protein tyrosine phosphatase containing SH2 domain. J Biol Chem 268:2816–2820

Function and regulation of somatostatin receptor subtypes

Agnes Schonbrunn, Yi-Zhong Gu, Patricia J. Brown and David Loose-Mitchell

Department of Pharmacology, University of Texas Medical School in Houston, P.O. Box 20708, Houston, TX77225, USA

Abstract. The five known somatostatin receptors serve unique biological roles by virtue of their tissue-specific expression and particular biochemical properties. However, the function of any individual receptor in its normal physiological milieu is not understood. Studies to address this problem have been difficult because tissues and cell lines often express multiple somatostatin receptors and, in the absence of receptor-selective somatostatin analogues, the actions of individual receptors cannot be identified. Moreover, the biological and biochemical actions of somatostatin receptors depend on their cellular environment, so that the behaviour of a receptor expressed in heterologous cells does not necessarily mimic that of endogenous receptors. We have developed two approaches to examine somatostatin receptors which circumvent these problems. Using a biotinylated somatostatin analogue for affinity purification, we isolated somatostatin receptors together with associated G proteins. Subsequent analysis of the purified complex with G protein-specific antibodies showed that the somatostatin receptors in AR42J cells preferentially couple with two pertussis toxin-sensitive G proteins: $G_i\alpha1$ and $G_i\alpha3$. To examine individual receptor types, we developed receptor-specific antibodies and used them to show that both sstr1 and sstr2 proteins were present in the GH_4C_1 pituitary cell line whereas AR42J cells contained sstr2 but not sstr1. Immunoprecipitation of receptor–G protein complexes from GH_4C_1 cells showed that sstr1 and sstr2 are both coupled to pertussis toxin-sensitive G proteins, in contrast to the results observed when these receptors are overexpressed in some non-endocrine cells. We also showed that the somatostatin receptors in GH_4C_1 cells are subject to both homologous and heterologous hormonal regulation. The mechanisms involved in the regulation of different receptor types are now being characterized using the receptor-specific antibodies to isolate the individual receptor proteins. Elucidating signal transduction by endogenous somatostatin receptors as well as their hormonal regulation will be critical for understanding the functions of these receptors in the different physiological targets of somatostatin.

1995 Somatostatin and its receptors. Wiley, Chichester (Ciba Foundation Symposium 190) p 204–221

Somatostatin was discovered over 20 years ago and subsequently shown to regulate a wide variety of endocrine, exocrine, gastrointestinal and neuronal functions (Reichlin 1983, Yamada & Chiba 1989, Epelbaum 1986). Within the next few years, the first high affinity somatostatin receptors were identified (Schonbrunn & Tashjian 1978) and these receptors were shown to modulate multiple signal transduction pathways in target cells including adenylyl cyclase, ion channels and phosphatases (Patel & Tannenbaum 1990, Melmed 1993). Now, five somatostatin receptor genes have been cloned (sstr1–5). When these receptors are expressed individually in mammalian host cells, they are able to regulate many of the signal transduction pathways previously implicated as mediating somatostatin's effects, as well as some new ones that had previously not been recognized (Table 1). These new developments have engendered enormous excitement in the field; however, the many possible permutations between the different receptor types and the potential down-stream effectors have greatly increased the difficulty of defining the mechanisms of somatostatin action in each of its physiological targets. Hence, recent research has focused on elucidating the physiological and biochemical functions of each of the individual receptor subtypes with the goal of answering three fundamental questions. (1) Which somatostatin receptor is responsible for eliciting each of the biological effects produced by somatostatin? (2) What signal transduction machinery does each somatostatin receptor activate and to what extent is its mechanism of action cell type dependent? (3) How are the individual receptor types regulated?

Two complementary strategies have been utilized to address these questions. Studies of endogenous somatostatin receptors ensure that the properties of the receptor under study are being characterized in their native environment, in the presence of all the cellular machinery required for normal function. However, such studies are difficult because most somatostatin-responsive tissues and many clonal cell lines (Patel et al 1994, Schonbrunn & Loose-Mitchell 1993, Brown et al 1993) express several types of somatostatin receptor. Specific agonists or antagonists are not available for the different receptors, so it is not possible to determine which receptor is responsible for a particular biological or biochemical action of somatostatin. Further, because endogenous somatostatin receptors are generally expressed at very low levels, biochemical studies are limited by the availability of material for study.

Many of the problems inherent in the study of endogenously expressed receptors can be overcome by transfecting receptor-negative cells with expression plasmids encoding individual somatostatin receptors. Such cells are engineered to express only a single receptor type, so there is no confusion about which receptor is being investigated. Further, the use of strong promoters ensures that receptor expression occurs at high levels, often 10 to 100 times that found in normal tissues. These high receptor densities facilitate biochemical studies by providing a ready source of receptor protein. Nonetheless, the host cells used for such studies provide a foreign environment for the expressed receptors and may not contain the same signal transduction machinery present in the normal target cells. For example, none of the cell types in which somatostatin receptors have been expressed to date would be expected to show an electrophysiological response to hormone stimulation because they lack the ion channels known to be regulated by somatostatin in electrically excitable cells. Moreover, the signal transduction pathways activated by receptors depend on receptor density (Milligan 1993). Quantitative analysis showed that the potency of β-adrenergic agonists to stimulate adenylyl cyclase varied from 200 nM to 0.2 nM as receptor density was increased from 5 to 5000 fmol/mg protein (Whaley et al 1994). Therefore, low affinity interactions between receptors and down-stream effectors, which are insufficient to produce changes in second messenger levels at normal receptor densities and agonist concentrations, can become functionally important in cells expressing high levels of receptor.

In this review, we summarize two different aspects of our studies, focusing on the properties of endogenously expressed receptors in secretory cells. Firstly, we have begun to characterize the G proteins coupled to endogenous somatostatin receptors in endocrine and exocrine cells. Secondly, we have examined both homologous and heterologous regulation of somatostatin receptors.

Signal transduction by somatostatin receptors

Early studies indicated that signal transduction by somatostatin receptors was mediated by G proteins, a family of ubiquitous GTPases (Spiegel et al 1992). These G proteins then act as transducers between the membrane somatostatin receptors and their intracellular effectors, leading either to inhibition, as with adenylyl cyclase and voltage-dependent Ca^{2+} channels, or to activation, as with K^+ channels (Schonbrunn 1990). The nature of the effector proteins regulated by particular receptors depends on the G proteins which they activate, with each G protein being able to recruit one or a limited number of effectors. Hence the G protein specificity exhibited by individual somatostatin receptors is a critical determinant in signal transduction.

Several independent lines of evidence support the involvement of G proteins in somatostatin action. Guanine nucleotides regulate the interconversion of somatostatin receptors between two states having either low or high affinity for hormone, and somatostatin stimulates the activity of high affinity membrane GTPases (Koch & Schonbrunn 1984, Offermanns et al 1989). Both these results

TABLE 1   Some second messengers regulated by somatostatin receptors

Second Messenger	Endogenous receptors	Transfected receptors
cAMP	↓	↔ or ↓
Cytosolic Ca^{2+}	↓	↑
IP_3 (inositol trisphosphate)	↔	↓
Diacylglycerol	↔	↑

show that a close association exists between the receptors and GTP binding proteins. Moreover, pertussis toxin, which covalently modifies and thereby inactivates the G_i/G_o family of G proteins, markedly decreases the binding of somatostatin to its membrane receptor (Koch et al 1985), inhibits somatostatin stimulation of GTPase activity (Offermanns et al 1989) and blocks the inhibitory effects of somatostatin on second messenger levels as well as on hormone secretion (Koch et al 1985, Yajima et al 1986). Finally, the deduced amino acid sequences of the cloned somatostatin receptors are highly homologous to that of other G protein-coupled receptors (Bell et al 1995, this volume). This receptor family shares characteristic structural features including seven putative transmembrane domains and three intracellular regions implicated in G protein coupling and receptor regulation. However, although pertussis toxin blocks some of somatostatin's actions, it does not inhibit them all. For example, neither somatostatin stimulation of Na^+–H^+ exchange nor somatostatin inhibition of pancreatic cell proliferation is blocked by pertussis toxin (Barber et al 1989, Viguerie et al 1989). Hence, endogenous somatostatin receptors do not act only via pertussis toxin-sensitive G proteins. Moreover, in some over-expression systems, it has not been possible to demonstrate G protein coupling by certain somatostatin receptors (Bell & Reisine 1993).

Identification of G proteins associated with somatostatin receptors by affinity purification of the somatostatin receptor–G protein complex

To elucidate the first step in the signal transduction pathway activated by somatostatin, we have begun to identify the G protein subunits which couple to endogenous somatostatin receptors. Our initial strategy, which is shown in Fig. 1, involved affinity purification of somatostatin receptors in association with G proteins (Brown & Schonbrunn 1993). In order to stabilize the complex between the membrane receptor and the G proteins that it activates, we saturated the receptors with agonist prior to solubilization. The fact that the receptor–G protein complex was formed while these proteins were in their native membrane environment rather than in detergent solution ensured that the complex reflected the normal specificity of the receptor for different G proteins. Instead of the native hormone, a biotinylated somatostatin analogue, (N-biotinyl -Leu[8],D–Trp[22],Tyr[25])somatostatin-28, was used as the receptor agonist because this peptide is able to bind simultaneously both to the receptor and to streptavidin with high affinity (Schonbrunn et al 1993). Following analogue binding, the ligand–receptor–G protein complex was solubilized with detergent using a procedure which removed essentially all of the somatostatin receptor from the membranes in an active and highly stable form (Brown & Schonbrunn 1993). The complex was then adsorbed to immobilized streptavidin. After extensive washing to remove non-specifically adsorbed proteins, we eluted the streptavidin column with guanine nucleotides in order to selectively dissociate the receptor

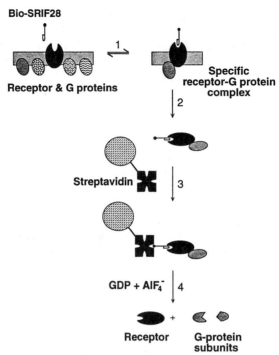

FIG. 1.   Affinity purification of a somatostatin receptor–G protein complex. Step 1: Membranes are incubated with the biotinylated somatostatin analogue, (N-biotinyl-Leu[8],D–Trp[22],Tyr[25])somatostatin-28 (abbreviated Bio-SRIF28), to occupy receptors and to stabilize the specific receptor–G protein complex. Step 2: The intact ligand–receptor–G protein complex is solubilized with detergent. Step 3: The detergent extract is applied to streptavidin agarose, which adsorbs the Bio-SRIF28–receptor complex as well as other biotinylated proteins. Step 4: Following extensive washing, application of GDP and AlF_4^- causes dissociation of G proteins from the receptor and conversion of the adsorbed somatostatin receptors to a low affinity state. Thus, specific elution of the receptors and G protein subunits occurs, whereas Bio-SRIF28 and other biotinylated proteins remain bound to the streptavidin agarose. Modified from Brown & Schonbrunn (1993).

from the biotinylated ligand. Thus, two activities of the receptor played an integral part in the purification of the receptor–G protein complex: the high affinity binding of agonist in the absence of guanine nucleotides and the decrease in ligand binding affinity produced when the receptor was uncoupled from G proteins by the addition of guanine nucleotides. The proteins eluted from the column were then subjected to SDS polyacrylamide gel electrophoresis and analysed by immunoblotting with specific G protein antibodies to identify the G protein subunits which co-purify with the receptor (Brown & Schonbrunn 1993).

The purification procedure was initially carried out with rat AR42J pancreatic acinar cells because they express high levels of somatostatin receptors.

Even though all three αi subunits are present in AR42J cells, the α_{i1} and α_{i3} subunits were preferentially associated with somatostatin receptors (Brown & Schonbrunn 1993). Furthermore, although the cells contain both $\beta 35$ and $\beta 36$ subunits, the receptors appeared to interact preferentially with $\beta 36$. Therefore, somatostatin receptors in this pancreatic cell line demonstrate specificity in their interactions with both the α and β subunits of G proteins. Luthin et al (1993) used the same general strategy to identify the G proteins associated with somatostatin receptors in the rat GH_4C_1 pituitary cell line and showed that those receptors interacted preferentially with α_{i2} and α_{i3}. To determine whether the difference in G protein coupling specificity in these two cell types could be explained by the presence of different somatostatin receptor types, we characterized the receptor mRNAs present in AR42J and GH_4C_1 cells using the reverse transcriptase polymerase chain reaction, followed by DNA hybridization with receptor-specific oligonucleotides. We found that AR42J cells express *sstr2* but not *sstr1* mRNA whereas GH_4C_1 cells express similar levels of both *sstr1* and *sstr2*. This analysis does not show that the receptor proteins are also expressed nor does it indicate the relative abundance of the different receptor proteins. Nonetheless, the presence of several different somatostatin receptors will complicate the interpretation of the affinity purification experiments because ligand-based affinity chromatography will purify G proteins associated with all somatostatin receptors present and cannot be used to characterize the G proteins which interact with specific somatostatin receptor types. Hence, alternative approaches were required to determine whether a single receptor type is coupled to the same G proteins in different target cells and whether different somatostatin receptors in the same cell couple to different G proteins.

Use of receptor-specific antibodies to identify and purify endogenous somatostatin receptors

To isolate individual somatostatin receptor types, we generated type-specific somatostatin receptor antibodies and developed procedures for the efficient immunoprecipitation of somatostatin receptors (Brown et al 1993, Gu et al, unpublished observations). We initially identified unique sequences within sstr1, 2 and 4, and then used KLH-coupled (keyhole limpet haemocyanin-coupled) polypeptides corresponding to these sequences as immunogens. The resulting rabbit polyclonal antibodies are each able to efficiently immunoprecipitate somatostatin receptors from CHO (Chinese hamster ovary) cells expressing the corresponding receptor type but not from cells expressing heterologous receptor types. For example, the sstr1 peptide antibody routinely precipitated 70% of the soluble ($[^{125}I]Tyr^{11}$)somatostatin–receptor complex prepared from CHO-K1 cells expressing *sstr1*, but precipitated less than 1% of the complex from cells expressing other somatostatin receptors. Similar results were obtained with the sstr2 and sstr4 antibodies. Preimmune sera were ineffective. Further, addition of antigen peptide during the incubation with the corresponding antiserum completely inhibited immunoprecipitation of the receptor, whereas unrelated peptides had no effect.

To characterize the nature of the receptor protein recognized by the different antibodies, we prepared membranes from untransfected CHO-K1 cells, as well as CHO cells expressing individual somatostatin receptor types, and affinity labelled the receptors with ($[^{125}I]Tyr^{11}$,ANB-Lys^4)somatostatin (Azido-somatostatin). We have previously shown that this photoaffinity analogue specifically labels high affinity somatostatin receptors (Brown & Schonbrunn 1990). We found that Azido-somatostatin covalently labelled broad bands at 60 and 85 kDa in sstr1- and sstr2-expressing CHO cells, respectively, but there was no specific labelling in the parental CHO-K1 cell line. Labelling was completely inhibited by 100 nM somatostatin. Further, both the apparent molecular weights of the specifically labelled proteins and their diffuse migration in SDS polyacrylamide gels were consistent with the presence of multiple glycosylation sites in these receptor types.

When photoaffinity-labelled receptors were immunoprecipitated, each antiserum precipitated only the corresponding receptor types. For example, using the sstr1 antiserum, no specifically labelled bands were observed following SDS polyacrylamide gel electrophoresis of the immunoprecipitates from photoaffinity-labelled CHO-K1 or sstr2-expressing CHO-K1 cells. However, a broad 60 kDa band was evident in the immunoprecipitate from CHO sstr1-expressing cells and immunoprecipitation of this protein was completely blocked by antigen peptide. These studies demonstrated that each somatostatin receptor peptide antibody could specifically precipitate the appropriate somatostatin receptor protein.

To our knowledge, this is the first time that antibodies have been successfully generated to sstr1 and sstr4. Moreover, the three antibodies that we have generated are the first to be able to immunoprecipitate any of the somatostatin receptors efficiently, since 60–80% of the soluble receptors are routinely purified. These antisera therefore provide unique and powerful new tools for studies of the individual somatostatin receptor proteins.

High affinity somatostatin receptors have been identified in both GH_4C_1 and AR42J cells by ligand binding (Schonbrunn & Tashjian 1978, Viguerie et al 1988). To determine the type of somatostatin receptor present in these cells, we incubated membranes from each cell type with ($[^{125}I]Tyr^{11}$)somatostatin, solubilized the ligand–receptor complex and immunoprecipitated it with sstr1 and sstr2 antibodies. The sstr1 antiserum precipitated about 45% of the total ($[^{125}I]Tyr^{11}$)somatostatin–receptor complex from GH_4C_1 cells, but less than 1% of the complex from AR42J cells. Hence, functional sstr1 protein is expressed only by the former. In contrast, sstr2 antiserum precipitated 55% of the ligand receptor complex from GH_4C_1 cells and 80% of the complex from AR42J cells. Thus, both cell types contain sstr2 protein, and this isoform represents most, if not all, of the receptors in AR42J cells. As would be expected, immunoprecipitation of the receptors from GH_4C_1 cells by the sstr1 and sstr2 antibodies was additive: over 90% of the receptor was precipitated when both antibodies were added simultaneously. Therefore, use of the receptor-specific antisera identified the functional receptor proteins present in GH_4C_1 and AR42J cells and allowed the separation of the individual receptor types in the cell line which expressed multiple somatostatin receptors.

To determine whether these antibodies coprecipitated the receptor with associated G proteins, we took advantage of the ability of GTP analogues to uncouple the receptor–G protein complex and thereby decrease receptor affinity for agonists. Membranes from CHO-K1 cells expressing either sstr1 or sstr2 were prelabelled with ($[^{125}I]Tyr^{11}$)somatostatin, solubilized and immunoprecipitated with receptor-specific antisera. Subsequently, the immunoprecipitate was resuspended in buffer and incubated with GTPγS. The observation that the guanine nucleotide-stimulated hormone dissociation from the immunoprecipitated receptor demonstrated that the receptor remained functionally associated with G proteins. Similarly, when the endogenous sstr1 and sstr2 proteins were immunoprecipitated from GH_4C_1 cells individually, hormone binding to the precipitated receptors was markedly inhibited by both GTPγS and pertussis toxin. These results show that sstr1 and sstr2 are both coupled to pertussis toxin-sensitive G proteins in the GH_4C_1 cell line. Similar experiments showed that sstr2 is also primarily coupled to pertussis toxin-sensitive G proteins in AR42J cells. Thus, the observation that these receptors are not G protein coupled in some expression systems probably reflects the foreign environment provided by those hosts rather than their normal signalling properties.

The antibodies we have described now allow the purification of individual receptor–G protein complexes from both normal cellular targets and from transfected cells. Therefore, the G proteins associated with these receptors in different cell types can be determined. Such studies, which are currently in progress, should further elucidate the signal transduction pathways utilized by these receptors in different cell types.

Somatostatin receptor regulation

In addition to somatostatin signalling, we have also been interested in the mechanisms involved in the hormonal regulation of cellular responsiveness to somatostatin. The recognition that multiple somatostatin receptors are expressed in most somatostatin-responsive tissues and cells has forced a re-evaluation of previous studies showing that somatostatin receptors, as measured by high affinity ligand binding, are subject to both homologous and heterologous regulation. In GH_4C_1 cells which, as indicated above, express both sstr1 and sstr2, total receptor number is known to be regulated by somatostatin itself

(Presky & Schonbrunn 1988) as well as by thyrotropin-releasing hormone, phorbol esters, glucocorticoids and thyroid hormone (Schonbrunn & Tashjian 1980, Hinkle et al 1981, Osborne & Tashjian 1982, Schonbrunn 1982). However, the specific receptors affected in each case remain to be characterized. Here we will briefly review our studies aimed at understanding whether somatostatin receptors become desensitized following chronic somatostatin treatment and discuss the mechanisms implicated in the homologous regulation of somatostatin receptors.

Most peptide hormones and growth factors undergo rapid, receptor-mediated internalization following binding to cell surface receptors. In contrast, in both GH_4C_1 pituitary cells and rat RINm5F insulinoma cells, receptor-bound somatostatin remains on the cell surface until it slowly dissociates (Presky & Schonbrunn 1986, Sullivan & Schonbrunn 1986). Therefore, to determine the effect of somatostatin pretreatment on the binding and functional properties of the receptor, the peptide bound to receptors during pretreatment must be removed without damaging the receptor. We have previously shown that pre-incubating cells for 30 min with a saturating concentration of non-labelled somatostatin inhibits the subsequent binding of ($[^{125}I]Tyr^1$)somatostatin (Presky & Schonbrunn 1988). This inhibition could be due either to the continued occupancy of cell surface receptors by the unlabelled hormone or to rapid receptor down-regulation. However, incubation of cells with pH 5 buffer for 10 min restores ($[^{125}I]Tyr^1$)somatostatin binding in the pretreated cells to control values indicating that receptor levels were not altered (Presky & Schonbrunn 1988). Washing non-pretreated cells with a pH 5 buffer does not affect receptor binding, demonstrating that the receptors are not damaged by this wash. Therefore, exposure of cells to low pH following somatostatin pretreatment results in the rapid release of prebound somatostatin from receptors. Similar studies showed that the pH 5 wash does not perturb coupling of the somatostatin receptors to adenylyl cyclase (Presky & Schonbrunn 1988). Hence this wash was used in all subsequent experiments to dissociate the somatostatin–receptor complex rapidly after chronic somatostatin pretreatment of cells.

To determine whether chronic somatostatin treatment results in receptor down-regulation, we incubated GH_4C_1 cells with saturating concentrations of somatostatin for various time periods at 37 °C, washed them with pH 5 buffer to dissociate prebound peptide and incubated them with ($[^{125}I]Tyr^1$)somatostatin to measure receptor binding. Somatostatin pretreatment produces a time-dependent increase in ($[^{125}I]Tyr^1$)somatostatin binding which at 20 h reaches a maximum value of 200% that of untreated cultures (Presky & Schonbrunn 1988). Scatchard analysis demonstrated that the number but not the affinity of the receptors is altered (Presky & Schonbrunn 1988). Furthermore, pretreatment of cells with $(-)-N^6$-2-(phenylisopropyl)-adenosine (PIA) and carbamyl choline chloride (carbachol), two other agents which mimic somatostatin's action of decreasing intracellular cAMP and Ca^{2+} concentrations by binding to adenosine and acetylcholine receptors on these cells, did not mimic the somatostatin-induced increase in receptor number (Presky & Schonbrunn 1988). Hence, occupancy of the somatostatin receptor appeared to be necessary for somatostatin receptor up-regulation.

To investigate whether somatostatin pretreatment increased receptor synthesis, we examined the effect of cycloheximide on somatostatin receptor regulation (Fig. 2). As expected, chronic incubation of cells with this protein synthesis inhibitor decreased ($[^{125}I]Tyr^1$)somatostatin binding, demonstrating that it inhibited the synthesis of new receptors effectively. However, cycloheximide did not affect the twofold increase in ($[^{125}I]Tyr^1$)somatostatin binding produced by somatostatin pretreatment (Fig. 2). Thus, the increase in receptor number could not be due to enhanced receptor synthesis, suggesting that somatostatin may either reduce receptor degradation or cause slow redistribution of pre-existing receptors to the plasma membrane (Presky & Schonbrunn 1988).

To determine whether desensitization to somatostatin occurred, we also examined the effects of chronic somatostatin pretreatment on the ability of somatostatin to inhibit adenylyl cyclase (Presky & Schonbrunn 1988). Preincubating GH_4C_1 cells with somatostatin increased the stimulatory effect of vasoactive intestinal peptide (VIP) and forskolin on cAMP formation. However, it did not alter either basal cAMP levels or the potency of somatostatin to inhibit cAMP accumulation in a subsequent challenge. Although the twofold increase in receptor levels produced by somatostatin pretreatment would indicate that cellular responsiveness might be correspondingly increased, it is not possible

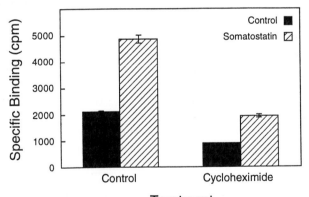

FIG. 2.   Effect of cycloheximide on somatostatin receptor modulation. GH_4C_1 cells were incubated in serum free medium containing 10 µg/ml cycloheximide for 60 min at 37 °C. Somatostatin 100 nM was then added to the dishes shown and the incubation was continued for an additional 24 h at 37 °C. The cells were subsequently rinsed with pH 5 buffer to remove receptor-bound somatostatin, and ($[^{125}I]Tyr^1$)somatostatin binding was then determined. In parallel dishes, quantitation of $[^3H]Leucine$ incorporation after the 60 min cycloheximide treatment showed that protein synthesis was inhibited by 82%. Modified from Presky & Schonbrunn (1988).

to detect a twofold change in somatostatin potency in the assays available (Presky & Schonbrunn 1988). Nonetheless, these studies clearly show that a 24 h somatostatin pretreatment does not desensitize GH_4C_1 cells to a subsequent somatostatin challenge (Presky & Schonbrunn 1988).

The observation that chronic somatostatin exposure does not cause homologous desensitization in GH_4C_1 cells and increases rather than decreases somatostatin receptor number differs markedly from the rapid receptor desensitization which has been reported for many other peptide, growth factor and neurotransmitter receptors. However, this behaviour mimics several of somatostatin's effects *in vivo*, which also do not desensitize (Hugues et al 1986, Tannenbaum et al 1989). Nonetheless, somatostatin receptor desensitization has been observed in some cells although in many cases this desensitization is unusual in that it occurs over the course of days or weeks, rather than over minutes or hours (Smith et al 1984, Heisler & Srikant 1985, Mahy et al 1988, Koper et al 1990). Hence, it will be interesting to determine whether different somatostatin receptors are subject to different regulatory mechanisms as well as whether the regulation of a particular receptor differs among target cells.

Our results indicate that somatostatin receptors are regulated post-translationally in GH_4C_1 pituitary cells. Studies presented by Berelowitz et al (1995, this volume) have demonstrated that receptor mRNA levels are also subject to hormonal regulation in the closely related GH_3 cell line. Therefore, multiple mechanisms are likely to be involved in somatostatin receptor regulation and elucidating these processes will require antibodies to quantitate receptor protein and specific assays to monitor receptor function as well as molecular tools to characterize transcriptional and translational effects.

Summary

We have described two approaches to elucidate the signal transduction mechanisms activated by somatostatin receptors. Receptor purification using either ligand based affinity chromatography or immunoprecipitation with receptor subtype specific antibodies demonstrated that both sstr1 and sstr2 are coupled to pertussis toxin-sensitive G proteins in pancreatic exocrine and pituitary endocrine cells. Identification of the G proteins associated with pancreatic somatostatin receptors demonstrated that these receptors interact with specific α and β G protein subunits. Continuing studies with the receptor-specific antibodies are aimed at comparing the G protein specificity exhibited by different somatostatin receptors in target cells. We also showed that somatostatin treatment of GH_4C_1 pituitary cells leads to an increase in somatostatin receptor levels independently of new protein synthesis. The nature of the somatostatin receptors involved, as well as the mechanisms by which the receptor proteins are regulated, are the subject of ongoing investigations.

Acknowledgements

We thank Drs Allen Spiegel and Andrew Shenker for providing G protein antibodies and Drs Philip Stork, Michael Berelowitz and Jack Bruno for providing sstr-expressing CHO cell lines. We also thank Fengyu Jiang, Anthony Garcia and Ellen McAllister who contributed their technical expertise to the projects described. This work was supported by Grant DK32234 from the National Institute of Arthritis, Diabetes, Digestive and Kidney Diseases.

References

Barber DL, McGuire ME, Ganz MB 1989 Beta-adrenergic and somatostatin receptors regulate Na^+-H^+ exchange independent of cAMP. J Biol Chem 264:21038–21042

Bell GI, Reisine T 1993 Molecular biology of somatostatin receptors. Trends Neurosci 16:34–38

Brown PJ, Lee AB, Norman MG, Presky DH, Schonbrunn A 1990 Identification of somatostatin receptors by covalent labeling with a novel photoreactive somatostatin analog. J Biol Chem 265:17995–18004

Brown PJ, Schonbrunn A 1993 Affinity purification of a somatostatin receptor–G protein complex demonstrates specificity in receptor–G protein coupling. J Biol Chem 268:6668–6676

Brown PJ, Gu Y-Z, Schonbrunn A 1993 Selective immunoprecipitation of somatostatin receptor subtype 1 with an antipeptide antibody. Proc 75th Meet Endocr Soc p 410 (abstr)

Epelbaum J 1986 Somatostatin in the central nervous system: physiology and pathological modifications. Prog Neurobiol 27:63–100

Heisler S, Srikant CB 1985 Somatostatin-14 and somatostatin-28 pretreatment down-regulate somatostatin-14 receptors and have biphasic effects on forskolin-stimulated cyclic adenosine $3'5'$-monophosphate synthesis and adrenocorticotropin secretion in mouse anterior pituitary tumour cells. Endocrinology 117:217–225

Hinkle PM, Perrone MH, Schonbrunn A 1981 Mechanism of thyroid hormone inhibition of thyrotropin-releasing hormone action. Endocrinology 108:199–205

Hugues JN, Enjalbert A, Moyse E et al 1986 Differential effects of passive immunization with somatostatin antiserum on adenohypophysial hormone secretions in starved rats. J Endocrinol 109:169–174

Koch BD, Schonbrunn A 1984 The somatostatin receptor is directly coupled to adenylate cyclase in GH_4C_1 pituitary cell membranes. Endocrinology 114:1784–1790

Koch BD, Dorflinger LJ, Schonbrunn A 1985 Pertussis toxin blocks both cyclic AMP-mediated and cyclic AMP-independent actions of somatostatin: evidence for coupling of Ni to decreases in intracellular free calcium. J Biol Chem 260:3138–3145

Koper JW, Hofland LJ, Van Koetsveld PM, den Holder F, Lamberts SWJ 1990 Desensitization and resensitization of rat pituitary tumor cells in long-term culture to the effects of the somatostatin analogue SMS201–995 on cell growth and prolactin secretion. Cancer Res 50:6238–6242

Luthin DR, Eppler CM, Linden J 1993 Identification and quantification of G_i-type GTP binding proteins that copurify with a pituitary somatostatin receptor. J Biol Chem 268:5990–5996

Mahy N, Woolkalis M, Manning D, Reisine T 1988 Characteristics of somatostatin desensitization in the pituitary tumor cell line AtT-20. J Pharmacol Exp Ther 247:390–396

Melmed S (ed) 1993 Molecular and clinical advances in pituitary disorders—1993. Endocrine Research and Education, Inc., Los Angeles, CA. p 1–316

Milligan G 1993 Mechanisms of multifunctional signalling by G protein-linked receptors. Trends Pharmacol Sci 14:239–244

Offermanns S, Schultz G, Rosenthal W 1989 Secretion-stimulating and secretion-inhibiting hormones stimulate high-affinity pertussis-toxin sensitive GTPases in membranes of a pituitary cell line. Eur J Biochem 180:283–287

Osborne R, Tashjian AH Jr 1982 Modulation of peptide binding to specific receptors on rat pituitary cells by tumor-promoting phorbol esters: decreased binding of thyrotropin-releasing hormone and somatostatin as well as epidermal growth factor. Cancer Res 42:4375–4381

Patel YC, Tannenbaum GS (eds) 1990 Somatostatin: basic and clinical aspects. Metabolism (suppl 2) 39:1–191

Patel YC, Panetta R, Escher E, Greenwood M, Srikant CB 1994 Expression of multiple somatostatin receptor genes in AtT-20 cells. Evidence for a novel somatostatin-28 selective receptor subtype. J Biol Chem 269:1506–1509

Presky DH, Schonbrunn A 1986 Receptor-bound somatostatin and epidermal growth factor are processed differently in GH_4C_1 rat pituitary cells. J Cell Biol 102:878–888

Presky DH, Schonbrunn A 1988 Somatostatin pretreatment increases the number of somatostatin receptors in GH_4C_1 pituitary cells and does not reduce cellular responsiveness to somatostatin. J Biol Chem 263:714–721

Reichlin S 1983 Somatostatin I. N Engl J Med 309:1495–1501

Schonbrunn A 1982 Glucocorticoids down-regulate somatostatin receptors on pituitary cells in culture. Endocrinology 110:1147–1154

Schonbrunn A 1990 Somatostatin action in pituitary cells involves two independent transduction mechanisms. Metabolism (suppl 2) 39:96–100

Schonbrunn A, Loose-Mitchell D 1993 Somatostatin receptor 1 is expressed in GH_4C_1 pituitary cells. Proc 75th Meet Endocrine Soc p 476 (abstr)

Schonbrunn A, Tashjian AH Jr 1978 Characterization of functional receptors for somatostatin in rat pituitary cells in culture. J Biol Chem 253:6473–6783

Schonbrunn A, Tashjian AH Jr 1980 Modulation of somatostatin receptors by thyrotropin-releasing hormone in a clonal pituitary cell strain. J Biol Chem 255:190–198

Schonbrunn A, Lee AB, Brown PJ 1993 Characterization of a biotinylated somatostatin analog as a receptor probe. Endocrinology 132:146–154

Smith MA, Yamamoto G, Vale WW 1984 Somatostatin desensitization in rat anterior pituitary cells. Mol Cell Endocrinol 37:311–318

Spiegel AM, Shenker A, Weinstein LS 1992 Receptor-effector coupling by G proteins: implications for normal and abnormal signal transduction. Endocr Rev 13:536–565

Sullivan SJ, Schonbrunn A 1986 The processing of receptor-bound [^{125}I-Tyr11]somatostatin by RINm5F insulinoma cells. J Biol Chem 261:3571–3577

Tannenbaum GS, Painson J-C, Lengyel MJ, Brazeau P 1989 Paradoxical enhancement of pituitary growth hormone (GH) responsiveness to GH-releasing factor in the face of high somatostatin tone. Endocrinology 124:1380–1388

Viguerie N, Tahiri-Jouti N, Estève J-P et al 1988 Functional somatostatin receptors on a rat pancreatic acinar cell line. Am J Physiol 255:G113–G120

Viguerie N, Tahiri-Jouti N, Ayral AM et al 1989 Direct inhibitory effects of a somatostatin analog, SMS201–995, on AR4-2J cell proliferation via pertussis toxin-sensitive guanosine triphosphate-binding protein-independent mechanisms. Endocrinology 124:1017–1025

Whaley BS, Yuan N, Birnbaumer L, Clark RB, Barber R 1994 Differential expression of the β-adrenergic receptor modifies agonist stimulation of adenylyl cyclase: a quantitative evaluation. Mol Pharmacol 45:481–489

Yajima Y, Akita Y, Saito T 1986 Pertussis toxin blocks the inhibitory effects of somatostatin on cAMP-dependent vasoactive intestinal peptide and cAMP-independent thyrotropin releasing hormone-stimulated prolactin secretion of GH3 cells. J Biol Chem 261:2684–2689

Yamada T, Chiba T 1989 Somatostatin. In: Marlouf G (ed) Handbook of Physiology, vol 2. American Physiological Society, Bethesda, MD p 431–453

DISCUSSION

Montminy: Does somatostatin pretreatment increase the response to VIP (vasoactive intestinal peptide) in terms of the accumulation of cAMP?

Schonbrunn: Yes. The proposed mechanism is by direct induction or stabilization of G_s because forskolin stimulation of cAMP formation is also increased.

Montminy: If you treat with somatostatin after pretreatment do you still see inhibition?

Schonbrunn: The magnitude of the stimulation by VIP increases, so the curve is shifted up, but the potency of somatostatin to inhibit is unchanged.

Berelowitz: When you immunoprecipitated labelled solubilized receptor and used GTPγS to dissociate the ligand, was that photoaffinity-labelled receptor, cross-linked receptor or ([^{125}I]Tyr^{11})somatostatin-bound receptor?

Schonbrunn: Noncovalently bound ligand was immunoprecipitated with the receptor. GTPγS stimulated the dissociation of the ligand–receptor complex in the immunoprecipitate, demonstrating that the receptor was still coupled to G proteins.

Berelowitz: Can you immunoprecipitate MK analogues with antibodies to sstr1 or are they only immunoprecipitated with sstr2 antibodies?

Schonbrunn: They are only precipitated using antibodies to sstr2.

Berelowitz: So the antibodies are specific?

Schonbrunn: Yes.

Berelowitz: Have you looked at any of the other somatostatin analogues?

Schonbrunn: We looked at both ^{125}I-labelled octreotide and [^{125}I]somatostatin-labelled receptors. However, the strongest evidence for antibody specificity comes from the experiments using CHO cell lines expressing different types of sstr—each antibody will only precipitate receptor from cells expressing the receptor to which the antibody was raised.

Epelbaum: Are the effects of GTPγS and pertussis toxin greater in the sstr2 immunoprecipitation than in the sstr1 immunoprecipitation?

Schonbrunn: Yes. GTPγS has a smaller effect on the binding affinity of sstr1 than on sstr2, as does pertussis toxin. This suggests that the conformational change, brought about by the G protein associating with the receptor, has a greater effect on the ligand-binding affinity of sstr1 than sstr2.

Epelbaum: Is this because the G proteins are different?

Schonbrunn: It could be a property of the receptor, or it could be a property of the G protein.

Eppler: Is there any evidence for cross-talk between somatostatin and VIP receptors so that binding of the ligand to the receptor results in a change of affinity of the other receptor for its ligand or the recruitment of receptors?

Schonbrunn: VIP does not have a significant effect on somatostatin receptors.

Eppler: Our GH_4C_1 cells have little if any $G_i\alpha1$, and a lot of $G_i\alpha2$ and G_o. We didn't find any G_o co-purifying with sstr2 and we would not expect to get co-purification of $G_i\alpha1$ (although we did find significant amounts of $G_i\alpha3$, which was also undetectable in our GH_4C_1 membranes). Our AR42J cells have $G_i\alpha1$ and $G_i\alpha2$, but only $G_i\alpha1$ is co-purified. So, it is possible that if sstr2 can choose between $G_i\alpha1$ and $G_i\alpha2$, it will choose $G_i\alpha1$. The observation that both $G_i\alpha2$ and $G_i\alpha3$ co-purified with the receptor might be explained if we had more than one type of somatostatin receptor in our preparation. Amino acid sequencing did not show anything except sstr2; but I've since shown (M. Eppler, unpublished results) that different digestion conditions are required to produce μ-opioid receptor peptides that can be used for sequencing, so another somatostatin receptor could have been present in our preparations which we might not have seen. *sstr1* mRNA is present in our GH_4C_1 cells (Bill Baumbach, American Cyanamid Co., personal communication), but I've shown that in membranes from these cells, the sstr1-selective ligand MK 678 competes for all the binding of $([^{125}I]Tyr^{11})$somatostatin-14 with an IC_{50} of less than 1nM, so there's no pharmacological evidence for sstr1 (M. Eppler, unpublished results).

Finally, how sure are you that the band you call $G_i\alpha2$ really is $G_i\alpha2$?

Schonbrunn: Our evidence is based on the specificity of the antibody and on the co-migration of the immunoreactive band with brain $G_i\alpha2$.

We've also done binding studies in our GH_4C_1 cells and we have found evidence for two somatostatin receptors—one that binds octreotide and one that does not. This is therefore consistent with our immunoprecipitation results.

Eppler: Michael Berelowitz showed earlier that the level of expression of receptors can change with different culture conditions (Berelowitz et al 1995, this volume). This might account for some of the differences in receptor types in GH_4C_1 cells in different labs.

Lamberts: Recently there was a paper by Vignolini et al (1994) in which they demonstrated the presence of both VIP receptors and somatostatin receptors on different human tumours. They presented evidence that somatostatin displaces VIP binding to the membrane, suggesting an interaction between the two receptors; however, we could not demonstrate such an interaction in our laboratory.

Taylor: I tried to repeat that work and didn't see any interaction.

Bruns: I investigated the binding affinity of VIP to sstr1–5 and found that VIP does not bind to any of them, so it is difficult to understand that paper by Vignolini et al (1994).

Taylor: You have to be careful with the specificity of the synthetic analogues because octreotide and lanreotide are both high affinity opiate ligands. In addition, we recently published a paper showing that many somatostatin peptides bind to neuromedin-B receptors (Orbuch et al 1993).

Humphrey: Agi Schonbrunn, do you have an explanation for the phenomenon that you described in GH_4C_1 cells of a somatostatin-induced doubling in receptor number?

Schonbrunn: My working hypothesis is that the binding of ligand to the receptor stabilizes it to degradation. This is the opposite of what you see with most other receptors where ligand binding stimulates receptor internalization and degradation. Stabilization has been shown for some receptors, however, e.g. the nicotinic acetylcholine receptor where binding of alpha bungarotoxin stabilizes that receptor to degradation (Gardner & Fambrough 1979).

Humphrey: What do you mean by stabilization? You showed that inhibiting protein synthesis by 85% didn't change the phenomenon of the increase in receptor number. You would expect to see agonist-induced receptor internalization in this situation, but it's quite possible that the agonist facilitates recycling of intracellular receptors back to the membrane rather than affecting degradation.

Schonbrunn: We showed that the somatostatin ligand is not internalized in GH_4C_1 cells—it stays on the surface of the cell (Presky & Schonbrunn 1986). These experiments were done in parallel with epidermal growth factor (EGF) which is one of the standard ligands that people use to look at receptor-mediated internalization. We were surprised that somatostatin does not stimulate this pathway.

Stork: The minimal effects of GTP analogues on somatostatin binding to sstr1 compared to sstr2 as well as the Mg^{2+}-independent binding of sstr1 in CHO cells suggest that non-classical G protein coupling of sstr1 occurs (Hershberger et al 1994). Would this affect the ability to elute sstr1 with GTPγS, since in some peoples' hands there is no GTP shift in binding of sstr1?

Schonbrunn: There's clearly a GTP effect on sstr1 binding, at least in GH_4C_1 cells. However, as this effect is not as large with sstr1 as with sstr2, it could explain why some people have had difficulty in seeing a GTP effect on sstr1. It could also explain why sstr1 might not be eluted off a ligand affinity column with GTP.

Bruns: In contrast to GH_4C_1 cells, Christiane Susini reported at this meeting (N. Viguerie & J. P. Estève, unpublished results) that pancreatic tumour cells may show receptor internalization, so we should be careful not to generalize somatostatin receptor internalization or desensitization for different cell types.

Schonbrunn: We were concerned that this was receptor specific or a peculiarity of the GH_4C_1 cell line. That is why we used EGF internalization as a control—to show that in the cells in which somatostatin is not internalized, EGF is. We also did parallel studies on RINm5F cells, a rat insulin-secreting cell line, and obtained exactly the same result (Sullivan & Schonbrunn 1986).

Patel: Is it sstr2 selective? You should be able to determine whether it is sstr2 that is recruited to the plasma membrane in your studies by immunoprecipitating this receptor with your receptor-specific antibodies.

Schonbrunn: We're in the process of doing that.

Patel: What's the time course of the effect?

Schonbrunn: It's a relatively slow increase which takes 24 h to reach a new steady state. This is consistent with receptors being stabilized to degradation.

Patel: During that time Mike Berelowitz showed that down-regulation of mRNA occurs.

Berelowitz: For *sstr2*, the levels of mRNA increase for the first 2 h, decrease to about 50% of the original message level at 6 h and at 48 h increase to 100%—a biphasic change (Berelowitz et al 1995, this volume). All the other somatostatin receptor mRNAs increase gradually reaching the maximum at about 24 h.

Patel: In the case of receptors expressed individually in CHO or COS cells, Terry Reisine's results suggest that *sstr1* is the only receptor that is not down-regulated in response to agonist exposure for two hours (Reino et al 1993).

Schonbrunn: Using the reverse transcriptase polymerase chain reaction, we have shown that GH_4C_1 cells express almost exclusively *sstr1* and *sstr2*. We can also immunoprecipitate essentially 100% of the ligand–receptor complex by a combination of the sstr1 and sstr2 antibodies. Therefore, those two receptors represent essentially all of the receptor protein present.

Berelowitz: But there are differences both between different cell lines and within the same cell line. We have GH3 cells that express only *sstr1* and *sstr2* and we have GH3 cells that have all five receptors.

Patel: Are all these differences just artifacts of different cell lines?

Schonbrunn: It is not known. That's why the development of specific antibodies is going to be important—to determine whether a particular cell line has the same receptor as the normal tissue from which it was derived.

References

Berelowitz M, Xu Y, Bruno JF 1995 Regulation of somatostatin receptor expression. In: Somatostatin and its receptors. Wiley, Chichester (Ciba Found Symp 190) p 111–126

Gardner JM, Fambrough DM 1979 Acetylcholine receptor degradation measured by density labeling: effect of cholinergic ligand and evidence of recycling. Cell 16:661–674

Hershberger RE, Newman BL, Florio T et al 1994 The somatostatin receptors SSR1 and SSR2 are coupled to inhibition of adenylyl cyclase in CHO cells via pertussis toxin sensitive pathways. Endocrinology 134:1277–1285

Luthin DR, Eppler CM, Linden J 1993 Identification and quantification of Gi type GTP binding proteins that copurify with a pituitary somatostatin receptor. J Biol Chem 268:5990–5996

Orbuch M, Taylor JE, Coy DH et al 1993 Discovery of a novel class of neuromedin-B receptor antagonists: substituted somatostatin analogues. Mol Pharmacol 44:841–850

Presky DH, Schonbrunn A 1986 Receptor-bound somatostatin and epidermal growth factor are processed differently in GH_4C_1 rat pituitary cells. J Cell Biol 102:878–888

Reino K, O'Carroll A-M, Hong H et al 1993 Characterization of cloned somatostatin receptors, sstr4 and sstr5. Mol Pharmacol 44:385–392

Sullivan SJ, Schonbrunn A 1986 The percentage of receptor-bound [^{125}I-Tyr^{11}]somatostatin by RINm5F insulinoma cells. J Biol Chem 261:3571–3577

Vignolini I, Yang Q, Shuren L et al 1994 Cross-competition between vasoactive intestinal peptide and somatostatin for binding to tumor cell membrane receptors. Cancer Res 54:690–700

Somatostatin receptors: clinical implications for endocrinology and oncology

S. W. J. Lamberts, W. W. de Herder, P. M. van Koetsveld, J. W. Koper, A. J. van der Lely, H. A. Visser-Wisselaar and L. J. Hofland

Department of Medicine, University Hospital Dijkzigt, 40 Dr. Molewaterplein, 3015 GD Rotterdam, The Netherlands

Abstract. Somatostatin receptors are present on most hormone-secreting tumours. They are the pathophysiological basis for the successful control of hormonal hypersecretion by pituitary adenomas, metastatic islet cell tumours and carcinoids during treatment with the long-acting somatostatin analogue octreotide. There is also evidence for inhibition of tumour growth in some of these patients. Visualization of somatostatin receptor-positive tumours is possible *in vivo* after the administration of ($[^{111}$In]diethylenetriaminepentaacetic acid)octreotide. Primary tumours are detected and often metastases that were previously unrecognized. Tumours that secrete growth hormone or thyroid-stimulating hormone and non-functioning pituitary adenomas, islet cell tumours, carcinoids, paragangliomas, phaeochromocytomas, medullary thyroid carcinomas and small-cell lung cancers are visualized in 70–100 % of cases. Meningiomas, renal cell cancers, breast cancers and malignant lymphomas are often somatostatin receptor positive, allowing their localization with this scanning procedure. In some of these tumours discrepancies have been noted between binding studies with somatostatin-14, somatostatin-28 and octreotide, which suggests the presence of somatostatin receptor subtypes on some tumours. Most hormone-secreting tumours react *in vitro* to octreotide with an inhibition of hormone release and growth. Cultured meningioma cells react to octreotide with a stimulation in growth, possibly by interference with the autocrine inhibitory growth control by interleukin 6. This suggests that the presence of somatostatin receptors on human tumours does not automatically imply a beneficial effect of somatostatin analogue therapy.

1995 Somatostatin and its receptors. Wiley, Chichester (Ciba Foundation Symposium 190) p 222–239

Somatostatin belongs to the rapidly expanding family of small peptides that play a regulatory role in the function of different organ systems throughout the body (Reichlin 1983). Somatostatin is present in and inhibits normal function of the central nervous system, the hypothalamus and the pituitary gland, the gastrointestinal tract, the exocrine and endocrine pancreas and the immune system (Reichlin 1983, 1993). In these different organs, somatostatin is involved in the inhibition of growth hormone (GH) and thyroid-stimulating hormone (TSH) release, of hydrochloric acid production, of intestinal contractility, of the secretion of most gastrointestinal and pancreatic hormones and of the function of activated mononuclear immune cells, respectively.

These different 'local' actions of somatostatin are mediated via specific, high affinity membrane receptors on the target cells (Reubi et al 1990a). Recently, it has become clear that there are at least five somatostatin receptors. These bind natural somatostatin with a high affinity, but differ in their binding characteristics to various long-acting analogues of the peptide (Bell et al 1995, Bruns et al 1995, this volume).

In view of its ability to affect so many physiological processes, somatostatin was expected to be of therapeutic value in clinical conditions involving hyperactivity of these systems (Guillemin 1978). These expectations were raised by the finding that high numbers of somatostatin receptors are present on most tumours that arise from tissues that express these receptors in the normal state (Reubi et al 1990a).

The practical use of native somatostatin was hampered by the multiple effects of the peptide, by the need for intravenous administration of the peptide, by its short duration of action (half-life in the circulation of less than three minutes after intravenous administration), and by the post-infusion rebound hypersecretion of hormones (Guillemin 1978). Many attempts have been made, therefore, to synthesize analogues of somatostatin that do not have the disadvantages of the native peptide. Step-by-step modification of the conformationally stabilized, central essential part of the somatostatin molecule resulted in the synthesis of octreotide (SMS 201-995, Sandostatin™), which is the first clinically widely used analogue (Bauer et al 1982). In rhesus monkeys octreotide inhibits the secretion of GH, glucagon and insulin 45, 11 and 1.3 times more potently, respectively, than does native somatostatin. After subcutaneous administration, the drug has an elimination half-life of two hours and no rebound hypersecretion of hormones is observed (Lamberts et al 1985a, Lamberts 1988). Two other somatostatin analogues with slightly different activity profiles, RC 160 and BIM 23014 (Hofland et al 1994), will probably be available for clinical use shortly.

Therapeutic applications of somatostatin for hormone-secreting tumours

Most hormone-secreting tumours originating from somatostatin target tissues have conserved somatostatin receptors which are often expressed homogeneously and in a high density over the tumour tissue. This is especially true for GH-secreting pituitary adenomas (Reubi & Landolt 1984), as well as for pancreatic islet cell tumours and carcinoids (Reubi et al 1990b). As hormonal hypersecretion in most patients with such tumours causes rather serious symptomatology, it was logical to start clinical investigations concerning the effect of somatostatin analogue treatment in patients with acromegaly and inoperable carcinoids and islet cell tumours.

Acromegaly

Acromegaly is virtually always caused by a GH-secreting pituitary tumour. It causes distressing disfigurements of the face, hands and feet in adulthood; in children it causes gigantism. The most important complaints include headache, perspiration, paresthesia in fingers and toes and tiredness. When treated insufficiently, persistent acromegaly approximately doubles mortality at all ages as a result of cardiovascular diseases associated with hypertension and diabetes mellitus, as well as an increased incidence of malignant diseases. Effective and safe long-term medical treatment could be of benefit in as many as 30–35% of patients with acromegaly in whom the pituitary tumour cannot be completely removed by trans-sphenoidal surgery.

In most patients with acromegaly, a single subcutaneous injection of 50 μg of octreotide suppresses circulating GH levels to virtually undetectable levels for 6–8 hours, after which the levels slowly return to pretreatment values (Fig. 1). Long-term therapy of acromegalic patients with octreotide in a dose of 100 μg subcutaneously two or three times daily resulted in most patients in a marked clinical improvement within days: excessive perspiration, headaches, paresthesias and fatigue diminished or disappeared in most individuals, while soft tissue swelling had clearly decreased after weeks (Lamberts et al 1985b, Lamberts 1988). The initially raised levels of circulating insulin-like growth factor 1 (IGF1) returned to normal in most patients, while the mean 24 h GH levels decreased by more than 80%. A slight, but significant, decrease in pituitary tumour size was observed during octreotide treatment in more than half of these patients. The mechanism of tumour shrinkage is probably a decrease in the size of individual tumour cells with an increased breakdown of GH (Lamberts 1988).

Octreotide effectively controls GH secretion in most patients with acromegaly. Patients who have undergone an unsuccessful trans-sphenoidal operation or who await the therapeutic effect of external pituitary irradiation are candidates for treatment with octreotide, but the observation that the pituitary tumours of elderly acromegalic people in general demonstrate a higher sensitivity to octreotide suggests that the drug can also be used as a primary treatment for these patients (van der Lely et al 1992).

Apart from transient crampy abdominal pains, flatulence, loose fatty stools and/or diarrhoea, no important side-effects were observed during octreotide treatment. In some of the patients, gallstones are formed, but these remain virtually always asymptomatic. Octreotide also causes a short-term inhibition and/or delay of insulin release in response to meals, (Fig. 1), but this is accompanied by only a slight decrease in glucose tolerance in some patients, without notable changes in levels of haemoglobin A_1C.

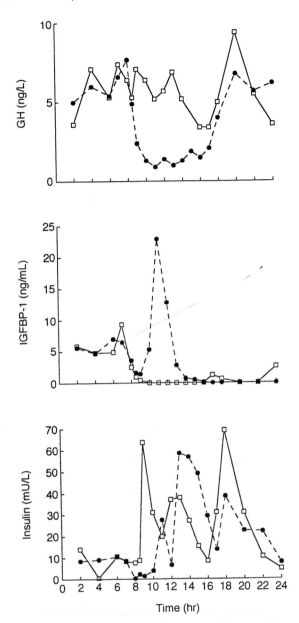

FIG. 1.   The effect of 50 μg octreotide (●) and a placebo (□) given subcutaneously at 8.15 am on circulating levels of growth hormone (GH), insulin-like growth factor binding protein-1 (IGF BP1) and insulin in a 42-year-old male patient with acromegaly.

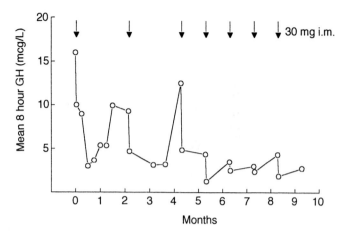

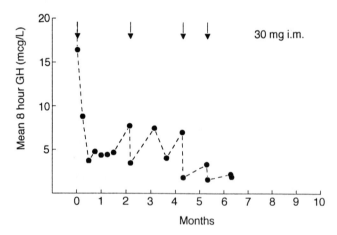

FIG. 2. The effects of repeated intramuscular (i.m.) administration of 30 mg of the long-acting depot formulation of octreotide in two acromegalic patients on mean levels of growth hormone (GH). Each point represents the mean of eight samples, each collected every hour between 8 am and 4 pm. Arrows indicate an injection.

At present, octreotide has to be administered subcutaneously twice or three times daily. The clinical efficacy of the drug is so impressive, however, that most patients accept this drawback. Recently, a long-acting application has been devised, in which 20–30 mg of octreotide is mixed within microspheres of

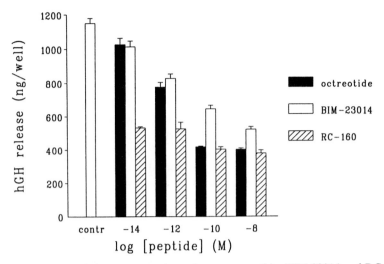

FIG. 3.   The effect of the somatostatin analogues octreotide, BIM 23014 and RC-160 on growth hormone (GH) release by cultured human GH-secreting pituitary adenoma cells. The cells were cultured for three days at a density of 10^5 cells per well per 1 ml MEM (modified Eagle's medium) + 10% fetal calf serum. On Day 3 of culture, the medium was changed and the cells were incubated for 24 hours with different concentrations of octreotide, BIM 23014 or RC-160.

DL–lactide-co-glycolide polymer: Sandostatin-LAR™ (long-acting repeatable). Preliminary studies indicate that this intramuscular depot preparation of octreotide is well tolerated and effectively controls hormonal hypersecretion in most acromegalic patients for 28–42 days. Initially, the drug was given at 60 day intervals. This resulted in a significant decrease in GH levels (Fig. 2) and IGF1 levels (not shown), as well as in clinical improvement. However, towards the end of the 60 days circulating GH levels tended to increase, indicating that the intramuscular depot of the drug was exhausted. After a change in the dose scheme to a monthly intramuscular injection, both the mean GH levels and IGF1 levels returned to normal in both patients. These observations are very promising with regard to the possibility that clinical symptomatology and hormonal hypersecretion can be controlled by 8–12 intramuscular injections of the long-acting octreotide application form per year in patients with acromegaly.

Another question regarding the future of medical treatment with somatostatin analogues of endocrine diseases, like acromegaly, is whether analogues other than octreotide have similar or better action profiles. Two analogues, RC 160 and BIM 23014 are currently under clinical investigation. We compared the *in vitro* effects of octreotide, RC 160 and BIM 23014 on GH release by cultured cells of human GH-secreting pituitary tumours, as well as in normal rat anterior pituitary cells (Hofland et al 1994). RC 160 significantly inhibited hormone release by cultured pituitary tumour cells in concentrations as low as 10^{-14} M, whereas at the same concentrations neither octreotide nor BIM 23014 exerted a significant effect (Fig. 3). At a concentration of 10^{-8} M, however, all three analogues showed a similar maximal inhibition. In normal rat anterior pituitary cell cultures, the IC_{50} values for inhibition of GH release were 0.1, 5.3, 47, 48, and 99 pM for RC 160, somatostatin-14, BIM 23014, octreotide and somatostatin-28, respectively. On the basis of these results, RC 160 appears to be about 500 times more potent than octreotide and BIM 23014 in inhibiting normal GH release *in vitro*. It is, however, at present unclear whether this will be reflected in a different sensitivity of normal and tumorous GH secretion *in vivo*, as the pharmacokinetic properties of these three octapeptide somatostatin analogues have not yet been compared.

Islet cell tumours and carcinoids

Islet cell tumours and carcinoids resemble the cells from which they originate in that they express somatostatin receptors in more than 80% of cases (Reubi et al 1990a,b). Most islet cell tumours and carcinoids are malignant and have already metastasized at the time of diagnosis. They are, in general, slow-growing tumours. The clinical distress in these patients is often related to hypersecretion of hormones which incapacitates them and necessitates repeated stays in hospital. Treatment with octreotide improves clinical symptomatology: watery diarrhoea, dehydration, flushing attacks, hypokaliaemia, peptic ulceration, life-threatening hypoglycaemic attacks and necrolytic skin lesions are well controlled during treatment with octreotide in most patients with metastatic carcinoids, vipomas (vasoactive intestinal peptide-secreting tumours), gastrinomas, insulinomas and glucagonomas (Kvols et al 1986, 1987). There is evidence that octreotide controls tumour growth in some of these patients: in about 20% of cases shrinkage of enlarged lymph nodes and liver metastases has been demonstrated (Kvols et al 1986). There is marked improvement in the quality of life of these patients during octreotide therapy; preliminary evidence also suggests prolonged survival.

Somatostatin and its analogues have been shown to inhibit the growth of a variety of transplantable tumours, like the Swarm chondrosarcoma, the Dunn osteosarcoma, acinar pancreatic carcinomas, ductal pancreatic adenocarcinomas and the Dunning prostate carcinoma (see Schally 1988). This might be achieved via one or more of the following mechanisms (Lamberts et al 1987, Schally 1988).

(1) Indirectly via inhibition of the secretion of hormones that are involved in the regulation of tumour growth, like GH, insulin or other hormones of the gastrointestinal tract.

(2) Direct or indirect (via GH) inhibition of the formation of IGF1 and/or other growth factors which stimulate tumour growth. Octreotide increases circulating levels of IGF binding protein 1 (Fig. 1), a serum protein that

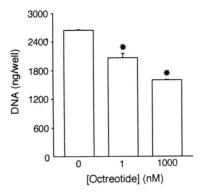

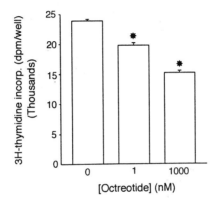

FIG. 4. The somatostatin analogue octreotide inhibits the growth of 7315b rat prolactinoma cells. Freshly dispersed prolactin-secreting 7315b rat pituitary tumour cells were cultured for seven days in MEM + 10% fetal calf serum (FCS). The cells were then harvested and re-seeded at a density of 40 000 cells per well in 1 ml MEM + 10% FCS and incubated for seven days without or with octreotide. After seven days, the cells were harvested and the DNA content (upper panel) or [^{3}H]thymidine incorporation (lower panel) was determined as described previously (Hofland et al 1990, Koper et al 1992). *p < 0.01 vs control (no octreotide).

modulates and inhibits the stimulatory effects of IGF1 at the cellular level (Ezzat et al 1992).

(3) Inhibition of angiogenesis. Octreotide exerts in several *in vitro* models a direct powerful inhibitory effect on angiogenesis.

(4) Direct inhibition of intracellular processes in tumour cells via specific somatostatin receptors. Somatostatin has a direct antiproliferative effect in several models, including gerbil fibroma, HeLa cells, the human breast cancer cell line MCF-7, the human pancreatic cancer cell line MIA PaCa-2, and in cultured prolactin-secreting rat pituitary tumour cells (Fig. 4). Hofland et al

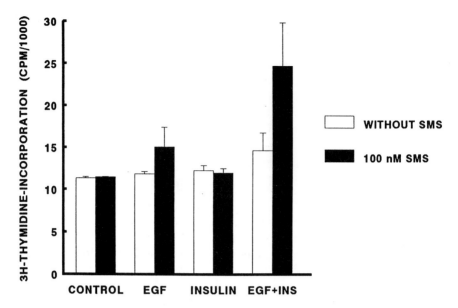

FIG. 5. The effect of octreotide on [³H] thymidine incorporation induced by epidermal growth factor (EGF) or insulin by cultured tumour cells from a patient with a meningioma. Meningioma cell cultures were preincubated for three days in serum-free medium. They were then incubated for 24 hours in fresh serum-free medium with 1.6 nmol/l EGF or 173 pmol/l insulin or both, with or without 100 nM octreotide (SMS). For the last four hours, 37 kBq [³H] thymidine was added to each well. Means ± SEM. For further details, see Koper et al (1992).

(1992) showed that a high number of somatostatin receptors on the tumour cells may be required for the antiproliferative effects of somatostatin analogues, whereas its antihormonal effect occurs in the presence of relatively low numbers of receptors. This suggests a dissociation of the antiproliferative and antihormonal effects induced by these compounds.

With regard to the intracellular effectors of somatostatin, several receptor-linked systems have been identified (see Patel et al 1990). These include reduction of the concentrations of two key intracellular mediators, cAMP and Ca^{2+}, through effects on membrane adenylyl cyclase and ionic K^+ and/or Ca^{2+} channels. Somatostatin also activates a tyrosine phosphatase, which is negatively regulated by EGF (epidermal growth factor) binding to the EGF receptor (Pan et al 1992).

In conclusion, these results suggest that somatostatin may be one of a class of hormones that act as endogenous tumour growth inhibitors or as modulators of growth stimulatory peptides. However, unexpected findings suggest a need of caution to translate these preclinical *in vivo* and *in vitro* studies to human pathophysiology. It was demonstrated that meningiomas, which are tumours arising from the leptomeninx, express high numbers of somatostatin receptors (Reubi et al 1987). Our studies indicated that somatostatin and octreotide inhibited adenylyl cyclase activity in cultured human meningioma cells, but unexpectedly stimulated the growth of these tumour cells *in vitro* in the presence of EGF (Fig. 5, Koper et al 1992). Octreotide probably interferes with the inhibitory autocrine growth control by interleukin 6 which is produced by these meningioma cells (Adams et al 1992). This suggests that octreotide should not be used in the treatment of patients with meningiomas.

Visualization of tumours by detection of somatostatin receptors

Because of the high density of somatostatin receptors on human neuroendocrine tumours, it has been possible to visualize somatostatin receptor-positive tumours *in vivo* after the injection of radionuclide-coupled somatostatin (Krenning et al 1989). ([^{111}In]DTPA)octreotide (DTPA, diethylenetriaminepentaacetic acid) scintigraphy is a very simple technique which allows the visualization of primary carcinoids and islet cell tumours; it also often detects unexpected distant metastases. Other tumours detected include paragangliomas, phaeochromocytomas, medullary thyroid cancers and small-cell lung cancers (for review see Lamberts et al 1991). Detection of a tumour by scintigraphy predicts a beneficial response to octreotide treatment. Some other tumour types, like those of the breast, the kidney and meningiomas, often express so many somatostatin receptors that their localization and spread can be demonstrated with somatostatin receptor scintigraphy. Finally, malignant lymphomas and tissues affected by granulomatous diseases and some autoimmune disorders are often somatostatin receptor positive (see Krenning et al 1993).

These developments allow a totally new diagnostic and therapeutic approach in patients with somatostatin receptor-positive tumours.

(1) The possibility of a beneficial effect of chronic treatment with somatostatin analogues of patients with inoperable somatostatin receptor-positive tumours can now be evaluated more precisely because the presence or absence of these receptors can be followed with receptor scintigraphy before and during treatment. Still, the unexpected effects of octreotide on *in vitro* meningioma tumour cell growth should be kept in mind.

(2) Radiolabelled somatostatin analogues were recently shown to be successful in radioguided surgery: with a hand-held radionuclide probe the primary and metastatic tumour deposits of somatostatin receptor-positive tumours could be detected intraoperatively after the injection of ([^{125}I]Tyr^3)octreotide (Schirmer et al 1993). This procedure is of potential benefit for optimizing surgical intervention in diseases like medullary thyroid cancer, carcinoids and islet cell tumours. Octreotide labelled with a low-energy γ photon-emitting radionuclide administered 24–48 h before operation might improve the radioguided surgery of somatostatin receptor-positive tumours.

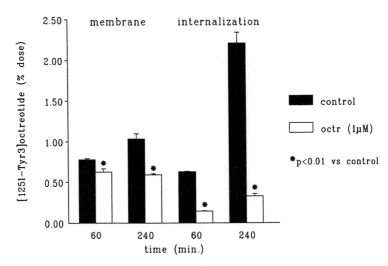

FIG. 6. Binding and internalization of ($[^{125}I]$ Tyr^3)octreotide by cultured cells from a human carcinoid tumour. Dispersed cells were cultured at a density of 10^6 cells per well in 3 ml culture medium. On Day 3 of culture the medium was changed and the cells were incubated for 60 or 240 min in DMEM (Dulbecco's modified Eagle's medium) + 30 mM HEPES + 0.1% BSA with approximately 0.1 nM ($[^{125}I]$ Tyr^3)-octreotide, with or without excess (1 μM) unlabelled octreotide to determine non-specific binding and internalization. At the end of the incubation period membrane-bound and cell-associated radioactivity were determined by acid stripping as described previously (Presky & Schonbrunn 1988). *p < 0.01 vs control.

(3) A further development is radiotherapy with radiolabelled, chelated somatostatin analogues. The accumulation of ($[^{111}In]$ DTPA)octreotide in the carcinoid and islet cell tumours of 11 patients was calculated to be $4.8 \pm 10.9\%$ (mean $\pm$ SD) of the injected dose (Pauwels et al 1992). The rapid decrease in blood radioactivity and the predominant renal clearance of ($[^{111}In]$ DTPA)octreotide are advantageous, although the amount of renal accumulation and the relatively long renal effective half-life at present limit the maximum radiation dose. Still these results are promising for the radiotherapy of inoperable somatostatin receptor-positive lesions.

Receptor-mediated endocytosis is a general process through which polypeptide hormones that bind to high affinity receptors on cell surfaces are internalized. Radiolabelled octreotide was taken up by cultured cells from a human carcinoid tumour such that after four hours more than 2% of the administered radioactivity was within the tumour cells (Fig. 6). The internalization was highly specific as the simultaneous administration of excess 'cold' octreotide prevented it virtually completely.

These recent developments raise the question of whether it might be feasible to manipulate the expression of somatostatin receptors by tumours *in vivo*, for

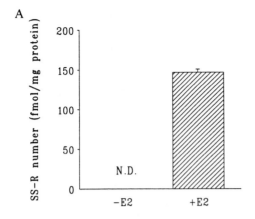

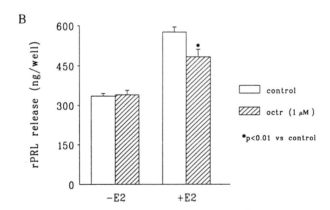

FIG. 7. (A)   Induction of somatostatin receptor (SS-R) expression by oestradiol. Freshly dispersed prolactin-secreting 7315b rat pituitary tumour cells were plated at a density of 10^6 cells/flask and cultured for six days in MEM + 10% horse serum without or with 17-β-oestradiol (1 nM). The cells were collected after six days and washed twice with saline. Membranes were isolated as described previously (Hofland et al 1992). The number of receptors for ($[^{125}I]$ Tyr^3)octreotide were determined by Scatchard analysis. (B) Octreotide inhibits prolactin release by cultured 7315b rat pituitary tumour cells. The cells were cultured for six days at a density of 40 000 cells per well per 1 ml DMEM + 10% horse serum without or with 17-β-oestradiol (1 nM). On Day 6 of culture, the medium was changed, then the cells were incubated for seven days with or without 1 μM octreotide. The culture medium was collected on Day 13 and the prolactin concentration was measured. *p < 0.01 vs control.

example in the hours before radioguided surgery and/or radiotherapy with radionuclide-coupled somatostatin analogues. There is clinical evidence in patients with Cushing's syndrome that hypercortisolism suppresses the expression of somatostatin receptors on some adrenocorticotropic hormone-secreting tumours. Pretreatment with a high dose of the glucocorticoid receptor blocker RU 486 reverses this process (see Lamberts et al 1994). We have preliminary evidence that it is possible to manipulate somatostatin expression at least *in vitro*. No somatostatin receptors can be detected on cells of the prolactin-secreting rat pituitary tumour 7315b cultured in the absence of oestradiol. Addition of 1 nM of the steroid rapidly induces high numbers of such receptors (Fig. 7A). Octreotide has no effect on prolactin secretion by these cells in the absence of oestradiol, but inhibits it in the presence of the steroid (Fig. 7B).

Conclusion

These studies have shown that the somatostatin analogue, octreotide, can be used to control hypersecretion of hormones by some tumours and may inhibit tumour growth in some patients. However, stimulation of meningioma cell growth *in vitro* has been observed, so future clinical studies should be pursued cautiously. Radiolabelled octreotide may be used to detect primary and metastatic tumours. Overall, somatostatin and its analogues show promise for clinical intervention in endocrinology and oncology.

References

Adams EF, Todo T, Rafferty B, Mower J, Fahlbusch R 1992 Human meningiomas secrete interleukin-6 in cell culture. J Endocrinol (suppl) 135:86

Bauer W, Briner U, Doepfner W et al 1982 SMS 201-995: a very potent and selective octapeptide analogue of somatostatin with prolonged action. Life Sci 31:1133–1141

Berelowitz M, Xu Y, Song J, Bruno JF 1995 Regulation of somatostatin receptor mRNA expression. In: Somatostatin and its receptors. Wiley, Chichester (Ciba Found Symp 190) p 111–126

Bruns Ch, Weckbecker G, Raulf F, Kaupmann K, Schoeffter Ph, Lübbert H, Hoyer D 1995 Characterization of somatostatin receptor subtypes. In: Somatostatin and its receptors. Wiley, Chichester (Ciba Found Symp 190) p 89–110

Ezzat S, Ren SG, Braunstein GD et al 1992 Octreotide stimulates insulin-like growth factor-binding protein-1 (IGFBP-1) expression in human hepatoma cells. Endocrinology 131:2479–2481

Guillemin R 1978 Peptide in the brain: the new endocrinology of the neurons. Science 203:390–402

Hofland LJ, van Koetsveld PM, Lamberts SWJ 1990 Percoll density gradient centrifugation of rat pituitary tumor cells: a study of functional heterogeneity within and between tumors with respect to growth rates, prolactin production and responsiveness to the somatostatin analog SMS 201-995. Eur J Cancer & Clin Oncol 26:37–44

Hofland LJ, van Koetsveld PM, Wouters N, Waaijers M, Reubi J-C, Lamberts SWJ 1992 Dissociation of antiproliferative and antihormonal effects of the somatostatin analog octreotide on 7315b pituitary tumor cells. Endocrinology 131: 571-577

Hofland LJ, van Koetsveld PM, Waaijers M, Zuyderwijk J, Lamberts SWJ 1994 Relative potencies of the somatostatin analogs octreotide, BIM 23014, and RC-160 on the inhibition of hormone release by cultured human endocrine tumor cells and normal rat anterior pituitary cells. Endocrinology 134:301-306

Koper JW, Markstein R, Kohler C et al 1992 Somatostatin inhibits the activity of adenylate cyclase in cultured human meningioma cells and stimulates their growth. J Clin Endocrinol & Metab 74:543-547

Krenning EP, Bakker WH, Breeman WA et al 1989 Localisation of endocrine-related tumours with radioiodinated analogue of somatostatin. Lancet I:242-244

Krenning EP, Kwekkeboom DJ, Bakker WH et al 1993 Somatostatin receptor scintigraphy with [^{111}In-DTPA-d-Phe1]- and [^{123}I-Thyr3]-octreotide: the Rotterdam experience with more than 1000 patients. Eur J Nucl Med 18:1-16

Kvols LK, Moertel CG, O'Connell MJ et al 1986 Treatment of the malignant carcinoid syndrome, evaluation of a long-acting somatostatin analogue. N Engl J Med 315: 663-666

Kvols LK, Buck M, Moertel CG et al 1987 Treatment of metastatic islet cell carcinoma with a somatostatin analogue (SMS 201-995). Ann Intern Med 107:162-168

Lamberts SWJ 1988 The role of somatostatin in the regulation of anterior pituitary hormone secretion and the use of its analogs in the treatment of human pituitary tumors. Endocr Rev 9:417-436

Lamberts SWJ, Oosterom R, Neufeld M, del Pozo E 1985a The somatostatin analog SMS 201-995 induces long-acting inhibition of growth hormone secretion without rebound hypersecretion in acromegalic patients. J Clin Endocrinol & Metab 60:1161-1165

Lamberts SWJ, Uitterlinden P, Verschoor L, van Dongen KJ, del Pozo E 1985b Long-term treatment of acromegaly with the somatostatin analogue SMS 201-995. N Engl J Med 313:1576-1580

Lamberts SWJ, Koper JW, Reubi J-C 1987 Potential role of somatostatin analogues in the treatment of cancer. Eur J Clin Invest 17:281-287

Lamberts SWJ, Krenning EP, Reubi J-C 1991 The role of somatostatin and its analogs in the diagnosis and treatment of tumors. Endocr Rev 12:450-482

Lamberts SWJ, de Herder WW, Krenning EP, Reubi J-C 1994 Editorial: A role of (labelled) somatostatin analogs in the differential diagnosis and treatment of Cushing's syndrome. J Clin Endocrinol & Metab 78:17-19

Pan MG, Florio T, Stork PJS 1992 G protein activation of a hormone-stimulated phosphatase in human tumor cells. Science 256:1215-1217

Patel YC, Murthy KK, Escher EE, Banville D, Spiess J, Srikant CB 1990 Mechanism of action of somatostatin: an overview of receptor function and studies of the molecular characterization and purification of somatostatin receptor proteins. Metabolism (suppl 2) 9:63-69

Pauwels S, Jamar F, Leners N, Fiasse R 1992 Efficacy of Indium-111-pentatreotide scintigraphy in endocrine gastro-entero-pancreatic (GEP) tumours. Gastroenterology 102:A387

Presky DH, Schonbrunn A 1988 Somatostatin pretreatment increases the number of somatostatin receptors on GH$_4$C$_1$ pituitary cells and does not reduce cellular responsiveness to somatostatin. J Biol Chem 263:714-721

Reichlin S 1983 Somatostatin. N Engl J Med 309:1495-1501, 1556-1563

Reichlin S 1993 Neuroendocrine–immune interactions. N Engl J Med 329:1246–1253

Reubi J-C, Landolt AM 1984 High density of somatostatin receptors in pituitary tumours from acromegalic patients. J Clin Endocrinol & Metab 59:1148–1153

Reubi J-C, Lang W, Maurer R, Koper JW, Lamberts SWJ 1987 Distribution and biochemical characterization of somatostatin receptors in tumours of the human central nervous system. Cancer Res 47:5758–5765

Reubi J-C, Kvols L, Krenning EP, Lamberts SWJ 1990a Distribution of somatostatin receptors in normal and tumor tissue. Metabolism 39:78–81

Reubi J-C, Kvols LK, Waser B et al 1990b Detection of somatostatin receptors in surgical and percutaneous needle biopsy samples of carcinoids and islet cell carcinomas. Cancer Res 50:5969–5977

Schally AV 1988 Oncological applications of somatostatin analogues. Cancer Res 48:6977–6985

Schirmer WJ, O'Dorisio TM, Schirmer TP, Mojzisik CM, Hinkle GH, Martin EW 1993 Intraoperative localization of neuroendocrine tumors with 125–I–TYR–octreotide and a hand held gamma-detecting probe. Surgery 114:745–752

Van der Lely AJ, Harris AG, Lamberts SWJ 1992 The sensitivity of growth hormone secretion to medical treatment in acromegalic patients: influence of age and sex. Clin Endocrinol 37:181–185

DISCUSSION

Lightman: What is the mechanism behind tumour shrinkage? You assume that tyrosine phosphatases are involved, but it is possible that somatostatin is affecting local IGF (insulin-like growth factor) production or local IGF binding protein (IGF BP) synthesis. I doubt that the circulating levels of growth hormone (GH) are relevant. Do you have any studies which give an insight into the local mechanism?

Lamberts: There are several mechanisms by which somatostatin analogue therapy may control tumour growth (Lamberts et al 1982, Schally 1988). The first is by a direct antimitotic action via the somatostatin receptors themselves. Indeed, it has been demonstrated that somatostatin and its analogues can directly inhibit the growth of primary cultures of pituitary and breast cancer cell lines (Hofland et al 1992, Setyono-Han et al 1987). Secondly, the tumour growth inhibitory effects might be mediated by changes in growth factor levels, since Melmed's group has shown that somatostatin can increase the circulating levels of IGF BP1 (Ezzat et al 1991). Presumably, IGF BP1 acts as an inhibitor of IGF1 action at the level of the IGF1 receptor.

Woltering et al (1991) showed inhibitory effects of somatostatin on angiogenesis. It is possible that angiogenesis is essential for metastases to grow in normal liver. If somatostatin analogues have such an effect on angiogenesis *in vivo*, one might expect a delaying effect of these drugs on the growth of liver metastases.

Berelowitz: What is the signal to noise ratio 24 h after a scan for breast cancer metastases, for example?

Lamberts: More than 90% of the isotope-coupled somatostatin analogue, ([^{111}In]DTPA)octreotide (DTPA, diethylenetriaminepentaacetic acid), is excreted after 24 h. The head and chest do not have significant background radioactivity after 24–36 h, but some remains in the upper abdomen and in the kidneys. In order to scan these areas you need computerized tomography and special measures, like emptying the gastrointestinal tract, to optimize the pictures (Krenning et al 1993).

Berelowitz: Is it possible to attach a porphyrin molecule to the analogue and photolyse the cells with a laser?

Lamberts: This is a possibility, but one of the major problems with porphyrin-sensitized photodynamic therapy is that it can only be directed to selective discrete tumour localizations and not to widespread metastases (Hill et al 1992).

Beaudet: The excellent penetration of the radioactivity is surprising. Presumably, the compound does not cross the blood–brain barrier. Have you ever observed a breakdown of this barrier in tumours?

Lamberts: Meningiomas are tumours of the blood–brain barrier. Their blood supply comes from systemic bloodflow. This is in sharp contrast to astrocytomas and glioblastomas which are localized within the blood–brain barrier. However, in many of these primary brain tumours, the blood–brain barrier is disrupted locally, allowing radioactivity to enter somatostatin receptor-positive tumours.

Patel: The visualization of more than 90% of neuroendocrine tumours, pituitary tumours and lymphomas with ([^{111}In]DTPA)octreotide implies that octreotide is detecting sstr2, sstr3 and sstr5. Some insulinomas do not have sstr2, but presumably have sstr1 and sstr4 (Lamberts et al 1991, Kubota et al 1994). Do you have any analogues that can detect these?

Lamberts: It is not feasible to use somatostatin-14 and somatostatin-28 in clinical medicine because of their short half-life after intravenous administration. We developed ([^{111}In]DTPA)RC 160 in the hope that RC 160 would have slightly different binding characteristics (Breeman et al 1993). Our preliminary observations, however, indicate that they are similar, and that ([^{111}In]DTPA)RC 160 scintigraphy will not visualize octreotide receptor-negative insulinomas.

Patel: Christian Bruns, does CGP 23996 bind to all five receptors?

Bruns: Yes, but CGP 23996 is degraded *in vivo* much quicker than the other somatostatin analogues, such as octreotide, so it would be difficult to use it for labelling studies in humans.

Reichlin: David Coy, do you know of any analogues which are specific for sstr1 or sstr2 and are not degraded after internalization?

Coy: No. CGP 23996 has the minimum chain size required to interact with all five receptors. It is closely related to the full somatostatin sequence, but lacks the N-terminal two residues outside the ring and has a covalent peptide bond structure instead of a disulphide bridge. If you simplify the structure any further, the affinity for sstr1 and sstr4 is removed.

Reisine: I disagree with Christian Bruns with regards to the stability of CGP 23996. This compound is more stable than either somatostatin-14 or somatostatin-28 (Czernik et al 1983).

Bruns: We compared the stabilities of somatostatin-28, somatostatin-14, CGP 23996 and short chain analogues such as octreotide in tissue homogenates *in vitro* and found no difference in stability between somatostatin-14, somatostatin-28 and CGP 23996—they were all relatively unstable compared with octreotide.

Lamberts: Coupling to DTPA might increase the bioavailability of CGP 23996 as it did when coupled to octreotide (Krenning et al 1993).

Reisine: The stability studies should be repeated if CGP 23996 is the only compound that can bind to all the receptors; however, this peptide is no longer available.

Eppler: Is it possible to detect sstr1 with antibodies? sstr1 is probably less heavily glycosylated than sstr2 as it has one less glycosylation site (Yamada et al 1992), so there might be more amino acid sequence in the internal region available for binding to antibodies.

Lamberts: Monoclonal antibody technology *in vivo* has been disappointing. The advantage of peptide receptor scintigraphy is the low quantity of the peptide (5–10 μg) necessary to couple to the radionuclide, in contrast to the higher amount of antibody required which is often milligrams (Fischman et al 1993).

Eppler: Is that because they have a high turnover rate?

Lamberts: No, biodistribution is the major problem.

Reichlin: What is the relative affinity of antibody–antigen binding versus receptor–ligand binding?

Eppler: A monoclonal antibody against the N-terminal 19 amino acid residues of the interleukin-8 receptor has a dissociation constant (K_d) for binding to the receptor of about 10^{-11} M (Chuntharapai et al 1994). This is higher than the affinity of the ligand for the receptor.

Cohen: Monoclonal antibodies, in general, have a much lower affinity for their antigens than polyclonals; therefore, a higher concentration of the former antibody would be required.

Lamberts: Is there a reliable and simple technique which you can use to measure the turnover of somatostatin receptors on tumour cells?

Cohen: You can do pulse–chase and immunoprecipitation experiments on human transfected cell lines.

Reisine: You can also apply non-selective cross-linking agents such as EEDQ (*N*–ethoxycarbonyl–2–ethoxy–1,2–dihydroquinoline) to the cell. It's a standard procedure in other receptor systems, e.g. the dopamine receptor system—you inactivate the receptor on the cell surface and then look at its reappearance. You can also use covalent ligands that are commercially available. These are non-selective, so if you have a cell line that has multiple receptors, you have the problem that they will all be modified.

Another way is to use antibodies to look at immunoprecipitated material. Antibodies against the somatostatin receptors now exist, so it should be possible to do this.

References

Breeman WAP, Hofland LJ, Bakker WH et al 1993 Radioiodinated somatostatin analogue RC-160: preparation, biological activity, *in vivo* application in rats and comparison with [^{125}I-Tyr^3]octreotide. Eur J Nucl Med 20:1089–1094

Chuntharapai A, Lee J, Burnier J, Wood WI, Hebert C, Kim KJ 1994 Neutralizing antibodies to human IL-8 receptor A map to the NH$_2$-terminal region of the receptor. J Immunol 152:1783–1789

Czernik AJ, Petrack B 1983 Somatostatin receptor binding in rat cerebral cortex. J Biol Chem 258:5525–5530

Ezzat S, Ren S-G, Braunstein GD, Melmed S 1991 Octreotide stimulates insulin-like growth factor binding protein-1 (IGFBP-1) levels in acromegaly. J Clin Metab 73:441–443

Fischman AJ, Babich JW, Strauss HW 1993 A ticket to ride: peptide radio-pharmaceuticals. J Nucl Med 34:2253–2263

Hill JS, Kahl SB, Kaye AH et al 1992 Selective tumor uptake of a boronated porphyrin in an animal model of cerebral glioma. Proc Natl Acad Sci USA 89:1785–1789

Hofland LJ, van Koetsfeld PM, Wouters N, Waaijers M, Reubi J-C, Lamberts SWJ 1992 Dissociation of antiproliferative and antihormonal effects of the somatostatin analog octreotide on 7315b pituitary tumor cells. Endocrinology 131:571–577

Krenning EP, Kwekkeboom DJ, Bakker WH et al 1993 Somatostatin receptor scintigraphy with [^{111}In–DTPA–d–Phe^1]– and [^{125}I–Tyr^3]–octreotide: the Rotterdam experience with more than 1000 patients. Eur J Nucl Med 20:716–731

Kubota A, Yamada Y, Kagimoto S et al 1994 Identification of somatostatin receptor subtypes and an implication for the efficacy of somatostatin analogue SMS 201-995 in treatment of human endocrine tumors. J Clin Invest 93:1321–1325

Lamberts SWJ, Koper JW, Reubi J-C 1987 Potential role of somatostatin analogues in the treatment of cancer. Eur J Clin Invest 17:281–287

Lamberts SWJ, Krenning EP, Reubi J-C, 1991 The role of somatostatin and its analogues in the diagnosis and treatment of cancer. Endocr Rev 12:450–482

Schally AV 1988 Oncological applications of somatostatin analogues. Cancer Res 48:6977–6985

Setyono-Han B, Henkelman MS, Foekens JA, Klijn JGM 1987 Direct inhibitory effects of somatostatin (analogues) on the growth of human breast cancer. Cancer Res 47:1566–1570

Woltering EA, Barrie R, O'Dorisio TM et al 1991 Somatostatin analogues inhibit angiogenesis in the chick chorioallantoic membrane. J Surg Res 50:245–251

Yamada Y, Post SR, Wang K, Tager HS, Bell GI, Seino S 1992 Cloning and functional characterization of a family of human and mouse somatostatin receptors expressed in the brain, gastrointestinal tract and kidney. Proc Natl Acad Sci USA 89:251–255

Somatostatin analogues and multiple receptors: possible physiological roles

David H. Coy and Wocjiech J. Rossowski

Peptide Research Laboratories, Department of Medicine, Tulane University Medical Center, New Orleans, LA 70112, USA

Abstract. The availability of transfected cell lines expressing the five cloned somatostatin receptors has allowed extensive families of synthetic analogues to be screened for their binding affinities to these receptors. Cyclic octapeptides in the octreotide series have high affinity for sstr2 and low affinity for sstr1 and sstr4. Modifications to these analogues can have profound effects on their residual affinity for sstr3 and sstr5. Several linear peptides, incorporating the critical core –D– Trp–Lys– sequence of all active analogues, exhibit selectivity for either sstr3, sstr5 or both, but little affinity for the other three. From these analogues we have assembled a panel of compounds with a useful degree of specificity for sstr2, sstr3 and sstr5 and these have been used to evaluate receptor involvement in various physiological processes. Inhibition of pituitary growth hormone release correlates completely with sstr2 affinity and it appears from preliminary results that inhibition of pancreatic glucagon release is also sstr2 mediated. Effects on gastric acid secretion show a high degree of dependence on sstr2 affinity. Inhibition *in vivo* by somatostatin of insulin release, amylase release and binding to rat pancreatic acinar cells is highly correlated with sstr5 receptor affinity. Gastric smooth muscle cells have particularly high affinity for sstr3 receptor ligands. The availability of receptor-specific ligands is proving to be of great value in elucidating the physiological roles of each receptor. It is anticipated that these and future generations of analogues will have distinct therapeutic advantages over somatostatin itself and presently available cyclic octapeptides.

1995 Somatostatin and its receptors. Wiley, Chichester (Ciba Foundation Symposium 190) p 240–254

At the time of writing, five principal receptors of the hypothalamic inhibitory peptide, somatostatin (Brazeau et al 1973), have been isolated and cloned (Yamada et al 1993, Yasuda et al 1993, Bruno et al 1992, O'Carroll et al 1993). Transfection of each of these receptors into various cell lines and their subsequent expression has provided a unique opportunity for placing the extensive pharmacology of somatostatin analogues on a rational basis. For instance, our laboratory had accumulated in excess of 100 compounds over 25 years, many of which, in collaboration with Terry Reisine's group, have recently been examined for their binding affinity to each receptor (Raynor et al 1993a,b).

We were surprised to discover several classes of agonist analogues with useful levels of specificity for three of the receptor types. Some of these analogues have been chosen for physiological studies aimed at delineating the involvement of individual receptors and some of these are described here.

It should be noted that receptor nomenclatures (sstr 1–5) in the two earlier papers of Raynor et al (1993a,b) have now been changed by broad consensus to reflect their chronological discovery. Thus, in this paper, the receptor of O'Carroll et al (1993) that prefers somatostatin-28 is referred to as sstr5 rather than sstr4 and the receptor of Bruno et al (1993) is referred to as sstr4.

Analogue binding characteristics for the five receptors

Cyclic analogues

Table 1 shows the binding affinities (taken from Raynor et al 1993a) of several representative cyclic octapeptides in the octreotide (SMS 201-995, Sandostatin™) series (Bauer et al 1982), including octreotide itself (analogue 1). A striking result of the truncation of the somatostatin sequence in this type of analogue is the complete loss of affinity for sstr1 and sstr4, largely present in the central nervous system, and the retention of extremely high affinity for sstr2. Indeed, it is apparent in the original paper (Raynor et al 1993a) that the full somatostatin tetradecapeptide sequence, including the extracyclic amino portion, is required for retention of full binding affinity for sstr1 and sstr4. It can also be seen that much higher affinity than the original octreotide analogue (1, Table 1) can be achieved: for the analogues in Table 1 this is done by incorporating L-3-(2-naphthyl)alanine (L-Nal) at the C-terminus. At the same time, affinities for sstr3 and sstr5 are not increased, or in some cases are actually decreased, which makes these analogues far more specific than their earlier counterparts. This may result in fewer side-effects in certain clinical situations using this type of compound. Analogue 7 is most interesting for its ability to bind to the neuromedin B receptor (Orbuch et al 1993), where it functions as a moderately potent (IC_{50} 40 nM) competitive antagonist. It is the only cyclic octapeptide to have significant affinity for sstr4.

The last analogue in Table 1 (analogue 8) is a cyclic hexapeptide supplied by Dr Daniel Veber, which possesses possibly unique properties. This compound has little affinity for all receptors apart from sstr5, for which it appears to have extraordinarily high affinity, apparently due to the additional Trp residue. Several linear analogues discussed in the next section also have high affinity and selectivity for this receptor, however, analogue 8 seems to be uniquely specific and may be a valuable tool for investigating the physiological roles of sstr5.

Linear octapeptide analogues

Some time ago, our laboratory developed a simplified approach to the design of analogues of somatostatin with potent inhibitory activity for the release of

TABLE 1 Receptor affinities of the octreotide series analogues relative to somatostatin

Analogue	Receptor type/affinities[a]				
	sstr1	sstr2	sstr3	sstr4	sstr5
1. D-Phe-[Cys-Phe-D-Trp-Lys-Thr-Cys]-Thr-ol	>10 000	8.6	38	>1000	0.66
2. D-Nal[b]-[Cys-Tyr-D-Trp-Lys-Val-Cys]-Thr-NH$_2$	>10 000	5.7	70	>1000	1.90
3. D-Phe-[Cys-Tyr-D-Trp-Lys-Val-Cys]-Nal-NH$_2$	>10 000	0.007	813	>1000	0.64
4. D-Phe-[Cys-Tyr-D-Trp-Lys-Thr-Cys]-Nal-NH$_2$	>10 000	0.043	218	>1000	0.10
5. D-Phe-[Cys-Tyr-D-Trp-Lys-Abu[c]-Cys]-Nal-NH$_2$	>10 000	0.001	162	>1000	6.9
6. D-Phe-[Cys-Tyr-D-Trp-Lys-Ser-Cys]-Nal-NH$_2$	>10 000	0.002	112	>1000	1.2
7. D-Nal-[Cys-Tyr-D-Trp-Lys-Val-Cys]-Nal-NH$_2$	>10 000	15	2588	119	6.0
8. c(Ahep[d]-Trp-D-Trp-Lys-Thr-Phe)	>10 000	104	375	52	0.006

[a] Affinities are all relative, with somatostatin = 1
[b] Nal, L-3-(2-naphthyl)alanine
[c] Abu, L-α-amino-n-butyric acid
[d] Ahep, 7-aminoheptanoic acid

growth hormone (GH). This generated a large series of linear octapeptides (W. A. Murphy, J. E. Taylor, J. P. Moreau, D. H. Coy, unpublished abstract no. 104, 72nd Endocr Soc Meet, 1990) that were still able to mimic the conformational properties of the more rigid cyclic octapeptides by virtue of electronic attractions between aromatic groups in place of covalently bonded Cys residues and which also retained quite potent biological activities. For example, analogue 6 (Table 2), in which non-aromatic Ala rather than Phe residues are used in place of Cys, has reduced affinity for sstr2. Many of these peptides were subsequently screened (Raynor et al 1993a,b) for their binding affinities to each of the five receptors and this gave perhaps the most surprising results of all of the families of analogues yet tested.

A small subset of these peptides is shown in Table 2, beginning with analogue 1 which has the highest potency for GH release inhibition and also twice the affinity for sstr2 than has somatostatin itself. Although the core D–Trp–Lys sequence must be present for affinity to any of the receptors to be retained, subtle alterations to the adjacent amino acid side chains can have major effects on receptor selectivity. Thus, the simple replacement of a Thr by a Val residue in analogue 2 results in considerable loss of sstr2 affinity. However, three of the most dramatic observations were the very high affinity that analogue 3 exhibits for sstr5, analogue 5 for sstr3, and analogue 4 for both sstr3 and sstr5. This was accompanied, particularly in the case of the last analogue, by a high degree of binding specificity and little affinity for sstr1, sstr2 and sstr4.

Physiological studies with receptor-selective analogues

Effects on GH release

Included in the original study (Raynor et al 1993a), in which analogues were screened for binding to sstr1, sstr2 and sstr3, was a comparison of the binding data with the potency of each analogue in an *in vitro* test for inhibition of GH release from cultured rat pituitary cells. There was complete correlation between binding to sstr2 and effects on GH release, and the most potent peptides in both systems were always in the cyclic octapeptides series. Several were considerably more potent than the first generation analogues in the series such as octreotide (Bauer et al 1982); many were also more specific for sstr2.

Effects on endocrine pancreatic secretions

It is obvious that the availability of a group of somatostatin analogues with individual receptor specificities should provide an important tool for delineating which receptors are involved in producing the many physiological effects of endogenous somatostatin peptides. From the binding results already discussed, a panel of analogues was chosen to provide maximum specificity for sstr2, sstr3

TABLE 2 Receptor affinities of the linear octapeptide series relative to somatostatin

Analogue	Receptor type/affinities[a]				
	sstr1	sstr2	sstr3	sstr4	sstr5
1. D-Phe-Cpa[b]-Tyr-D-Trp-Lys-Thr-Phe-Thr-NH$_2$	>10 000	0.5	10	433	11
2. D-Phe-Cpa-Tyr-D-Trp-Lys-Val-Phe-Thr-NH$_2$	>10 000	8.7	60	254	9.4
3. D-Phe-Phe-Phe-D-Trp-Lys-Thr-Phe-Thr-NH$_2$	230	114	5.3	20.9	0.002
4. D-Phe-Phe-Phe-D-Trp-Lys-Val-Phe-Thr-NH$_2$	>10 000	164	0.5	89.5	1.6
5. D-Phe-Phe-Tyr-D-Trp-Lys-Val-Phe-D-Nal-NH$_2$	>10 000	>10 000	0.3	184	51
6. D-Phe-Ala-Phe-D-Trp-Lys-Val-Ala-Nal-NH$_2$	>10 000	22	1900	>1000	35

[a]See Table 1.
[b]Cpa, L-p-chlorophenylalanine

and sstr5 (all having little affinity for sstr1 and sstr4). These compounds were the following:

Somatostatin	Ala–Gly–[Cys–Lys–Asn–Phe–Phe–Trp–Lys–Thr–Phe–Thr–Ser–Cys]
DC-25-100 (sstr2)	D–Nal–[Cys–Tyr–D–Trp–Lys–Val–Cys]–Thr–NH_2
NC-8-12 (sstr2)	D–Phe–[Cys–Tyr–D–Trp–Lys–Abu–Cys]–Nal–NH_2
DC-25-12 (sstr3, sstr5)	D–Phe–Phe–Phe–D–Trp–Lys–Val–Phe–Thr–NH_2
DC-23-99 (sstr5)	D–Phe–Phe–Phe–D–Trp–Lys–Thr–Phe–Thr–NH_2
DC-25-20 (sstr3)	D–Phe–Phe–Tyr–D–Trp–Lys–Val–Phe–D–Nal–NH_2

The first *in vivo* experiment (Rossowski & Coy 1993) in which these five peptides were employed to track possible receptor involvement compared their effects on glucose-stimulated insulin release with that of somatostatin. The peptides were infused at four dose levels and the responses are shown in Fig. 1. EC_{50} values calculated from the data are shown in Table 5. The most striking observation was the virtual equivalent potencies of somatostatin and sstr5-selective linear analogue, DC-23-99. The two sstr2-selective cyclic octapeptides, particularly NC-8-12 which is a very potent and selective sstr2 ligand, had lower potency; the sstr3-selective analogue, DC-25-20 (not shown in Fig. 1) was devoid of activity. These results clearly demonstrated for the first time that complete dissociation of effects on GH and insulin release was possible. We believe this reflects a direct effect on receptors present on the β-cells of the rat endocrine pancreas, although an indirect effect on another factor(s) controlling *in vivo* insulin release cannot be ruled out.

In preliminary experiments, we have performed a similar experiment on the same group of analogues, only this time measuring effects on release of glucagon in fasted rats as detected by radioimmunoassay. At the time of writing only the results from infusion of 100 nm/kg per hour of each analogue and somatostatin were available (Fig. 2). Despite the lack of full dose–response data, it is apparent that the structure–activity relationships in this experiment were completely different from those obtained in the insulin experiment. In Fig. 2, the most potent inhibitors of glucagon release are the sstr2-selective cyclic octapeptides, DC-25-100 and NC-8-12. Analogues selective for sstr3 and sstr5, DC-25-20 and DC-23-99, were not active or displayed lower activity, respectively. In a control experiment, possible effects of altered insulin levels on glucagon release were eliminated by infusion of insulin antibodies and no effect was seen on the peptide responses.

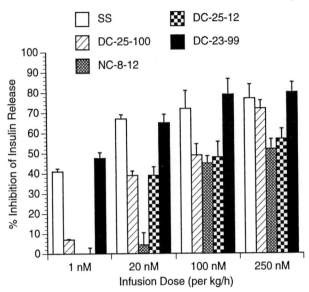

FIG. 1. Effects of infusion of various doses of somatostatin (SS) and four receptor-selective somatostatin analogues on glucose-stimulated insulin release in rats.

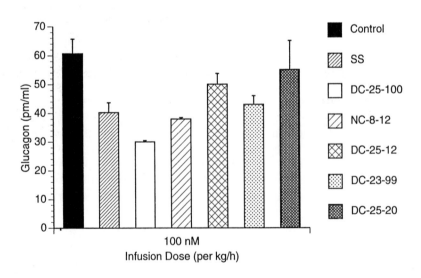

FIG. 2. Effects of infusion of somatostatin (SS) and five receptor-selective somatostatin analogues on glucagon levels in fasted rats.

TABLE 3 Somatostatin and analogue dose-dependent inhibition of bombesin-stimulated pancreatic amylase release

Peptide	Dose (nM/kg per hour)	% Inhibition
Somatostatin	0.5	29.5 ± 2.4
	1.0	54.6 ± 2.5
	5.0	79.0 ± 5.1
	10.0	81.5 ± 3.4
DC-25-100	0.5	48.0 ± 3.5
	1.0	63.5 ± 5.1
	5.0	78.6 ± 2.4
	10.0	81.5 ± 1.2
NC-8-12	0.5	27.0 ± 5.1[a]
	1.0	52.9 ± 3.6
	5.0	78.9 ± 2.4
	10.0	80.3 ± 3.8
DC-25-12	5.0	55.0 ± 4.8
	10.0	62.9 ± 4.0
	100.0	74.3 ± 2.6
	500.0	81.7 ± 3.4
DC-23-99	0.1	24.0 ± 4.8[a]
	1.0	66.5 ± 3.8
	10.0	71.1 ± 3.8
	100.0	76.7 ± 3.5
DC-25-20	10.0	22.3 ± 2.5[a]
	100.0	15.4 ± 1.3[a]

[a]Not significantly different from control

Although there has been some indication in the past that certain analogues displayed distinct effects (Meyers et al 1977) on insulin and glucagon release, the present results offer the first proof that somatostatin inhibition of release of the two hormones is under the control of separate receptor subtypes.

Effects on exocrine pancreatic secretions

Somatostatin is also a potent inhibitor of pancreatic enzyme secretion and we recently examined the same panel of analogues for their effects on amylase release (stimulated by bombesin infusion) in anaesthetized rats. Somatostatin

TABLE 4 **Somatostatin and analogue dose-dependent inhibition of pentagastrin-stimulated gastric acid release**

Peptide	Dose (nM/kg per hour)	% Inhibition
Somatostatin	0.5	29.5 ± 7.3
	1.0	50.1 ± 8.6
	5.0	56.3 ± 6.8
	10.0	79.9 ± 2.9
DC-25-100	1.0	43.1 ± 5.9[a]
	5.0	70.1 ± 3.6
	10.0	91.1 ± 3.1
NC-8-12	0.5	30.7 ± 8.3
	1.0	45.9 ± 3.2
	10.0	94.9 ± 2.7
DC-25-12	10.0	29.3 ± 10.5[a]
	100.0	49.6 ± 6.1
	250.0	79.3 ± 1.6
	500.0	91.2 ± 3.2
DC-23-99	10.0	36.6 ± 7.8[a]
	100.0	25.3 ± 9.0[a]
DC-25-20	10.0	34.8 ± 5.2
	100.0	58.0 ± 8.6
	250.0	57.1 ± 2.1

[a]Not significantly different from control

and the two cyclic octapeptides, DC-25-100 and NC-8-12, potently inhibit amylase secretion in the 0.5–10 nM infusion dose range (Table 3). EC_{50} values are shown in Table 5.

As for insulin, the sstr5-selective linear analogue, DC-23-99, was more potent, followed by the mixed sstr3- and sstr5-selective compound, DC-25-12. The sstr5-selective ligand, DC-25-20, was not active at the doses tested. From these observations, we concluded that this process is probably largely under sstr5 control, although the lack of total specificity of some of the analogues does not allow for complete elimination of sstr2 involvement.

The analogues were also tested for binding to rat pancreatic acinar cells; IC_{50} values from this experiment are shown in Table 5. The results closely parallel peptide effects on *in vivo* amylase release in that somatostatin and the sstr5-selective DC-23-99 display the highest affinity. However, again the two

TABLE 5 Comparison of binding affinities of linear somatostatin octapeptide analogues for five somatostatin receptors on transfected cells, pancreatic acinar cells, gastric smooth muscle cells and *in vivo* inhibition of rat insulin, gastric acid and amylase release

| Peptide | IC_{50} (nM) Somatostatin receptor | | | | | Acinar cells | Muscle cells | EC_{50} (nM) for inhibition in vivo of release of: | | |
	sstr1	sstr2	sstr3	sstr4	sstr5			Insulin	H^+	Amylase
Somatostatin	0.1	0.28	0.07	0.86	1.2	4.0	—	2.9	1.65	1.08
DC-25-100	789	1.60	5.6	0.1	>1000	4.7	2.6	55.7	1.44	0.47
NC-8-12	>1000	<0.01	11	6.0	>1000	5.3	0.04	183	1.26	1.19
DC-25-12	>1000	46	0.03	1.2	77	—	—	109	52.0	8.04
DC-23-99	23	32	0.42	0.002	18	3.8	0.008	1.5	n.a.	1.08
DC-25-20	>1000	>1000	0.02	43	158	41.0	0.9	>1000	63.0	>1000

cyclic octapeptides were also potent binders, so an sstr2 component on acinar cells cannot be ruled out.

Effects on gastric acid secretion

The results from experiments measuring analogue inhibition of pentagastrin-stimulated gastric acid release in conscious rats were more straightforward (Table 4) in that there was a clear return to sstr2 involvement. The only potent analogues were the cyclic octapeptides, which had EC_{50} values (Table 5) quite similar to somatostatin itself. DC-23-99, DC-25-12 and DC-25-20 were at least 50 times less potent, thus clearly eliminating sstr3 and sstr5 from this process.

sstr2 control of gastric acid release is supported by the recent finding (Prinz et al 1994) that sstr2 is preferentially expressed on rat enterochromaffin-like cells and that only sstr2 ligands inhibit the gastrin-induced Ca^{2+} signal in these cells. DC-25-12 and DC-25-20 were 100–1000 times less effective in inhibiting this response or gastrin-stimulated histamine release from these cells.

In all of the animal or cell models thus far described, it is noteworthy that few effects could be attributed to the sstr3-selective analogue, DC-25-20. Thus far, the only system where this peptide acts positively is in guinea-pig gastric smooth muscle cells (Gu et al 1994). The IC_{50} values of the selective analogues for binding to these cells are shown in Table 5; both compounds selective for either sstr3 or sstr5 have high affinity for these cells. Given the almost full selectivity of DC-25-20 and some affinity of DC-23-99 for sstr3, it is tempting to assume that these cells possess primarily sstr3. Gu et al (1994) have shown that the biological activity with which these binding data correlate best is inhibition of relaxation of these cells induced by vasoactive intestinal peptide.

Conclusions

Given the multitude of biological actions of somatostatin, it has been a long-standing objective of pharmacological studies to develop synthetic analogues selective for subsets of some of them. This was achieved unwittingly with several families of peptides, notably the cyclic octreotide peptides which we now know inadvertently lost most of their affinity for sstr1 and sstr4. The availability of the five transfected cell lines has certainly placed the field on a far more rational basis and it has been gratifying that examination of the synthetic analogue base has yielded analogues with useful specificity for additional receptor types.

Much work remains to be done to correlate these findings with potential therapeutic benefits. Certainly, the specificity results need to be confirmed for the full set of five human receptors (the original work described in this paper has largely involved rat receptor clones). Some differences in absolute selectivity values for some of the same analogues when tested on human receptors have recently been reported (C. B. Srikant, A. Warszynska, M. Greenwood,

R. Panetta, Y. C. Patel, unpublished abstract no. 674, 76th Endocr Soc Meet, 1994). They also used ^{125}I-labelled somatostatin-28, known to have particularly high affinity for sstr5 (O'Carroll et al 1993), as the tracer rather than the various labelled receptor-specific ligands employed by Raynor et al (1993a,b). A comparison of the results achieved using transfected cell lines with those using natural cell types expressing each of the subtypes also needs to be made.

We have found that a potential technical problem associated with some, but not all, of these analogues lies in their extreme 'stickiness' for glass and plastic. Thus, for instance, serial dilutions carried out with the same pipette tip will result in elution of excess peptide at each dose level and the subsequent creation of artificially shifted dose–response curves and calculation of overly optimistic binding constants. Conversely, the same peptides left in dilute solution will be rapidly lost onto the containing surfaces.

Despite these problems, it appears safe to say that each of the receptor subtypes possesses ligand recognition requirements sufficiently different to be readily exploited using synthetic analogues of somatostatin. In the future, this should result in the development of therapeutic agents beyond the currently clinically available octreotide analogues with fewer side-effects and greater efficacy in many areas of medicine.

References

Bauer W, Briner U, Doepfner W et al 1982 SMS 201-995: a very potent and selective octapeptide analogue of somatostatin with prolonged action. Life Sci 3l:1133–1140

Brazeau P, Vale W, Burgus R, Ling N, Butcher M, Guillemin R 1973 Hypothalamic polypeptide that inhibits the secretion of immunoreactive pituitary growth hormone. Science 179:77–79

Bruno JF, Xu Y, Song JF, Berelowitz M 1993 Molecular cloning and functional expression of a brain-specific somatostatin receptor. Proc Natl Acad Sci USA 89:11151–11155

Gu Z-F, Mantey SA, Coy DH, Maton PN, Jensen RT 1994 The SSTR-3 subtype of somatostatin receptor mediates the inhibitory action of somatostatin on gastric smooth muscle cells. Am J Physiol, in press

Meyers C, Arimura A, Gordin A et al 1977 Somatostatin analogs which inhibit glucagon and growth hormone more than insulin release. Biochem Biophys Res Commun 74:630–636

O'Carroll A-M, Lolait SJ, König M, Mahan LC 1993 Molecular cloning and expression of a pituitary somatostatin receptor with preferential affinity for somatostatin-28. Mol Pharmacol 42:939–946

Orbuch M, Taylor JE, Coy DH et al 1993 Discovery of a novel class of neuromedin B receptor antagonists, substituted somatostatin analogues. Mol Pharmacol 44:841–850

Prinz C, Sachs G, Walsh JH, Coy DH, Wu V 1994 The somatostatin receptor subtype on rat enterochromaffinlike cells. Gastroenterology 107:1067–1074

Raynor K, Murphy WA, Coy DH et al 1993a Cloned somatostatin receptors: identification of subtype-selective peptides and demonstration of high affinity binding of linear peptides. Mol Pharmacol 43:838–844

Raynor K, O'Carroll A-M, Kong HY et al 1993b Characterization of cloned somatostatin receptors SSTR4 and SSTR5. Mol Pharmacol 44:385–392

Rossowski WJ, Coy DH 1993 Potent inhibitory effects of a type four receptor-selective somatostatin analog on rat insulin release. Biochem Biophys Res Commun 197:366–371

Yamada Y, Post SR, Wang K, Tager HS, Bell GI, Seino S 1993 Cloning and functional characterization of a family of human and mouse somatostatin receptors expressed in brain, gastrointestinal tract, and kidney. Proc Natl Acad Sci USA 89:251–255

Yasuda K, Rens-Domiano S, Breder CD et al 1993 Cloning of a novel somatostatin receptor, SSTR3, coupled to adenylyl cyclase. J Biol Chem 267:20422–20428

DISCUSSION

Cohen: This is direct evidence that receptor selectivity is mediated by the sequence and conformation of the hormonal peptide or of its analogues. In certain derivatives, there is a Phe/Tyr–D–Trp–Lys–Val/Thr–Cys consensus sequence. The D–Trp in position 2 or 3 or the Phe–Phe bond of the natural tetradecapeptide is essential to determine the type of reverse turn and the proportion of peptides that assume a β-turn conformation in solution. This conformation also minimizes cleavage of Phe–Phe or Phe–Trp bonds. We have found proteases that selectively cleave somatostatin (de Morais Carvalho et al 1992, Delporte et al 1992) depending on the β-turn or the two hydrophobic side chains. These chains are the locus for quite a number of thermolysin-like proteases.

The conformation of the peptide in solution is therefore important and an equilibrium exists between these different peptide conformations. Particular amino acid chains make contact with certain sites on different receptors and binding affinities of different analogues may differ by one to three orders of magnitude as a result of making additional contacts between the peptide and the receptor. On the basis of all these considerations, can you speculate what the ideal agonist sequence would be?

Coy: Analogues for the somatostatin receptors require the conserved pharmacophoric region. Receptor specificity may arise from additional side chains on either side of that region or increased flexibility of linear peptides.

We have a reasonably good molecular model for sstr2 analogues where the side chains are flexible and cannot be pinned down by NMR (nuclear magnetic resonance) or computer studies. You would have to make complicated analogues to force the side chains into rigid positions as a structural probe.

Cohen: The linear peptides do not have a disulphide bridge, but they may have a similar conformation in solution.

Coy: Yes, but unlike residues in a cyclic structure, individual residues in a linear structure are free to rotate.

Reisine: Murray Goodman's lab in San Diego broke every peptide bond in MK 678 and put the side groups back in every possible orientation. We tested these isomers for sstr2 binding, and from the results a 3D peptide structure which binds sstr2 was developed (Huang et al 1992). It should now be possible to design non-peptides with a similar 3D structure. This will also be possible for sstr5 because we now have a ligand that binds and are beginning to understand which residues and side groups are required for receptor selectivity. It's more problematic for sstr1 and sstr4 because there are no selective ligands.

It is quite possible that some of these peptides are non-competitive because their affinity is higher than the femtomolar range. We iodinated these peptides to test for sstr2 binding, but it is difficult to iodinate and retain the binding affinity; for instance, iodinated NC 8-12 does not bind. This may either be due to inappropriate iodination or some aspect of the molecule that is unstable.

Patel: Most of these octapeptide analogues begin with the D–Trp modification. Is this D–Trp modification responsible for the loss of sstr1 and sstr4 binding?

Coy: No, because there are several analogues in the Raynor studies which contained D–Trp and had a good affinity for all five receptors (Raynor et al 1993a,b).

Berelowitz: It seems that small analogues have a very high affinity for sstr2, sstr3 and sstr5. Is there any evidence that the somatostatin molecule has a preference for sstr1 and sstr4 because it is larger?

Coy: Somatostatin-28 has a higher affinity for sstr4 than somatostatin-14, but it is not known what structural feature is responsible for this. It is possible that the additional portion of the extended chain affects the conformation of the pharmacophoric region at a distance.

Reisine: CGP 23996 is the only compound of smaller size than somatostatin-14 that can interact with all the receptors, but it is no longer available.

Schonbrunn: We have an N-terminal biotinylated somatostatin-28 analogue which can be coupled to streptavidin. This coupling results in only a threefold decrease in binding affinity. This suggests that the dramatic difference in the N-terminus of this analogue produces only a subtle change in binding affinity (Schonbrunn et al 1993).

Susini: Yes, I agree. We compared the binding affinities of somatostatin-28 with somatostatin-28 in the presence of somatostatin-28 (1-14) antiserum and found no change.

Bruns: There seems to be a good correlation for somatostatin analogues between their inhibition of growth hormone (GH) release and binding affinity for sstr2. Would it be possible to distinguish between the inhibitory effect on insulin secretion, glucacon secretion and GH secretion if we had sstr2, sstr3 or sstr5 selective analogues?

Coy: You would hope that selective peptides would determine which receptors are involved with the effects of particular hormones, but there are probably additional factors involved in the *in vivo* secretion of hormones.

Patel: We find that human sstr2, sstr3 and sstr5 bind hexapeptide and both cyclic and linear octapeptide analogues. Human sstr1 and sstr4 form a separate class because they bind these analogues poorly.

In a systematic comparison of the agonist selectivity of 32 of David Coy's synthetic somatostatin analogues for human sstr1–5 stably expressed in Chinese hamster ovary CHO-K1 cells, we found that the analogues exhibited about a 50-fold increase in binding potency compared with somatostatin-14 for only sstr2 and sstr3, and relative selectivity for only sstr2 which is at best only 35-fold (Patel & Srikant 1994). The potency and degree of selectivity of these analogues are several orders of magnitude lower than those reported by David Coy and Terry Reisine (Raynor et al 1993a).

These discrepancies may be due to methodological differences or species differences because we studied human receptors whereas Raynor et al (1993a) studied rodent receptors. Until this is resolved, caution should be exercised when using these compounds as superagonists or receptor-selective agonists.

Reisine: We tested 60–70 peptides on different somatostatin receptors, not to determine the actual K_i of each peptide, but to find their relative potencies. We have also studied receptor selectivity in tissues with more than one receptor type; for example, the effect of BIM 23052 in mouse pituitary tumour AtT-20 cells. Some peptides are selective for sstr2 versus sstr5 and may be able to discriminate between GH release and insulin release. They may not all have high discriminating affinities because some interactions may be non-competitive.

References

Delporte C, de Morais Carvalho K, Leseney AM, Winand J, Christophe J, Cohen P 1992 A new metallo-endopeptidase from human neuroblastoma NB-OK1 cells which inactivates atrial natriuretic peptide by selective cleavage at the Ser[123]–Phe[124] bond. Biochem Biophys Res Commun 182:158–164

de Morais Carvalho K, Joudiou C, Boussetta H, Leseney AM, Cohen P 1992 A peptide hormone inactivating endopeptidase (PHIE) in Xenopus laevis skin secretion. Proc Natl Acad Sci USA 89:84–88

Huang Z, He Y-B, Raynor K, Tallent M, Reisine T, Goodman M 1992 Main chain and side chain chiral methylated somatostatin analogs: syntheses and conformational analyses. J Am Chem Soc 114:9390–9410

Patel YC, Srikant CB 1994 Subtype selectivity of peptide analogues for all five cloned human somatostatin receptors (hsstr1–5). Endocrinology 135:2814–2817

Raynor K, Murphy WA, Coy DH et al 1993a Cloned somatostatin receptors: identification of subtype-selective peptides and demonstration of high affinity binding of linear peptides. Mol Pharmacol 43:838–844

Raynor K, O'Carroll A-M, Kong H et al 1993b Characterization of cloned receptors sstr4 and sstr5. Mol Pharmacol 44:385–392

Schonbrunn A, Lee AB, Brown PJ 1993 Characterization of a biotinylated somatostatin analog as a receptor probe. Endocrinology 132:146–154

Final discussion

Reichlin: Mark Eppler, what are the veterinary applications of this work?

Eppler: Basically, we want to make animals grow bigger and/or faster so they will be worth more money for the same amount of feed. Somatostatin is the major inhibitor of growth hormone (GH) release. Immunoneutralization of somatostatin in rats by injection of somatostatin antibodies (Wehrenberg et al 1983) produces increased release of GH and increased sensitivity to growth hormone-releasing hormone (GHRH). Immunoneutralization of somatostatin in rats and farm animals by active immunization sometimes produces increased growth rates, but the results have been inconsistent, suggesting that the situation is more complicated. So, one of our aims has been to identify which of the five somatostatin receptors is involved specifically in the regulation of GH release. Also, somatostatin is effective in the gastrointestinal tract, so it may be possible to develop selective somatostatin antagonists that have positive effects on nutrient uptake.

Epelbaum: Do any of the somatostatin analogues bind to opiate receptors?

Reisine: Octreotide and octapeptide analogues of octreotide are antagonists of the μ opiate receptor, but they do not interact with the $\varkappa$ or δ receptor. In contrast, the hexapeptide compounds, the linear peptides and somatostatin don't interact with any of the opiate receptors.

Epelbaum: Do opiate-derived analogues bind to sstr1 or sstr4?

Reisine: No.

Berelowitz: Do GH-releasing peptides (Bowers et al 1991) bind to somatostatin receptors?

Reisine: They do not interact with somatostatin receptors or opiate receptors.

Reichlin: Does this μ receptor binding explain the analgesic actions of octreotide?

Reisine: No, because it would be a μ antagonist. It's also controversial whether octreotide actually has analgesic properties. When somatostatin is injected intrathecally into rats, it has some analgesic effects, but it also causes paralysis and other side effects.

Taylor: Several people have injected somatostatin intrathecally and found that in rodents, somatostatin may, under certain conditions, induce analgesia and neurotoxicity (Meynadier et al 1985, Chrubasik et al 1985, Mollenholt et al 1988, 1990, Gaumann et al 1989, Penn et al 1990). It is possible that analgesia occurs in humans, but with low toxicity. The mechanism is uncertain, but it definitely does not involve the opiate receptor.

Reichlin: It has been reported that somatostatin prevents tissue injury from a variety of damaging factors; for example, alcohol-induced gastric damage (Usadel 1988). What is the basis of this so-called 'cytoprotective' action of somatostatin?

Taylor: It is not clear. Several papers also showed that somatostatin has a cytoprotective action in the liver (Kessler et al 1986, Ziegler et al 1985, 1986, 1991, Raper et al 1991, 1993, Fricker et al 1994). There's a low affinity, high volume transport site in the liver which excretes somatostatin and may also be involved in the cytoprotective effect. The mechanism is not known, but some evidence indicates that it is not related to the somatostatin receptors.

In collaboration with Nicholas Russo at The Mayo Clinic, we are working with cultured cholangiocytes from the gall bladder duct. We have shown that somatostatin inhibits the proliferation of cholangiocytes and clonal tumour cell lines from cholangiocarcinomas. This is an interesting phenomenon because most somatostatin peptides are brought through the gall bladder and are excreted unchanged or unmetabolized, so you have a sort of target delivery phenomenon.

Reichlin: Usadel (1988) finds that some somatostatin analogues which do not inhibit GH release are cytoprotective.

Taylor: Yes, some of these analogues are cyclic peptides that have a similar structure to cholic acid, although cholic acid does not bind to somatostatin receptors.

Lamberts: Continuous infusions of natural somatostatin reduces gastric acid production and have been used in the treatment of haemorrhages in the stomach. This protective action of somatostatin might be related to both its inhibitory action on pepsin and hydrochloric acid production.

References

Bowers CY, Sartar AO, Reynolds GA, Badger TM 1991 On the actions of the growth hormone-releasing hexapeptide, GHRP. Endocrinology 128:2027–2035

Chrubasik J, Meynadier J, Scherpereel P, Wunsch E 1985 The effect of epidural somatostatin on postoperative pain. Anesth Analg 64:1085–1088

Fricker G, Dubost V, Schwab D, Bruns C, Thiele C 1994 Heterogeneity in hepatic transport of somatostatin analog octapeptides. Hepatology 20:191–200

Gaumann DM, Yaksh TL, Post C, Wilcox GL, Rodriguez M 1989 Intrathecal somatostatin in cat and mouse studies on pain, motor behaviour and histopathology. Anesth Analg 68:623–632

Kessler H, Gehrke M, Haupt A, Klein M, Müller A, Wagner K 1986 Common structural features for cytoprotection activities of somatostatin, antamanide and related peptides. Klin Wochenschr (suppl VII) 64:7–78

Meynadier J, Chrubasik J, Dubar M, Wunsch E 1985 Intrathecal somatostatin in terminally ill patients. A report of two cases. Pain 23:9–12

Mollenholt P, Post C, Rawal N, Freedman J, Hokfelt T, Paulsson I 1988 Antinociceptive and 'neurotoxic' actions of somatostatin in rat spinal cord after intrathecal administration. Pain 32:95–105

Mollenholt P, Post C, Paulsson I, Rawal N 1990 Intrathecal and epidural somatostatin in rats: can antinociception, motor effects and neurotoxicity be separated? Pain 363–370

Penn RD, Paice JA, Kroin JS 1990 Intrathecal octreotide for cancer pain. Lancet I:738
Raper SE, Kothary PC, Kokudo N, DelValle J, Eckhauser FE 1991 The liver plays an important role in the regulation of somatostatin-14 in the rat. Am J Surg 151:184–189
Raper SE, Kothary PC, Kokudo N, DelValle J 1993 Hepatectomy impairs hepatic processing of somatostatin-14. Am J Surg 165:89–95
Usadel KH 1988 Hormonal and nonhormonal cytoprotective effect by somatostatins. Hormone Res 29:83–85
Wehrenberg WB, Ling N, Brazeau P, Esch F, Bohlen P, Guillemin R 1982 Physiological roles of somatostatin and somatocrinin in regulation of growth hormone secretion. Biochem Biophys Res Comm 109:562–567
Ziegler K, Frimmer M, Kessler H et al 1985 Modified somatostatins as inhibitors of a multispecific transport system for bile acids and phallotoxins in isolated hepatocytes. Biochem Biophys Acta 845:86–93
Ziegler K, Frimmer M 1986 Molecular aspects of cytoprotection by modified somatostatins. Klin Wochenschr (suppl VII) 64:87–89
Ziegler K, Lins W, Frimmer M 1991 Hepatocellular transport of cyclosomatostatins: evidence for a carrier system related to the multispecific bile acid transporter. Biochim Biophys Acta 1061:287–296

Summary

Seymour Reichlin

Division of Endocrinology, Tufts University, New England Medical Center, Box 268, 750 Washington Street, Boston, MA 02111, USA

Driven by dramatic new advances in the understanding of the molecular processes behind the control of somatostatin synthesis, secretion and action, this symposium has achieved a surprisingly seamless integration of the most basic aspects of somatostatin biology with physiological, neurological and clinical implications. The timing of this symposium is near perfect. The two years that have elapsed since the first report of the structure of a somatostatin receptor have witnessed the identification of four more somatostatin receptors (discussed by Graeme Bell) and the association of these receptors with previously recognized and pharmacologically distinct somatostatin-related ligands (discussed by Christian Bruns, Terry Reisine and David Coy). From an evolutionary viewpoint, it is staggering that the seven or eight somatostatin receptor variants, all homologous and all derived from a single ancestral gene, are localized (in humans) to five different chromosomes, and that the somatostatin gene is located on yet another chromosome. In the anglerfish, separate genes encode the two active forms of naturally occurring somatostatin, somatostatin-14 and somatostatin-28; whereas in higher vertebrates, including humans, tissue-specific prosomatostatin processing (discussed by Yogesh Patel) determines which form of somatostatin is secreted.

Recombinant receptor probes have been used to identify second messenger systems including several G proteins that regulate different intracellular processes (discussed by Terry Reisine, Agnes Schonbrunn and Christiane Kleuss). Selective binding of G protein subtypes provides a welcome molecular explanation for the tissue-specific actions of somatostatin. Using antisense knockout studies, Christiane Kleuss demonstrated that the coupling of a receptor to a Ca^{2+} channel was mediated by the G protein $G_o\alpha2\beta1\gamma3$. Christiane Susini described the coupling of the receptors to a tyrosine phosphatase which was proposed to be the basis for the antiproliferative action of somatostatin. Somatostatin as an antiproliferative agent was a recurring theme. Its potential as a treatment for cancer (discussed by Steven Lamberts) is bolstered by the finding that somatostatin receptors are present on many tumours, the evidence for tumour regression in model systems and the development of analogues with selectively high antitumour activity (discussed by David Coy). The clinical use of

258

radiolabelled somatostatin analogue imaging of neuroendocrine, lymphatic and mesodermal tumours and of localized activated monocytes/macrophages (discussed by Steven Lamberts) utilizes the observation that most of these abnormal cells express somatostatin receptors. A molecular explanation for the failure of some tumours with receptors for somatostatin-14 to bind octreotide has emerged—octreotide is receptor specific for sstr2, sstr3 and sstr5, and will not label tumours with sstr1 or sstr4, e.g. many insulinomas that are notorious for being resistant to octreotide therapy. The development of new analogues, possessing greater receptor selectivity and stability is clearly a challenge for peptide engineers (discussed by David Coy).

One of the happy outcomes of this symposium is that principal workers in the field have arrived at a consensus of terminology for the somatostatin receptors (see Appendix).

Receptor selectivity was also shown to be biologically relevant (discussed by Mike Berelowitz and Alain Beaudet). Receptor types are distributed in distinct but overlapping patterns in the brain and in peripheral tissues, may be coexpressed in individual cells and may, under some circumstances, be regulated independently. For example, Mike Berelowitz found that starvation reduces the amounts of *sstr1, sstr2* and *sstr3* mRNA in the pituitary, whereas *sstr4* and *sstr5* are unaffected. In contrast, experimental diabetes mellitus in rats lowers *sstr5* mRNA in the pituitary whereas *sstr4* is unaffected. Alain Beaudet described that somatostatin receptors are distributed selectively on subpopulations of hypothalamic hypophysiotropic cells and may be regulated differently by changes in target gland function. Clearly, the area of receptor regulation is one marked for an explosion of new information.

Little is known about the control or regulation of receptor expression which is important for understanding brain development. Patterns of receptor expression appear and then disappear at age-specific times in the fetus, usually in parallel with regional somatostatin gene expression (discussed by Philippe Leroux). These developmental patterns are under genetic control and can serve as a model of regulation of morphogenesis in higher vertebrates.

Understanding the role of somatostatin in brain disease will require the integration of these newly emerging techniques of receptor characterization with the more established methods for identifying somatostatin mRNA and protein in the brain. Marie-Françoise Chesselet reviewed a number of factors that increase the expression of somatostatin mRNA; for example, striatal somatostatin mRNA is activated by stimulation of NMDA (*N*–methyl–D–aspartate) receptors and by ischaemia (a situation in which NMDA receptors are known to be activated). Somatostatinergic interneurons appear to be resistant to many factors that damage brain tissue and are preserved selectively in Huntington's disease. Marie-Françoise Chesselet also showed that central dopaminergic pathways in the basal ganglia probably exert a tonic stimulatory effect on their targeted somatostatinergic neurons. The functional significance of regional changes in neuronal somatostatin content was brought out in discussion by Richard Robbins. The temporal lobe in patients with idiopathic epilepsy is depleted of somatostatin. The resultant loss of neuroinhibitory actions of somatostatin may contribute to epileptogenic hyperexcitability.

The culmination of many years' work on the mechanism of cAMP regulation of somatostatin synthesis by Marc Montminy, Richard Goodman and their respective groups (Kwok et al 1994, Arias et al 1994, Nordheim 1994) was discussed. The model emerging from their work has biological significance beyond the somatostatin field, for it provides a molecular mechanism for alteration of nuclear function in any cell type in which ligand–membrane binding triggers adenylyl cyclase activation. Moreover, the demonstration that the same mechanism can be influenced by the phosphatidylinositol pathway describes a final common path by which ligand binding to cell membranes can affect gene function through multiple second messengers. These mechanisms suggest potential new oncogenic pathways for inquiry.

References

Arias J, Alberts AS, Brindle P et al 1994 Activation of cAMP and mitogen responsive genes relies on a common nuclear factor. Nature 370:226–229
Kwok RPS, Lundblad JR, Chrivia JC et al 1994 Nuclear protein CBP is a coactivator for the transcription factor CREB. Nature 370:223–226
Nordheim A 1994 CREB takes CBP to tango. Nature 370:177–178

Appendix

Somatostatin receptor nomenclature consensus

It is now evident that there are at least five types of somatostatin receptor (Yamada et al 1992a,b, Kluxen et al 1992, Meyerhof et al 1992, Yasuda et al 1992, Bruno et al 1992, Rohrer et al 1993, O'Carroll et al 1992). The recent cloning of the receptor genes did not resolve, but rather compounded, the controversy surrounding the nomenclature for receptors that were previously characterized using available ligands (see Hoyer et al 1994). Thus the SSTR4 receptor cloned by Bruno et al (1992) was called SSTR5 by some authors (e.g. Bell & Reisine 1993, Raynor et al 1993a,b) and the SSTR4 receptor cloned by O'Carroll et al (1992) was called SSTR5 by others (e.g. Yamada et al 1993, Kaupmann et al 1993, Patel et al 1994).

It was agreed by the interested parties at this Ciba Foundation symposium to name the receptor reported by Bruno et al (1992) as SSTR4 and the receptor identified by O'Carroll et al (1992) as SSTR5. In keeping with current convention recommended by IUPHAR (International Union of Pharmacology), it was also agreed to use lower case letters to signify that the receptors are recombinant. This represents a significant step forward in resolving the somatostatin receptor nomenclature controversy and will aid understanding of the scientific literature. This system of nomenclature was therefore used throughout this volume.

Subsequent to the symposium, it was pointed out by IUPHAR that the letter 'r' for receptor is redundant in receptor nomenclature systems and that the approved appellations should be sst_1, sst_2, sst_3, sst_4 and sst_5. It is anticipated that scientists will readily adopt this new IUPHAR nomenclature system for somatostatin receptors once it is published.

Pat Humphrey and Terry Reisine*

*Glaxo Institute of Applied Pharmacology, University of Cambridge, Cambridge, UK and *Department of Pharmacology, University of Pennsylvania, Philadelphia, USA*

References

Bell GI, Reisine T 1993 Molecular biology of somatostatin receptors. Trends Neurosci 16:34–38

Bruno JF, Xu J, Song J, Berelowitz M 1992 Molecular cloning and functional expression of a brain-specific somatostatin receptor. Proc Natl Acad Sci USA 89:11151–11155

261

Hoyer D, Lübbert H, Bruns C 1994 Molecular pharmacology of somatostatin receptor subtypes. Naunyn-Schmiedeberg's Arch Pharmacol 350:441–453

Kaupmann K, Bruns C, Hoyer D, Seuwen K, Lubbert H 1993 mRNA distribution and second messenger coupling of four somatostatin receptors expressed in the brain. FEBS Lett 331:53–39

Kluxen FW, Bruns C, Lübbert H 1992 Expression cloning of a rat brain somatostatin receptor cDNA. Proc Natl Acad Sci USA 89:4618–4622

Meyerhof W, Wulfsen I, Schonrock C, Fehr S, Richter D 1992 Molecular cloning of a somatostatin-28 receptor and comparison of its expression pattern with that of a somatostatin-14 receptor in rat brain. Proc Natl Acad Sci USA 89:10267–10271

O'Carroll A-M, Lolait SJ, König M, Mahan LC 1992 Molecular cloning and expression of a pituitary somatostatin receptor with preferential affinity for somatostatin-28. Mol Pharmacol 42:939–946

Patel YC, Greenwood MT, Warszynska A, Panetta R, Srikant CB 1994 All five cloned human somatostatin receptors (hSSTR1–5) are functionally coupled to adenylate cyclase. Biochem Biophys Res Commun 198:605–612

Raynor K, Murphy WA, Coy DH et al 1993a Cloned somatostatin receptors: identification of subtype-selective peptides and demonstration of high affinity binding of linear peptides. Mol Pharmacol 43:385–392

Raynor K, O'Carroll AM, Kong H et al 1993b Characterisation of cloned somatostatin receptors. Mol Pharmacol 44:838–844

Rohrer L, Raulf F, Bruns C, Buettner R, Hofstaedter F, Schuele R 1993 Cloning and characterisation of a novel human somatostatin receptor. Proc Natl Acad Sci USA 90:4196–4200

Yamada Y, Post SR, Wang K, Tager HS, Bell GI, Seino S 1992a Cloning and functional characterisation of a family of human and mouse somatostatin receptors expressed in brain, gastrointestinal tract and kidney. Proc Natl Acad Sci USA 89:251–255

Yamada Y, Reisine T, Law SF et al 1992b Somatostatin receptors, an expanding gene family: cloning and functional characterisation of human SSTR3, a protein coupled to adenylyl cyclase. Mol Endocrinol 6:3236–2142

Yamada Y, Kagimoto S, Kubota A et al 1993 Cloning, functional expression and pharmacological characterisation of a fourth (hSSTR4) and fifth (hSSTR5) human somatostatin receptor subtype. Biochem Biophys Res Commun 195:844–852

Yasuda K, Rens-Domiano S, Breder CD et al 1992 Cloning of a novel somatostatin receptor, SSTR3, coupled to adenylyl cyclase. J Biol Chem 267:20422–20428

Index of contributors

Subject index